ECOLOGICAL HEALTH: SOCIETY, ECOLOGY AND HEALTH

ADVANCES IN MEDICAL SOCIOLOGY

Series Editor: Barbara Katz Rothman

Series Editor for Volumes 5–6: Gary L. Albrecht

Series Editor for Volumes 7–8: Judith A. Levy

Recent Volumes:

Volume 5:	Quality of Life in Health Care – Edited by Ray Fitzpatrick
Volume 6:	Case and Care Management – Edited by Gary L. Albrecht
Volume 7:	Emergent Issues in the Field of Drug Abuse – Edited by Judith A. Levy, Richard C. Stephens, and Duane C. McBride
Volume 8:	Social Networks and Health – Edited by Bernice A. Pescosolido and Judith A. Levy
Volume 9:	Bioethical Issues, Sociological Perspectives – Edited by Barbara K. Rothman, Elizabeth M. Armstrong and Rebecca Tiger
Volume 10:	Patients, Consumers and Civil Society – Edited by Susan M. Chambré and Melinda Goldner
Volume 11:	Understanding Emerging Epidemics: Social and Political Approaches – Edited by Ananya Mukherjea
Volume 12:	Sociology of Diagnosis – Edited by P. J. McGann and David J. Hutson
Volume 13:	Sociological Reflections on the Neurosciences – Edited by Martyn Pickersgill and Ira van Keulen
Volume 14:	Critical Perspectives on Addiction – Edited by Julie Netherland

ADVANCES IN MEDICAL SOCIOLOGY VOLUME 15

ECOLOGICAL HEALTH: SOCIETY, ECOLOGY AND HEALTH

EDITED BY

MAYA K. GISLASON

University of Sussex, Sussex, UK

United Kingdom – North America – Japan
India – Malaysia – China

Emerald Group Publishing Limited
Howard House, Wagon Lane, Bingley BD16 1WA, UK

First edition 2013

British Library Cataloguing in Publication Data
A catalogue record for this book is available from the British Library

ISBN: 978-1-78190-323-0
ISSN: 1057-6290 (Series)

CONTENTS

LIST OF CONTRIBUTORS *ix*

ACKNOWLEDGMENTS *xiii*

INTRODUCTION *xv*

PART I: CRITICAL THEORETICAL ENGAGEMENTS WITH SOCIAL-ECOLOGICAL APPROACHES TO HEALTH RESEARCH

EXPANDING THE SOCIAL: MOVING TOWARDS THE ECOLOGICAL IN SOCIAL STUDIES OF HEALTH
Maya K. Gislason *3*

TOWARDS A CRITICAL APPROACH TO ECOHEALTH RESEARCH, THEORY AND PRACTICE
Crescentia Dakubo *23*

EXPLORING ABORIGINAL PEOPLE'S CONNECTION TO COUNTRY TO STRENGTHEN HUMAN–NATURE THEORETICAL PERSPECTIVES
Jonathan Yotti Kingsley, Mardie Townsend and Claire Henderson-Wilson *45*

PART II: SOCIAL-ECOLOGICAL HEALTH INJUSTICES AND PATHS TO JUSTICE

TIBETAN PROTEST SELF-IMMOLATION IN CHINA: REFLECTIONS ON ECOLOGY, HEALTH AND POLITICS
Colin D. Butler *67*

ECOHEALTH THROUGH AN ABILITY STUDIES AND DISABILITY STUDIES LENS
Gregor Wolbring 91

STRUCTURAL VULNERABILITY AND NARRATIVE: SENSITISING CONCEPTS FOR UNDERSTANDING THE HEALTH IMPACTS OF CLIMATE CHANGE
Rebecca Moran, Julie Hollenbeck and Cassandra Phoenix 109

PART III: RESEARCHING HEALTH IN THE NEXUS BETWEEN THE SOCIAL AND THE ECOLOGICAL

HEALTH AND ENVIRONMENTAL POLITICS IN THE UNITED STATES: A HISTORICAL PERSPECTIVE
Jennifer Thomson 127

EXPLORING THE LINKS BETWEEN HIV/AIDS AND FORESTS IN MALAWI: MORBIDITY, MORTALITY, AND CHANGING DEPENDENCE ON FOREST RESOURCES
Joleen Timko 147

DROPS AND HOT STONES: TOWARDS INTEGRATED URBAN PLANNING IN TERMS OF WATER SCARCITY AND HEALTH ISSUES IN LEH TOWN, LADAKH, INDIA
Daphne Gondhalekar, Adris Akhtar, Pascal Keilmann, Jenny Kebschull, Sven Nussbaum, Sonam Dawa, Phuntsok Namgyal, Lobzang Tsultim, Tsering Phuntsog, Stanzin Dorje, Tsering Mutup and Phunchok Namgail 173

THE ECOLOGY OF DYING: COMMODITY CHAINS, GOVERNANCE, AND THE MEDICALIZATION OF END-OF-LIFE CARE
Christine Vatovec, Laura Senier and Michael Mayerfeld Bell 195

WHY IS AN INTEGRATED SOCIAL-ECOLOGICAL SYSTEMS (ISES) LENS NEEDED TO EXPLAIN CAUSES AND DETERMINANTS OF DISEASE? A CASE STUDY OF DENGUE IN DHAKA, BANGLADESH
Parnali Dhar Chowdhury and C. Emdad Haque *217*

PERPETUATING A REDUCTIONIST MEDICAL WORLDVIEW: THE ABSENCE OF ENVIRONMENTAL MEDICINE IN THE AMERICAN ADHD CLINICAL PRACTICE GUIDELINES
Manuel Vallée *241*

PART IV: SOCIAL–ECOLOGICAL APPROACHES TO STUDYING INSTITUTIONAL CHANGE FOR HEALTH

A SUSTAINABLE DEVELOPMENT AGENDA FOR THE UK NATIONAL HEALTH SERVICE (NHS): AN ORGANIZATIONAL LEARNING MODEL FOR DEFINING AND SUPPORTING GOALS
Claire Marsh *267*

ENVIRONMENTAL HEALTH RISK GOVERNANCE IN PRACTICE: LESSONS LEARNED FROM A FLEMISH CASE STUDY APPROACH
Kristien Stassen and Pieter Leroy *289*

OCEANS AND HUMAN HEALTH IN THE CARIBBEAN REGION
Alana Malinde S. N. Lancaster and Lyndon F. Robertson *311*

LIST OF CONTRIBUTORS

Adris Akhtar	Center for Development Research (ZEF), University of Bonn, Bonn, Germany
Michael Mayerfeld Bell	Department of Community and Environmental Sociology, University of Wisconsin-Madison, WI, USA
Gillian A. Bendelow	Department of Sociology, School of Law, Politics and Sociology, University of Sussex, Brighton, UK
Colin D. Butler	Faculty of Health, University of Canberra, Canberra, Australia
Parnali Dhar Chowdhury	Natural Resources Institute, University of Manitoba, Winnipeg, Canada
Crescentia Dakubo	Northern Ontario School of Medicine, Lakehead University, Thunder Bay, Canada
Sonam Dawa	Ladakh Ecological Development Group (LEDeG), Karzoo, Leh, Ladakh, India
Stanzin Dorje	Ladakh Ecological Development Group (LEDeG), Karzoo, Leh, Ladakh, India
Maya K. Gislason	Centre for Global Health Policy, University of Sussex, Brighton, UK
Daphne Gondhalekar	Center for Development Research (ZEF), University of Bonn, Bonn, Germany

C. Emdad Haque	Natural Resources Institute, University of Manitoba, Winnipeg, Canada
Claire Henderson-Wilson	Faculty of Health, Deakin University, Melbourne, Australia
Julie Hollenbeck	European Centre for Environment and Human Health, University of Exeter Medical School, Truro, UK
Jenny Kebschull	Center for Remote Sensing of Land Surfaces (ZFL), University of Bonn, Germany
Pascal Keilmann	Dritter Orden Clinic and MediHimal, Munich, Germany
Jonathan Yotti Kingsley	Faculty of Health, Deakin University, Melbourne, Australia
Alana Malinde S. N. Lancaster	Faculty of Law, University of the West Indies, Cave Hill Campus, Bridgetown, Barbados
Pieter Leroy	Institute for Management Research, Political Sciences of the Environment, University Nijmegen, Radboud, The Netherlands
Claire Marsh	Bradford Institute for Health Research & Sustainability Research Institute, University of Leeds, Leeds, UK
Rebecca Moran	European Centre for Environment and Human Health, University of Exeter Medical School, Truro, UK
Tsering Mutup	Block Medical Office, Karzoo, Leh, Ladakh, India
Phunchok Namgail	Ladakh Ecological Development Group (LEDeG), Karzoo, Leh, Ladakh, India

Phuntsok Namgyal	Ladakh Ecological Development Group (LEDeG), Karzoo, Leh, Ladakh, India
Sven Nussbaum	Center for Remote Sensing of Land Surfaces (ZFL), University of Bonn, Bonn, Germany
Tsering Phuntsog	Ladakh Ecological Development Group (LEDeG), Karzoo, Leh, Ladakh, India
Cassandra Phoenix	European Centre for Environment and Human Health, University of Exeter Medical School, Truro, UK
Lyndon F. Robertson	Caribbean EcoHealth Programme, Caribbean Institute for Meteorology and Hydrology Husbands, St. James, Barbados
Laura Senier	Department of Sociology and Anthropology and Department of Health Sciences, Northeastern University, Boston, MA, USA
Kristien Stassen	Centre for Economics and Corporate Sustainability (CEDON), University College Brussels (HUB), Brussels, Belgium
Jennifer Thomson	Bucknell University, Lewisburg, PA, USA
Joleen Timko	Faculty of Forestry, Africa Forests Research Initiative on Conservation and Development (AFRICAD), University of British Columbia, Vancouver, Canada
Mardie Townsend	Faculty of Health, Deakin University, Melbourne, Australia
Lobzang Tsultim	Ladakh Ecological Development Group (LEDeG), Karzoo, Leh, Ladakh, India

Manuel Vallée Department of Sociology, Faculty of Arts, The University of Auckland, Auckland, New Zealand

Christine Vatovec College of Medicine and Rubenstein School of Environment and Natural Resources, University of Vermont, Burlington, VT, USA

Gregor Wolbring Department of Community Health Sciences, Community Rehabilitation and Disability Studies Stream, Faculty of Medicine, University of Calgary, Calgary, Alberta, Canada

ACKNOWLEDGMENTS

The community of people who have made this edited collection possible are due a heartfelt thank you. I would like to thank Barbara Katz, editor of the Advances in Medical Sociology series, for her vision and support as well as those at Emerald who have facilitated its publication, in particular Cristina Irving Turner, Nikki Chapman, Claire Swift, and Jaya Chowdhury of MPS Limited. A heartfelt expression of gratitude also goes out to a large number of very kind anonymous reviewers, you know who you are, who gave generously of their time and contributed thoughtful reflections to this interdisciplinary project. To the authors, all of whom persevered even when life presented many unexpected turns and challenges, I thank you! Colleagues in both the Ecohealth and Medical Sociology communities, I am grateful for your support of a project which aims to ark across a range of conceptual spaces and empirical preoccupations. Finally, thank you to my family and friends – all of you – and very much to my daughter, Arwyn, to whom I dedicate my efforts to build health for all to enjoy today and for generations to come.

INTRODUCTION

Biomedical advances have undoubtedly bestowed benefits of increased life expectancy and protection from infectious disease, which enhance the quality as well as the length of life, at least in wealthier societies. Scientific and medical knowledge has also advanced significantly since the 17th century when the turn toward logical and scientific thought was made in Western philosophy. Since this time, a positivistic worldview has generated nuanced insights into science and medicine, through specialisms such as biology, genetics, and infectious disease medicine. What was lost, however, was a holistic paradigm within science and medicine which employed a whole-systems approach to the study of health and disease. Lost too, was a medical cosmology where studying the interplay between social and environmental determinants of health made sense (Jewson, 2009). In the 21st century, diagnostic categories have expanded greatly alongside the social patterning of health and illness associated with longevity and emotional/mental health, demanding multifactorial aetiologies which require that traditional categories, formulations and management strategies be reevaluated.

Recent sociopolitical (the rise of the risk society), epidemiological (the resurgence of infectious disease epidemics), and ecological (reminders of the fundamental dependency of humans on ecosystem services) events are precipitating a conceptual and practical reorganization of medical ideologies and practices (Bendelow, 2009; Benton & Redcliff, 2002; Samson, 1999). Within public health and health promotion the tradition of focusing on short-term health risks is also being extended to include the development of longer-term prevention based strategies which seek to not only contain and control public health problems but to also address their causes. Thus, while biomedicine may dominate current health care and public health practice, this is not the only story, nor the only explanatory paradigm.

The revitalization of concepts of integration and interdependence is occurring within medicine and public health. One concept that is re-emerging within health research and practice is holism and this is enabling radical changes in what is meant by health, illness, and disease. Holism is premised on the view that parts of a whole are intimately interconnected and can neither exist nor be understood independent of the whole (Samson, 1999, pp. 3–5). Medical sociologists working with the principles of holism observe that Western philosophy generally, and the biomedical cosmology specifically, have been built upon notions of coeternal binary oppositions (Samson, 1999, pp. 3–4). Through reawakening concepts such as holism, it is possible to challenge the tacit view of humans as distinct from animals, culture from nature, mind from body, scientific logic from subjective experience, and human health as separate from natural environmental and ecosystem health. From this vantage point, we can also set our sights much higher than simply on the avoidance of illness but can aim to build health and well-being at the scale of social and earth systems.

Approaching health in an integrated way also extends the purview of who holders of medical knowledge are and who can contribute to the project of generating health and well-being. Holistic and integrated concepts of health and illness remain axiomatic to ancient systems such as traditional Chinese medicine and Ayurveda, to Earth Medicine as practiced for millennia by Aboriginal and tribal peoples the world round. Holism is a principle that has also been pervasive as well as throughout the history of medicine in the so-called 'developed' world which has its roots in Hippocratic medicine. While there has always been pluralism in health and healing (Cant & Sharma, 1999, 2001); some literature on complementary and alternative medicine suggests that the rise in interest in reviving traditional health and healing concepts and practices is a reflection of people's increasing interest in being treated as biopsychosocial beings who are attempting to cultivate more harmony in their own lives and between themselves and their lived (social and biological) environments (see for instance Pietroni, 1992; Samson, 1999).

Although holistic, and in particular, integrated health care models purport to embrace the mind/body/society link, much remains to be done to understand the connections between human and animal bodies, natural ecologies and states of health and illness in contemporary medical models and public health practices as well as in the explanatory power of social science (Gislason, 2007). Increasingly the inability to understand the connections between the social and the ecological is coming to be seen as an "intellectual obstacle" to meeting a range of contemporary social challenges, including those in the realm of health and illness. In particular the mind/body, individual/social, ecology/society, human/animal, and clinical medicine/public health splits are seen to obstruct a more integrated theoretical approach to health and illness and impede investigation of the relationship between these abstractly counterposed domains (Bendelow, 2009;

Williams & Bendelow, 1998). Social scientists in general, and medical sociologists in particular, by tackling the thorny and complex issues surrounding biopsychosocial and integrative approaches to health, are making visible the biological and systems-based aspects of human health and forging spaces within medical sociology to study the interplay between social and environmental determinants of health.

Building bridges between theories, disciplines, methods, practices and mandates requires boundary crossing. Thinking ecologically about health and equally approaching health as a social-ecological issue is at the crux of new work which is trying to simultaneously address issues of social justice, ecological integrity and social, physical, and mental health and well-being. This book breaks ground by bringing insights from medical sociology into conversation with the newly emerging field of ecosystem based approaches to health research or *Ecohealth*. Foregrounded are myriad ways 'the social' can be imagined, researched and brought to life in work that links social and environmental determinants of health. Insights from critical theory, principles of social equity and social justice, research from biology and earth systems science, and social principles of sustainable development that link international governance with local realities are brought into conversation in the chapters of this edited collection.

The key concepts foregrounded in this book are health, ecology, and society. Health in this context refers to human, animal and ecosystem health and understands that health is produced in the nexus between human, animal (including microbial), and ecological activity. Health is seen as relational, interdependent, and cumulative as it is not only produced and sustained through genetics but also through how individual's treat their own bodies, how society's care for the medical and public health of its people and how earth systems are impacted by human activity. Relatedly, ecological health is produced through the ways in which humans across the planet care for or neglect animals (both wild and domestic), the biological communities within which they live, work, and play and the Earth systems which regulate and sustain our home planet. In this context, ecological health is a term denoting the interconnection between the social and the ecological as it produces health and illness. Another way of expressing this dynamic is through the term social-ecological health issues. Ecohealth is often used in this book to refer specifically to the principles and pillars of ecosystem based approaches to health which take a systems based approach to working on health at the intersection between human, animal (including microbial), and ecological well-being. This is a medical scientific approach based on the principles of interdependence and interrelationships.

A social approach to the study of ecology and health, as imagined in this book, is one that embraces disciplines that range from medicine, law and urban planning through to history, disability studies, and Aboriginal health. Each author seeks to translate knowledge across disciplines to offer some unique insights into how health and illness are produced in the interplay between social and ecological factors. While the authors in this book hail from a variety of disciplines, cultures, countries, and points of entry into this project, all share a strong commitment to advancing the study of health as a social-ecological issue. Working beyond traditional conceptualizations of "the environment" and embracing a complexification of the social-natural interface by using ecological principles are hallmarks of the boundary pushing that is presented in this book. As a reader of this text, you are both scholar and witness to this important frontier work within medical sociology, social epidemiology, and health studies. We invite you to move on from here: to stimulate, refine and move to action conceptual and applied work which brings together scholars, medical, public health, and veterinary practitioners and community stakeholders to conduct work that is intersectoral, policymakers, multistakeholder, and transdisciplinary and which is oriented to social justice, ecological conservation, and sustainability.

The book is organized into four interrelated parts that explore the interfaces between the social and the ecological discussed in this introduction. Part I is titled "Critical Theoretical Engagements with Social-Ecological Approaches to Health Research" and in which the author's use critical social theory to stimulate thoughtful conceptualizations of the project of studying health as a social-ecological issue. Gislason writes broadly about moving toward a social-ecological approach within medical sociology. Dakubo extends the project by describing how some of the key concepts of the ecohealth approach, such as community, participation, and transdiciplinarity, can be enriched through the application of critical theoretical perspectives. Examples of what this critical ecohealth studies approach looks like are drawn from Dakubo's research in rural Northern Ghana. The section is further enhanced by Kinglsey, Townsend, and Henderson-Wilson who draw upon insights offered from Aboriginal people's connection to Country in Australia in order to suggest ways that this knowledge can be used to strengthen human–nature theoretical perspectives, in particular the concepts of biophelia, solastalgia, topophilia, and theories of place.

"Social-Ecological Health Injustices and Paths to Justice" is the title for Part II of this book. Commitments within both medical sociology and ecohealth to not only identify social injustices but to also apply insights to the project of shifting power relations and ameliorating injustices are underscored in this section. The three contributing authors bring together biomedical and philosophical theories, knowledge gained from lived experience and a commitment to redressing social-ecological injustices. Butler opens this section with a chapter on social justice and ecology in Tibet and suggest that Tibetan self-immolation is a reflection of profound social and ecological suffering. Butler argues that this social-ecological issue is fundamentally about health and well-being and should be of great concern for scholars working in the nexus between the social and the ecological. Wolbring then highlights issues of disability and environmental degradation and argues that while these are matters of import themselves, lessons learned from disability health studies can be abstracted to shed light on the disabling practices that anthropocentric capitalistic societies are creating for people globally. This section concludes with Moran, Hollenbeck, and Phoenix who suggest that tools from critical social theory and methodology, namely narrative research, can enrich studies on climate change and health and underscore that issues of equity and justice are central to the project of "giving climate change a face."

Part III of the book focuses, as the title suggests, on "Researching Health in the Nexus between the Social and the Ecological." The chapters in this section link the theoretical project of understanding health as a social-ecological issue with methodological and disciplinary-specific approaches to research. To illustrate the range of possibility for applying this framework to social studies of health, this section offers chapters written by an historian and a medical sociologist through to urban planners and scholars working on issues of natural resource management and forestry. First, Thomson offers a critical historical look at how social movements in the United States over the last several decades have constructed the interrelationship between health and the environment and how these frameworks have been put to use as political tools. Timko then explores the links between HIV/AIDS and forests in Malawi. In particular, she reviews research which shows that forests – as ecosystems and as generators of ecosystem services – in rural Malawi are becoming increasingly important to individuals and communities who are living with and caring for people with HIV/AIDS. Shifting geographical and disciplinary focus, an international team of urban planners, GIS specialists, doctors, and community stakeholders with Gondhalekar as lead author, take an integrated approach to urban planning in Leh Town, Ladakh, India in order to address growing issues of water scarcity and diarrhoeal disease which is on the rise in part due to increases in tourism. Vavotec, Senier, and Bell then look at the health system through the lens of the ecological implications of palliative care for end-of-life cancer patients. This study focuses on the social and ecological constructions of commodity chains within medicalized care settings. Moving to a focus on medical research practice, Chowdhury and Haque reflect on the development of theories of disease causation. Drawing from a case study in Dhaka, Bangladesh, the authors argue that an integrated social-ecological systems approach is needed in order to understand the interplay of factors that cause diseases to emerge. Finally, medical sociologist Vallee puzzles over the lack of integration of environmental health insights into clinical practice guidelines relating to toxicant exposure, such as lead poisoning. The links between ADHD and lead poisoning in children in United States are the focus of Vallee's research.

Part IV, the final section in this book, is entitled "Social-Ecological Approaches to Studying Institutional Change for Health." This section is premised on the view that when studying social-ecological systems understanding the interplay between agents and social structures is an integral component of change making. The section opens with research highlighting the importance of institutional learning. Marsh introduces an organizational learning framework which she has used to think about change within the NHS Sustainable Development agenda in the United Kingdom and offers insights gathered from this research journey. Reflecting on the role of risk governance in the elaboration of a formal environmental health policy arena in Europe, Stassen and Leroy present the tool of Environmental Health Risk Governance. Drawing examples from a Flemish case study, the two authors show how taking a complexity embracing approach to understanding the links between health, the environment and risk can help to develop new policy arenas and governance practices. Finally, Lancaster and Robertson draw our attention to the legal dimensions of water governance through their study of how the Caribbean Region is addressing the links between human and ocean health through the use of legal and policy tools, mandates, and international conventions.

As you will observe, all of these chapters seek to show how health and disease can be studied at the nexus between cultural practices, biological activities, social contexts, and medical and public health systems. A question this book begins to answer is "what are benefits to the social studies of health and illness when 'health' is studied at the nexus between the natural and the social world?" Integrated social-ecological work on health, like the scholarship shared in this book, is important because current trends in illness and disease show that the distances in time and space between human activity, environmental degradation, and illness and disease outcomes are compressing and disease pathways are becoming more direct. As the gap closes, the issues intensify and there is an onus on social thought to address the issues borne of an era where "humanity's use of the biosphere is no longer sustainable" and both the natural and the social worlds are losing resilience which is "the capacity of a system to absorb disturbance and still retain its basic function and structure" (Walker & Salt, 2006, pp. viii–xiii). Pragmatically, this project matters because how health is constructed as a social-ecological issue has significant implications for people's access to health care services, for the choice and range of treatments available and for the sorts of interventions undertaken by health activists, health care and public health practitioners, veterinarians, policy-makers, governmental agencies, and international global health organizations.

Maya K. Gislason
Gillian A. Bendelow

REFERENCES

Bendelow, G. (2009). *Health, emotion and the body*. Cambridge: Polity Press.

Benton, T., & Redcliff, M. (Eds.). (2002). *Social theory and the global environment*. London: Routledge.

Cant, S., & Sharma, U. (1999). *A new medical pluralism: Alternative medicine, doctors, patients and the state*. London: UCL Press.

Gislason, M. (2007). Biophelia and passionate sociology: Tools for making sense of health, illness and disease. *Conference Presentation to the 6th Global Conference on Making Sense of Health, Illness and Disease*, July 9–12, University of Oxford.

Jewson, N. D. (2009). The disappearance of the sick man from medical cosmology, 1770-1870. *International Journal of Epidemiology*, *38*, 622–633.

Pietroni, P. (1992). Beyond the boundaries: Relationship between general practice and complementary medicine. *British Medical Journal*, *305*, 564–566.

Samson, C. (Ed.). (1999). *Health studies: a critical and cross-cultural reader*. Oxford: Blackwell Publishers.

Walker, B., & Salt, D. (2006). *Resilience thinking: sustaining ecosystems and people in a changing world.* Washington, DC: Island Press.

Williams, S. J., & Bendelow, G. (1998). *The lived body: Sociological themes, embodied issues.* London: Routledge.

PART I
CRITICAL THEORETICAL ENGAGEMENTS WITH SOCIAL-ECOLOGICAL APPROACHES TO HEALTH RESEARCH

EXPANDING THE SOCIAL: MOVING TOWARDS THE ECOLOGICAL IN SOCIAL STUDIES OF HEALTH

Maya K. Gislason

ABSTRACT

Purpose – *A crucial contemporary public health issue is the construction and contestation of the relevance of the natural world to human health.*

Approach – *Taking a critical approach, this chapter examines how the natural environment as a health determinant is positioned in relation to the 'social' within social theory generally and social epidemiological studies of health, illness and disease specifically.*

Findings – *This study shows how current constructions of social and natural environmental health drivers contour social approaches to the study of health and proposes an integrated social-ecological approach for generating new contributions of social epidemiology to research on environmentally driven health injuries.*

Originality – *The research breaks ground for further social scientific studies of health and the environment and in particular substantiates the call for an extended notion of the 'environment' using ecological*

Ecological Health: Society, Ecology and Health
Advances in Medical Sociology, Volume 15, 3–22

ISSN: 1057-6290/doi:10.1108/S1057-6290(2013)0000015004

principles. Methodologically, the interdisciplinary reach of this research draws attention to the tensions that arise when working across the medical, natural and social sciences.

Keywords: Health; medical sociology; ecological–social health issues; social epidemiology; dualisms

INTRODUCTION

I know that the molecules in my body are traceable
to phenomena in the cosmos
We are all connected;
To each other, biologically
To the earth, chemically
To the rest of the universe atomically.

Neil deGrasse Tyson (2011)

The social sciences have generated bountiful insights into the ways humans inhabit their world, which range from the sublime through to the horrific. Given the amount of work yet to be done to make the world a better place, for social studies of health it may seem counterintuitive to shift focus away from issues of social inequality and suffering in order to study the interplay between social and natural systems. But then it is also becoming increasingly clear that misery grows in contexts of environmental degradation with profound implications for human health and wellbeing. The fact is that humans inhabit living environments as one species amongst millions bound together within social and ecological webs of life-giving interdependence. An important unit of survival, therefore, is always organism and environment (Bateson, 2000).

Within public health a view that is gaining strength is the importance of maintaining a whole health picture, one where the health of the Earth is considered a public health issue. Proponents argue that losing sight of this perspective can lead to practices which mortgage "the wellbeing of future generations against the greed of present generations, measured in terms of current trends in drawing down natural capital through overconsumption, population growth and growth in the abuse and/or inequitable use of technology" (Soskolne & Bertollini, 1999, p. 21). Concerted appeals for action in the public health sector are coupled with predictions that in the face of declining resources and collapsing natural life support systems current levels of public health will become difficult, if not impossible, to maintain (Corvalan, Hales, & McMichael, 2005; OHI, 2013; WHO/Europe, 2004; WHO, 2007). To move away from producing sickness and towards generating and sustaining health, the natural environment will need to become a core consideration of national health agendas and an organising principle of public health systems (Aguirre, Ostfeld, Tabor, House, & Pearl, 2002; Aron & Patz, 2001; Lang, 2009; Rayner, 2009; Soskolne & Bertollini, 1999).

Grounded firmly in the terrain of critical social theory, this chapter considers how sociological thought broadly, and medical sociological approaches to the study of health and illness specifically, can embrace the co-constitutionality of the natural and the social. This chapter begins by suggesting that sociological research on health and illness needs to address the importance of the ecological dimensions of medical and public health harms if, as a discipline, we are going to make meaningful contributions to research, policy and practice intended to improve states of health and wellbeing globally. Unpacking and harnessing the concepts of the social, the natural and the notion of health *in* the environment are the focus of the three sections that follow. The chapter concludes with thoughts on how to move towards a social-ecological approach to studying health, illness and disease within the context of social studies of health.

TOUCHING THE EARTH: UNSETTLING THE SOCIAL

When health issues leak out of binary frameworks and flow across policy sectors and disciplinary divides, they illuminate the complexity, interactivity and co-determinacy of humanity's relationships with the natural world (Bateson, 2000; Dubos, 1959; Moss & Teghtsoonian, 2008; Samson, 2008). Particularly evocative are those health issues which clearly demonstrate that anthropogenic activities drive environmental events which in turn produce human disease and social suffering. These same issues are also motivating as they underscore that environmentally driven health harms are actually those that humans have the greatest chance of ameliorating as their sources are often human perceptions and practices.

Despite the power people have to change the story of environmental health issues, the processes and systems that produce them are complex and within them issues of social equity and social justice abound. For example, worldwide 13 million deaths could be prevented each year by making the environment healthier and in the least developed countries, one third of death and disease is thought to be a direct result of environmental causes. In material terms, according to the World Health Organisation (WHO) this means that:

- In children under the age of five, one third of all disease is caused by environmental factors such as unsafe water and air pollution.
- Globally, natural disasters such as droughts, floods and storms kill more women than men, and tend to kill women at a younger age.
- Better environmental management could prevent 40% of deaths from malaria, 41% of deaths from lower respiratory infections, and 94% of deaths from diarrhoeal disease – three of the world's biggest childhood killers.
- Water scarcity affects one in three people on every continent of the globe. The situation is getting worse as needs for water rise along with population growth, urbanization and increases in household and industrial uses.

Research from the WHO 'The Health and Environment Linkages Initiative' (HELI) also shows that these trends are likely to intensify as

> rapid, unplanned and unsustainable patterns of urban development are making developing cities focal points for many emerging environmental and health hazards. As urban environments grow, the quality of the urban environment will play an increasingly important role in public health with respect to issues ranging from solid waste disposal, provision of safe water and sanitation, and injury prevention, to the interface between urban poverty, environment and health. (HELI, 2013)

While there is an unequal toll that environmental drivers are having on human life and wellbeing in absolute terms in the developing world, in the developed world and in countries with more robust health care systems and resources, producing healthier environments could also significantly reduce the incidence of cancers, cardiovascular diseases, asthma, lower respiratory infections, musculoskeletal diseases, poisonings and drowning (Prüss-Üstün & Corvalán, 2006; Soskolne & Lee, 2002). In the European region, for example, 20%–24% of all deaths are considered to have the environment as a major contributing factor, which has led to EU-wide initiatives to study health in relation to air quality, chemical safety, environment and health information systems, housing, noise and occupational health, with a focus on children's health (WHO/Europe, 2010).

Both within social theory as well as in medicine and public health, three conceptual constructs that help organise thinking about health in the interplay between the social and the environmental (including their interconnectivity) are the social, the natural and health. Studying the interplay between these three arenas also helps assign responsibility for their prevention and control, particularly in relation to health risks. An unpacking of these concepts is the task of the rest of this section. In this work I have sought to deconstruct these notions by viewing each as a conceptual container so as to highlight how ideas about the relationship between humans and environments have been constructed and contested. Focusing on knowledge formation (via construction and contestation) is particularly relevant when trying to lay a foundation for studying health as a social-ecological issue epistemological and ontological changes are afoot. Increasingly the divide between nature and humans is considered a fiction and the interplay between them is seen to not only constitute but to also transcend social reality (Latour, 1991; Lee, 1997).

The Social

In sociology, 'the social' demarcates a phenomenon created through individual or collective human activity. Typically, the social is imagined in relation to human society and its modes of organisation and draws attention to their roles in producing the social world. Following the atomised view of the world cultivated within Western thought, the honing of 'the social' has involved a scything of the organistic and the natural from conceptualisations of the human world and produced anthropocentric approaches to building social theory. The hegemony of 'purified approaches' within science and education has relegated the natural and the social to discrete theoretical and disciplinary domains (Hinchliffe & Woodward, 2000). However, the boundaries between disciplines and methods have always been somewhat permeable and within sociology the natural world has never completely disappeared. Rather the natural, the biological, and the environmental persist – often as an absent presence and an undeclared force against which the social is constantly being imagined (Nash, 2006). Yet, when the solidity, purity and boundedness of concepts are questioned, space between the binaries is opened up and the relationships and practices that are already and always active within these spaces are brought into the light.

In that 'the social' is constituted through a series of iterative and interactive processes and actors, across the ages social scholars have drawn attention to the material and discursive processes that produce the social world. Latour cautions therefore that 'the social' should be thought of as referring to:

> a stabilized state of affairs, a bundle of ties that, later, may be mobilized to account for some phenomena or another. There is nothing wrong with this use of the word as long as it designates what is *already* assembled together. Problems arise however, when 'social' begins to mean a type of material, as if the adjective was roughly comparable to other terms like 'wooden', 'steely', 'biological', 'economical', 'mental', 'organizational', or 'linguistic'. (Latour, 2007, p. 1)

In contrast to a purified notion of the social used to demarcate the space of humans within positivistic cosmologies, a linguistic interlude shows that the Latin *socius* refers broadly to the interactivity and co-existence of organisms (irrespective of their awareness of interconnection or whether or not their interactions are voluntary) within communities. Relatedly, a group of organisms (humans, plants and animals) sharing common resources (derived from nature, culture and society) constitutes a community (Barnes, 2000). While some social scientists continue to argue that such an integrated definition of 'the social' or 'community' is unwieldy and that the social sciences need to develop and defend their expertise in relation to a bounded notion of the social (Fuller, 2006), this view is becoming outmoded. In addition, within sociology a purified notion of the social has never been a universally accepted approach.

Environmental sociology, for example, is one arena where the environment is treated as integral to the social world. This sub-discipline draws on work by some of sociology's founding thinkers, in particular Karl Marx and Emile Durkheim, who theorised the social without removing it from its larger earthly context. Not only *where* (the space between the social and the natural environment) but also *how* the social is produced is important for these social thinkers. To illustrate, Karl Marx, in collaboration with Friedrich Engels, developed conflict social theory around the notion of interdependence (Foster, 1991; Foster, 2000). Importantly, the concept of interdependence helped them highlight the necessity of human–environmental interactions to the building and maintenance of the social world. Terms they developed, such as dialectical materialism (used to study the phenomena of nature through discourse), helped them express this relationship theoretically. Marx and Engles argued, for example, that 'nature' is an integrated whole which connects phenomena organically and is also an arena within which things are dependent and co-determinate. Interdependence, therefore, is a concept that connects people to the Earth not only emotionally and empathically but also economically, morally and biologically. Stalin, a Marxist scholar, has shown that in fact Marx and Engels drew examples from Darwinian science through to chemistry and medicine when developing their theoretical works. Stalin writes, 'the dialectical method requires that phenomena should be considered not only from the standpoint of their interconnection and interdependence, but also from the standpoint of their movement, their change, their development, their coming into being and going out of being' (1940). Stalin also underscores that in *The Communist Manifesto* (1848), Marx and Engels described the universal interdependence of nations and the alienation of human labour as connected to the estrangement of human beings from nature. For example, in *The Grundrisse* Marx laments how humans and nature are brought into association through relations of production:

> It is not the unity of living and active humanity with the natural, inorganic conditions of their metabolic exchange with nature and hence their appropriation of nature, which requires explanation or is the result of a historic process, but rather the separation between these inorganic conditions of human existence and this active existence, a separation which is completely posited only in relation of wage labor and capital. (Marx in Foster, 2000, p. 1)

The notions of interdependence, connection and associations presented through dialectical materialism, are useful to the project of rethinking the social. One reason is that the notion of interdependence is derived from a holistic observation that through industrialised (now globalised) economic systems the relationship between humans and the natural world is being increasingly fragmented and is leading to various forms of disaffection for people. Certainly a continued concern for health is a clear indicator of the link between rampant economic development, its erosion of ecological integrity and the resilience of earth systems and human health in the face of this onslaught.

Marx was not the only 'founding father' of sociology to think about the interface between the social and the environmental. Durkheim's works is also important when rehearsing the value of 'the social' as a theoretical tool for studying how societies maintain coherence and integrity vis-a-vis the natural world. Durkheim promoted the discipline of sociology as a holistic method for studying societies as large integrated wholes (Durkheim, 1950). Durkheim observed that society is more than the sum of its parts and that it functions as an organic whole, a 'thing-like' entity with its own life and logic. Such an approach opens up possibilities for theorising society as does the notion that change (in social beliefs, actions and architecture) is a 'social fact'; making fluidity and growth fundamental qualities of the social world.

In part due to Durkheim, a functionalist view of society as a system or, in other words, the view of society as an organism, actively invokes the natural world as metaphor but also as a presence (not an absence) shaping the human world. Looking to the future, using ecological principles to expand upon Durkheim's notions of holism, organicity and integrity could strengthen social approachces to studying communities as organic wholes.

The Natural Environment

Presently, however, arguments about the proper way to frame the relationship between the social and the natural abound in social theory. On one hand, and as has already been noted, purified notions of the separateness of the natural from the social have permeated classical social theory and have come to be treated as relatively stable 'social facts'. On the other hand, 'the social' and 'nature' do not describe a domain of reality; rather these concepts operate as 'two collectors that were invented together, largely for polemical reasons' (Latour, 2007, p. 110). Efforts to empty these concepts of meaning are also confounding as is the case within contemporary social theoretical debates which suggest that nothing is really 'natural' anymore and, if so, where does nature end and the social begin? For example, green spaces, parks, gardens, forests and even areas of re-wilded wilderness are shown to be manufactured through human activity in tandem with natural forces (Greening, 2009, p. 164). Also, but for a few noteworthy exceptions, the 'wilderness' or notions of rewilding social spaces are rarely theorised (Benton & Redclift, 2002). In addition, when social thought is 'greened' it is often through the use of natural metaphors ('the natural turn'). While this 'natural turn' enlivens language through references to the natural world it does not necessarily herald the inclusion of biota into conceptualisations of the social to the degree that thinking socially would lead to 'immersion thinking' which is rooted in observations such as the view that:

> We are immersed in life. We breathe it in, we walk on it, we touch it. Each footstep on a fertile lawn or forest mat will send tremors to trillions of bacteria, millions of algae, fungi, and protozoa, and hundreds of insects and worms. The skin on our bodies, when viewed microscopically, is a teaming matrix of tiny caverns filled with bacteria, viruses, and mites ... Life abounds most everywhere inhabited by humans. Life thrives on the nutrients in the soil and water, the oxygen and carbon dioxide in the air, and on the sunlight that ultimately powers most life.

In effect, these theorisations can ensnare 'the natural' within metaphor, putting natural images to the task of thinking about the social (yet again as a purified space) (Code, 2006). Overall, these efforts do little to destabilise the anthropocentrism of much social theory which conceives of the environment as either separate from the social or as a 'setting' for the main event: the human drama.

Theoretical movements calling for the natural environment to be placed centrally within the social imagination have done the most for removing purified notions of the social. Examples include work conducted in the 1960s and 1970s when environmental sociology was revitalised and offered fresh frameworks for theorising the social and the environmental as interactional. For example, Catton and Dunlap offered to social theory 'The New Ecological Paradigm' (NEP) as an alternative to what they termed the 'Human Exemptionalist Paradigm' (HEP) – the purified view of humans as 'exempt' from environmental forces because of their social, cultural, economic and technological prowess (Catton & Dunlap, 1978, 1979, 1980). In the midst of social uprisings, Eco-Marxism was also developed by using social conflict theories to conduct materialist analyses of environmental conflicts in the 1970s. The societal–environmental dialectic proposed by Schnaiberg was a substantive contribution of this neo-Marxist movement as it observed that governments and industries continued to privilege economic growth over environmental integrity and even over the health and wellbeing of the populace, unless and until economic and political commitments to sustainable development were seriously called into question (Schnaiberg, 1980).

In the 1980s and 1990s, an integrationist approach was developed, which Buttel and Humphrey (2002) refer to as 'the sociology of environmental reform'. Departing from more radical, anti-establishment theoretical movements of the 1970s and 1980s, this approach engendered collaboration. Ulrich Beck (1995) is a social theorist whose work reflects the ethos of his time. He develops a project of reflexive modernisation emphasising the external 'environmental' character of hazards as well as their 'suppressed sociality' – the social drivers behind them. He cautions that *post-histoire* thinking, the view of the immortality of human societies, is in fact a thinking that produces an 'end-of-societal-history' thesis because it fails to consider the role of humans in producing the hazards of the time. He posits that the survival of the social habitus (but not the natural world *per se*) will be dependent upon overcoming the principle of 'organised non-liability' which denies the presence of hazards ('makes them mute') and also makes it difficult to assign responsibility for ameliorating the risks they pose. Rallying optimism, Beck (1995) argues that the way forward is to build an ecological democracy based on principles of accountability.

A quintessential approach of environmental social theory emerging out of the 1990s and 'noughties', however, was Ecological Modernization Theory (EMT). Described as 'the social scientific interpretation of environmental reform processes at multiple scales in the contemporary world' (Mol, Sonnenfeld, & Spaargaren, 2009, p. 1), EMT focused on how 'environmental interests have become incorporated into more and more aspects of social relations and institutions, as well as into contemporary human values, cultures and everyday practices' (Mol et al., 2009, p. 1). EMT, its proponents argue, replaces the concept of sustainable development (SD) because to SDs political and economic preoccupations it adds analytical and sociological awareness (Spaargaren, Mol, & Buttel, 2000, p. 333). Appreciating that movements ebb and flow and therefore ideologies and social practices change across uneven trajectories, critics of EMT echo a question posed to earlier social and theoretical movements: can EMT change structures in ways that earlier movements could not? Their concern is that in the 'new green wave' of the 21st century incorporation will continue to be *ad hoc*, even in contexts where there is the capacity to respond comprehensively, such as in advanced capitalist societies (Coffey & Major, 2005; Goldsmith, Allen, Allaby, Davoll, & Lawrence, 1972; Lafferty & Hovden, 2002).

While only a decade has passed since the most recent of the environmental–theoretical discourses discussed above were presented (perhaps not enough time for social theory to be significantly revised) pressing real world issues suggest it is time again for sociology to consider the relationship between humans and the natural world. Global environmental change, the links between ecological degradation and civil conflicts, increasing environmental illnesses, the exhaustion of natural resources and the rise of industrialised meat production are all traceable causes of human illness and suffering. In fact, it is increasingly difficult to see human beings as standing outside of the natural world and as disconnected from non-human bodies (Cochrane, 2010):

> in a world where risk and uncertainties seem to be piling up on top of one another, if we are to make improvements to people's lives, not to mention other species lives, then any attempts to understand nature without society, or to understand society without nature, will prove insufficient to the task. (Hinchliffe & Woodward, 2000, pp. 3–4)

There is work waiting for medical sociologists, environmental sociologists and social theorists in other disciplines to undertake, tasks that could be informed by thinking about health as embedded within the natural world (Chivian & Bernstein, 2008; IPCC, 2013; Corvalan et al., 2005) as described in the following section. Rethinking the interdependence of the social, the natural and the biological is already being treated as a pertinent project. These frameworks are being called for within emerging health and environment policy arenas, they are needed to catalyse social equity movements and to reformulate philosophical and humanist assumptions about humans' place on the planet.

Health in the Environment

As is the preoccupation of this chapter, an explicit reason to rethink 'the social' and 'the natural' within social theory is to rethink health as a social-ecological construct. Like 'the social' and 'the natural,' 'health' (as well as related concepts such as wellbeing, illness and disease) is 'as much a social construct as a biological characteristic. [Health] is the product of a complex interaction of different factors: this is true at both individual and population levels' (Sengupta, 2009, p. 19). Health is also different in that it is contoured not only by cultural frameworks, social forces and disciplinary contexts replete with their respective traditions and values (Rose, 2006; Shilling, 2008; Turner, 2008) but also by personal biology and experiences which are lived through the intimacy of one's own body (Moss & Teghtsoonian, 2008; Williams and Bendelow, 1998).

Making the theoretical connection between an 'environment' out there and 'health' in the body (or groups of bodies if a population is affected) can be difficult. Thinking biomedically about the intersection between environmental drivers, disease emergences, injury events and anthropogenically induced environmental destruction can also seem a rather contradictory undertaking (Kroll-Smith, Brown, & Gunter, 2000; Moss & Teghtsoonian, 2008). One source of the difficulty is that understanding health in the environment requires working in the tension between abstraction and specificity and at the interface between forces traditionally divvied up as the turf of medicine and public health, the natural sciences and, the social sciences and humanities. Even the WHO's universal definition of health, through its declaration, somewhat confounds the clear demarcation of what health is and who is responsible for it as it is imagined as a 'state of complete physical, mental and social wellbeing and not merely the absence of disease and infirmity' which requires that both an art and a science approach to public health is cultivated (WHO, 1946). Reminiscent of the Bateson observation of the environment–organism interface, the WHO definition also highlights that health can be supported, improved or detracted from depending on the setting (Stewart & Jarvis, 2009, p. 168).

As might be expected, social health settings and relatedly social determinants of health tend to be the focus of social studies of health and are defined as:

> The conditions in which people are born, grow, live, work and age, including the health system. These circumstances are shaped by the distribution of money, power and resources at global, national and local levels, which are themselves influenced by policy choices. The social determinants of health are mostly responsible for health inequities – the unfair and avoidable differences in health status seen within and between countries. (CSDH, 2008)

According to the WHO's approach to health promotion, two other public health milieu, the built and the natural environment, are also important health settings and correspondingly each are described as a health determinant. Typically, environmental health determinants are used to study the natural environment–human health interface and it is the concept of 'environmental health' which has emerged as a principal framework through which public health work links human health to environmental determinants. However, as I showed with the concepts of the social and the environmental, what 'environmental health' means is also a highly variable construct. At the broadest level, environmental health determinants are understood to be

> all the physical, chemical, and biological factors external to a person, and all the related factors impacting behaviours. [Environmental health] encompasses the assessment and control of those environmental factors that can potentially affect health. It is targeted towards preventing disease and creating health-supportive environments. This definition excludes behaviour not related to the environment, as well as behaviour related to the social and cultural environment, and genetics. (WHO, 2007)

Environmental health determinants have been effectively used to show how environmental degradation can be linked to human health injuries. The WHO, for example, holds environmental hazards responsible 'for as much as a quarter of the total burden of disease worldwide, and more than one-third of the burden among children' (Prüss-Üstün & Corvalán, 2006). Recent studies show that environmental factors influence 85 out of the 102 categories of diseases and injuries listed in the World Health Report (WHO, 2007). What is clear when studying both social and environmental determinants of health is that within their theoretical definitions, in the field when using epidemiological instruments to measure them and in the knowledge an analysis of their data produce, the social and the environmental are shown to be interactional. There tends to be gaps, however, between reality as it is unfolding and the tools and structures available for expressing the interactivity between health issues that cross boundaries. Events which catalyse change by highlighting the interplay between the social and the ecological in relation to health issues are, however, inspiring a rethinking of the frameworks and methods used to study health issues emerging in the nexus between social and ecological systems.

Exactly how social–environmental health phenomena are becoming important (again) also matters. Produced through science and interpreted within biomedical frameworks, research on environmental health injuries is being cited as a significant health phenomenon of the 21st century (Corvalan, et al., 2005; IPCC, 2013; McMichael et al., 2003; Soskolne, 2008). However, it is when framed as health hazards (potential threats which can be sources of a health injury) and risks ("a quantified estimation of that threat") (Stewart & Jarvis, 2009, p. 169) that the environment often comes to matter the most. In part the notion of the 'environment' as risk has been easy to fold into contemporary public health discourses because risk thinking is increasingly central to the Weltanschauung (worldview) of modern culture (Beck, 1992; Giddens, 1999). As is the case with risk society, modern health systems also organise themselves around ideas about risks, with environmental concerns becoming more pertinent to issues of safety or preparing for the future (Beck, 1995; Giddens, 2009). Myriad initiatives are trying to develop measurements for 'environmental' risks or hazards and how the environment is being operationalised within them is an important aspect of this work. Making sense of both the direct and indirect mechanisms which link the social and the environmental through health phenomena is integral to these projects.

Of course, these issues also have explicit public health implications (Eisenberg et al., 2007). Within the context of public health, practitioners look for empirical things such as diseases and syndromes 'that might have environmental causes or which are modified, either positively or negatively, by an environmental factor' (Stewart & Jarvis, 2009, p. 170). The 'source-pathway-receptor' link is typically the formula used to ascertain if there is an environmental driver at work and the rule is that if one or more of the three is missing there can be no stated threat to health. Public health also has a practical mandate when it comes to health responses as it is directed to ameliorate health damages. Therefore, a definition of the environment must synchronise with the mandates and constraints of public health interventions (see Gislason, 2010), which means that often the environment is defined pragmatically. The criteria can be, for example, that a natural environment be 'amenable to change' and 'reasonably modifiable' so that a public health intervention can be rationalised (Prüss-Üstün & Corvalán,

2006, p. 23). Such an approach is realistic but is also a compromise as it facilitates the folding of the natural into the social in a way that maintains a focus on the 'natural' environment within the purview of health based on the criteria that it be 'fit for (social) purpose' as a site of intervention. An effect of this approach is that within the context of public health, 'health in the environment' becomes something more than a biological, natural or social phenomena – it becomes an enviro-social-biological administrative construct (Gislason, 2013). To study health at the nexus between the social and the natural environment, therefore, is also to study administrative, disciplinary, and socio-cultural activities as they are produced in relation to natural processes and assembled together to generate a framework through which humans' dependence on the natural environment is understood.

Moving beyond the 'cult of humanity' (Dew, 2007) within public health and medicine as well as within the study of the social world is challenging. When the goal is to improve health and wellbeing within both social and ecological systems, however, instruction can be taken from environmental epidemiology which has developed a rigorous approach to both identifying the environment as an external factor and measuring how it impacts people where they live, work and play. Granted, social factors mediate the potential impact of the natural environment on human health as the environment affects people's general states of vulnerability and susceptibility (measured through socioeconomic status, for example). However, the environmental epidemiological oeuvre is that it considers two environmental vectors: (1) proximate (downstream) determinants of health – those closely related in time and space to the injuries they produce – and (2) distant (upstream) determinants, which are far apart in time and space from the harms they produce. Specifically, (1) proximate environmental health determinants can be biological agents in the air, water and soil while (2) distant health determinants are often social in origin, such as:

> policies that drive current levels of population growth, consumption and waste issues, and the uses of technology. For example, the environmental, transboundary transport of contaminants through the food chain has resulted in global chemical contamination. Other transboundary issues include acid precipitation, ozone, greenhouse gasses, and hazardous wastes. Global ecological integrity (i.e., the ability of life-support systems to sustain themselves in the presence of polluting forces) and global change (including concerns about climate change from global warming, ozone depletion, and the loss of biodiversity) are also distant health determinants. (Soskolne & Lee, 2002)

Thinking about the environment as a vector is useful to the project of thinking about the environment in a novel way within social studies of health. Bringing the natural environment centrally into the health equation as an agent and amplifying its range by considering both proximate and distant scales of time and space are effective ways to draw attention to the social forces driving environmental health injuries. Comparing this dynamic view of the environment and health to static ones often at work within social epidemiology raises questions about how notions of the social, the environment and health are produced within specific social practices and disciplinary frameworks.

MOVING TOWARDS A SOCIAL-ECOLOGICAL APPROACH

Constructing the environment through frameworks which juxtapose the social with the biological or the natural produce approaches which cannot easily study interconnections and interdependencies between the social and the ecological. Such dualistic, frameworks are antiquated in the current age of complexity, as is poignantly illustrated by health issues emerging in relation to global environmental change. Here, ecological concepts emerge as a potent conceptual framework for social theorising. Concepts such as ecosystems extend the view of the social by drawing attention to interrelationships and exchanges between humans and nature and in this way connect back to the Latin notion of the '*socius*', which is community. Thinking about the social as an assemblage, a community of biotic and abiotic components, enriches the metaphors available within social thought. More to the point, however, it improves capacity to analyse the contexts within which the social is produced and to engender a realistic accounting of the materials upon which the social is built, which include not only ideologies and discourses but also biotic matter, chemical compounds, and natural elements such as air and water, all housed along with humans in the meta-context of the ecosphere. In terms of theory, this approach requires working at the interplay between the material and the discursive and in the nexus between social and environmental determinants of health.

Integrated thinking about health as a social-ecological issue breaks ground for further social scientific studies of health and the environment and substantiates the call for an extended notion of 'the social' as well as 'the environment' through using ecological concepts. Calling for a rethinking of the social within the context of social theory should extend to health research and, therefore, to a rethinking of the 'social' within social epidemiological studies of health, illness and disease. Of course, the marginalisation of the environmental within the social sphere and the subjugation of animal health to human health is maintained by relations of power which produce a distorted notion of the autonomy of social communities from natural ones and humans from animals. 'Rethinking the prevailing definition of what counts as sociology [and] enlarging that definition considerably' (Agger, 1991, p. 126) is essential to the project of studying concepts traditionally disconnected from one another within the social sciences, veterinary sciences, public health and medicine.

This integrated approach to health research is emerging at a time when 'modern life has taught us that both nature and humankind are more complicated than the dialectical notions of the nineteenth century supposed' (Raskin in Lee, 1997, p. 17). Taking research and practice paths that reject modern theoretical preoccupations contoured by deterministic, linear explanations within the individual and society matrix remind us there was also a time when 'science was not easily divided between the "natural" and the "social" [and] when there was a curiosity about [nature] and systematic inquiry into all aspects of the world in which [people] lived was conducted, generally without specialization' (Lee, 1997, p. 16). Transdisciplinary research, hybrid disciplines and creolized scholars are among those reaching across disciplinary boarders in order to work within the richness and plurality of perspectives available to all of us (Czarniawska, 2007). When 'health' is placed at the centre of inquiry the myriad ways in which the social and the natural are co-constitutive can become more visible to a range of knowledge producers and users. Pragmatically, researching health in the society-ecology nexus is a response to the view that "only collaboration between different sectors can protect human health from risks from a hazardous or contaminated environment" (World Health Organization Regional Office for Europe, 2013). How to work together to achieve these goals is a real and pressing set of challenges for social, medical and public health organisations around the world.

ACKNOWLEDGEMENTS

Thank you to professors Jackie Cassell and Gillian Bendelow as well as to Dr. Cameron Duder and Grant Sheppard for helping me think through and express the work in this chapter. This research was conducted through the generous support of a doctoral fellowship from the Commonwealth as well as from the Social Sciences and Humanities Research Council of Canada.

REFERENCES

Agger, B. (1991). Critical theory, poststructuralism, postmodernism: Their sociological relevance. *Annual Review of Sociology*, *2011*(August), 105–131.

Aguirre, A. A., Ostfeld, R. S., Tabor, G. M., House, C., & Pearl, M. C. (2002). *Conservation Medicine: Ecological health in practice*. Oxford: Oxford University Press.

Aron, J. L., & Patz, J. (2001). *Ecosystem change and public health*. Baltimore: John Hopkins University Press.

Barnes, P. (2000). *Pie in the sky*. Washington, DC: Corporation for Enterprise Development.

Bateson, G. (2000). *Steps to an ecology of mind*. Chicago, IL: University of Chicago Press.

Beck, U. (1992). *Risk society: Towards a new modernity*. New Delhi: SAGE Publications.

Beck, U. (1995). *Ecological politics in an age of risk*. Cambridge: Polity Press. Retrieved from http://books.google.co.uk/books?hl=en&lr=&id=H9QxLtkfNhEC&oi=fnd&pg=PP8&dq=ulrich+beck+risk+society+ecology&ots=FAZOTelTFK&sig=0O1cUOt2CUZPzuPK8ul2DZF2LQE#v=onepage&q&f=false

Benton, T., & Redclift, M. (Eds.). (2002). *Social theory and the global environment*. London: Routledge.

Buttel, F. H., & Humphrey, C. R. (2002). Sociological theory and the natural environment. In R. E. Dunlap & W. Michelson (Eds.), *Handbook of environmental sociology* (pp. 33–69). Westport: Greenwood Press.

Catton, W., & Dunlap, R. E. (1978). Paradigms, theories, and the primacy of the HEP-NEP distinction. *The American Sociologist*, *13*(November), 256–259.

Catton, W., & Dunlap, R. E. (1979). Environmental sociology. *Annual Review of Sociology*, *5*, 243–273.

Catton, W., & Dunlap, R. E. (1980). A New Ecological Paradigm for post-exuberant sociology. *American Behavioral Scientist*, *24*(September/October), 15–47.

Chivian, E., & Bernstein, A. (Eds.). (2008). *Sustaining life: How human life depends on biodiversity*. New York, NY: Oxford University Press.

Cochrane, A. (2010). Environmental ethics. In J. Frieser & B. Dowden (Eds.), *Internet encyclopedia of philosophy*. Retrieved from http://www.iep.utm.edu/envi-eth/. Accessed on July 31, 2013.

Code, L. (2006). *Ecological thinking: The politics of epistemic location*. New York, NY: Oxford University Press.

Coffey, B., & Major, A. (2005). Towards more integrated natural resource management in Victoria: Possible elements of an integrated state-wide policy framework. *Australasian Journal of Environmental Management*, *12*(September/Supplementary Issue), 29–38.

Corvalan, C. F., Hales, S., & McMichael, A. (2005). *Ecosystems and human well-being health synthesis*. A Report of the Millennium Ecosystem Assessment, World Health Organization, Geneva, Switzerland.

CSDH. (2008). *Closing the gap in a generation: Health equity through action on the social determinants of health*. Final Report of the Commission on Social Determinants of Health. Geneva, World Health Organization.

Czarniawska, B. (2007). *On creole researchers, hybrid disciplines and pidgin writing* (pp. 59–67). Presented at the Inter: A European Cultural Studies Conference, Göteborg University, Sweden: Linköping University Electronic Press, Linköpings universitet.

deGrasse Tyson, N. (2011). *Symphony of science: We are all connected*. Retrieved from http://www.youtube.com/watch?v=XGK84Poeynk%3e. Accessed on September 28.

Dew, K. (2007). Public health and the cult of humanity: A neglected Durkheimian concept. *Sociology of Health & Illness*, *29*(1), 100–114. doi:10.1111/j.1467-9566.2007.00521.x

Dubos, R. (1959). *Mirage of health: Utopias, progress and biological change*. New Brunswick: Rutgers University Press.

Durkheim, E. (1950). *The rules of sociological method*. Glencoe: The Free Press.

Eisenberg, J. N. S., Desai, M. A., Levy, K., Bates, S. J., Liang, S., Naumoff, K., & Scott, J. C. (2007). Environmental determinants of infectious disease: A framework for tracking causal links and guiding public health research. *Environmental Health Perspectives*, *115*(8), 1216–1236.

Foster, J. B. (1991). *The vulnerable planet: A short economic history of the environment*. New York, NY: Monthly Review Press.

Foster, J. B. (2000). *Marx's ecology: Materialism and nature*. New York, NY: Monthly Review Press. Retrieved from http://tifcss.org/graphics/marx_ecology.pdf

Fuller, S. (2006). *The new sociological imagination*. London: SAGE Publications.

Giddens, A. (1999). Risk and responsibility. *Modern Law Review*, *62*(1), 1–10.

Giddens, A. (2009). *The politics of climate change*. Cambridge: Polity Press.

Gislason, M. (2013). West Nile virus: The production of a public health pandemic. *Sociology of Health and Illness*, *35*(2), 188–199.

Gislason, M. K. (2010). Sounding a public health alarm: producing West Nile Virus as a newly emerging infectious disease epidemic. In A. Mukherjea (Ed.), *Understanding Emerging Epidemics: Social and Political Approaches* (pp. 77–100). Bingley: Emerald Group Publishing Ltd.

Goldsmith, E., Allen, R., Allaby, M., Davoll, J., & Lawrence, S. (1972). *A blueprint for survival*. Harmondsworth: Penguin.

Greening, K. (2009). Public health and the natural environment. In F. Wilson & M. Mabhala (Eds.), *Key concepts in public health* (pp. 163–167). London: Sage.

Hinchliffe, S., & Woodward, K. (Eds.). (2000). *The natural and the social: uncertainty, risk, change*. London: Routledge.

IPCC. (2013). *Intergovernmental panel on climate change*. Retrieved from http://www.ipcc.ch/working_groups/working_groups.shtml. Accessed on July 31.

Kroll-Smith, S., Brown, P., & Gunter, V. J. (2000). *Illness and the environment: A reader in contested medicine*. New York, NY: New York University Press.

Lafferty, W., & Hovden, W. (2002). *Environmental policy integration: Towards an analytical framework* (No. 72). (P. for R. and D. for a S. Society, Trans.). Oslo: University of Oslo.

Lang, T. (2009). Reshaping the food system for ecological public health. *Journal of Hunger and Environmental Nutrition*, *4*(3), 315–335.

Latour, B. (1991). The impact of science studies on political philosophy. *Science, Technology & Human Values*, *16*(1), 3–19.

Latour, B. (2007). *Reassembling the social: An introduction to actor-network theory*. Oxford: Oxford University Press.

Lee, M. (1997). From enlightenment to chaos: toward nonmodern social theory. In R. Eve, S. Horsfall & M. Lee (Eds.), *Chaos, complexity and social theory: Myths, models and theories* (pp. 15–29). London: SAGE Publicatios.

McMichael, A. J., Campbell-Lendrum, D. H., Corvalan, C. F., Ebi, K. L., Githeko, A. K., Scheraga, J. D., & Woodward, A. (2003). *Climate change and human health: From risks to responses*. Geneva: World Health Organisation.

Mol, A. P. J., Sonnenfeld, D. A., & Spaargaren, G. (2009). *The ecological modernisation reader: Environmental reform in theory and practice*. London: Routledge.

Moss, P., & Teghtsoonian, K. (2008). *Contesting illness: Processes and practices*. Toronto: University of Toronto Press.

Nash, L. (2006). *Inescapable ecologies: A history of environment, disease and knowledge*. Berkeley, CA: University of California Press.

One Health Initiative (OHI). (2013). Missions Statement. Retrieved from http://www.onehealthinitiative.com/mission.php. Accessed on July 31, 2013.

Prüss-Üstün, A., & Corvalán, C. (2006). *Preventing disease through healthy environments: Towards an estimate of the environmental burden of disease* (pp. 86–89). France: World Health Organization. Retrieved from http://www.who.int/quantifying_ehimpacts/publications/preventingdisease.pdf

Rayner, G. (2009). Conventional and ecological public health. *Journal of Public Health*, *123*(9), 587–591.

Rose, N. (2006). *Politics of life itself: Biomedicine, power and subjectivity in the twenty-first century*. Princeton: Princeton University Press.

Samson, C. (1999). *Health studies: A critical and cross-cultural reader*. Oxford: Blackwell Publishers.

Samson, C. (2008). The rule of terra nullius and the impotence of international human rights for indigenous peoples. *Essex Human Rights Review*, *5*(1).

Sengupta, S. (2009). Determinants of health. In F. Wilson & M. Mabhala (Eds.), *Key concepts in public health* (pp. 15–20). London: Sage.

Schnaiberg, A. (1980). *The environment: From surplus to scarcity*. New York, NY: Oxford University Press. Retrieved from http://media.northwestern.edu/sociology/schnaiberg/1543029_environmentsociety/index.html. http://www.socialresearchmethods.net/Gallery/Neto/Envsoc1.html

Shilling, C. (2008). *Changing bodies: Habit, crisis and creativity*. London: SAGE Publications.

Soskolne, C. L. (2008). *Sustaining life on earth: environmental and human health through global governance*. Plymouth, MA: Lexington Books.

Soskolne, C. L., & Bertollini, R. (1999). *Global ecological integrity and "sustainable development": Cornerstones of public health: A discussion document*. Geneva: World Health Organization. Retrieved from http://www.colinsoskolne.com/documents/WHO-1999_Discussion_Document.pdf

Soskolne, C. L., & Lee, E. S. (2002). Environmental determinants of health. In L. Breslow (Ed.), *Encyclopedia of public health* (Vol. 2). Gale Cengage. Retrieved from http://www.enotes.com/environmental-determinants-health-reference/. Accessed on July 31, 2013.

Spaargaren, G., Mol, A. P. J., & Buttel, F. H. (Eds.). (2000). *Environment and global modernity*. London: Sage.

Stalin, J. (1940). *Dialectical and historical materialism*. In J. Series Stalin (Ed.), Berkley, CA: International Publishers. Retrieved from http://books.google.com/books/about/Dialectical_and_historical_materialism.html?id = wvS4AAAAIAAJhttp://www.marxists.org/reference/archive/stalin/works/1938/09.htm

Stewart, A. G., & Jarvis, R. (2009). Environment and public health. In F. Wilson & M. Mabhala (Eds.), *Key concepts in public health* (pp. 168–173). London: SAGE Publications Inc.

The Health and Environment Linkages Initiative (HELI). (2013). *The urban environment*. World Health Organization, Geneva Switzerland. Retrievable from http://www.who.int/heli/risks/urban/urbanenv/en/. Accessed on July 31.

Turner, B. S. (2008). *The body and society: Explorations in social theory* (3rd ed.). London: SAGE Publications.

WHO. (1946). *Preamble to the constitution of the World Health Organization as adopted by the international health conference*. Presented at the International Health Conference, World Health Organization. Retrieved from http://www.who.int/suggestions/faq/en/index.html

WHO. (2007). *The world health report 2007 – A safer future: Global public health security in the 21st century*. (Series WHO, Ed.). Geneva: World Health Organization. Retrieved from http://www.who.int/whr/2007/en/index.html

WHO/Europe. (2004). *Fourth ministerial conference on environment and health: Children's environment and health action plan for Europe*. Budapest, Hungary, June 23–25. Retrieved from http://www.euro.who.int/. Accessed on July 31, 2013.

WHO/Europe. (2010). *The journey to Parma: A tale of 20 years of environment and health action in Europe*. WHO Regional office for Europe, Copenhagen. Retrieved from http://www.euro.who.int/__data/assets/pdf_file/0016/104515/E93535.pdf. Accessed on July 31, 2013.

Williams, S., & Bendelow, G. (1998). *The lived body: Sociological themes, embodied issues*. London: Routledge.

World Health Organization Regional Office for Europe. (2013). *Environment and health*. Retrieved from http://www.euro.who.int/en/what-we-do/health-topics/environment-and-health. Accessed on July 31.

TOWARDS A CRITICAL APPROACH TO ECOHEALTH RESEARCH, THEORY AND PRACTICE

Crescentia Dakubo

ABSTRACT

Purpose – *Ecosystem-based approaches to human health, also known as* Ecohealth approaches, *fall in an emerging field of study that focuses on understanding how social, political and economic factors shape people–environment interactions and consequently influence human health. The approach integrates scientific and local perspectives to develop interventions aimed at improving human health through sustainable ecosystem management.*

Three principles are central to traditional ecohealth approaches: (1) transdisciplinarity which integrates the expertise of professionals from the natural, health and social sciences with those of other stakeholders; (2) community participation; and (3) social equity and gender relations.

However, the extent to which these concepts benefit from rigorous theoretical analysis is still very limited. As an emerging field of study, the use of poststructuralist, feminist, postcolonial, and political ecology perspectives is still limited. This chapter illustrates the application of critical theory to the conceptual and methodological tools used in ecohealth research and practice.

Ecological Health: Society, Ecology and Health
Advances in Medical Sociology, Volume 15, 23–43

ISSN: 1057-6290/doi:10.1108/S1057-6290(2013)0000015005

Approach – *Critical theory is used to examine people–environment interactions, transdisciplinarity, community participation, and social equity and gender relations. The discussion is informed by an ecohealth project conducted in the Upper West region of Ghana.*

Research implications – *As an emerging field of study, ecohealth is yet to benefit from rigorous critical theoretical frameworks that are used to examine health and environment issues in other subdisciplines. This chapter illustrates the application of these perspectives to ecohealth.*

Originality/value – *This is the first chapter that outlines a critically informed approach to ecohealth research and practice.*

Keywords: Critical theory; ecohealth; community participation; transdisciplinarity; gender and social equity; Ghana

INTRODUCTION

Ecosystem-based approaches to human health, also known as *ecohealth approaches to human health*, is an emerging field of study that bridges thinking in the social, health and natural sciences. These approaches recognize the interdependencies between human health and ecosystem health, and seek to understand the social, political, and economic factors that influence such interaction and the resulting health implications (Dakubo, 2010; Forget & Lebel, 2001). Ecohealth-based studies use participatory and transdisciplinary processes to work with various stakeholders and in the process develop interventions that aim to simultaneously improve human health and promote sustainable ecosystem management practices.

There are a number of conceptual elements that underlie ecohealth. First, is the recognition that human beings are integral to ecosystems, and ecosystem management decisions must always take human health concerns into account. Second, ecohealth calls for the integration of expertise from the natural, social, and health sciences with traditional knowledge systems to allow for a comprehensive understanding of environment and health phenomena. Such transdisciplinary integration of knowledge systems is intended to contribute to a better understanding of the problem at hand and increase the likelihood of developing interventions that are socially and ecologically sound. Third, ecohealth encourages the use of participatory research approaches that involve beneficiaries of a study in all aspects of the research process. Such active involvement allows for a better understanding of the causal processes through which human health is influenced by environmental conditions. Lastly, the ecohealth approach calls for the adoption of social equity and gender lens in its investigations, with the realization that exposure to environmental hazards is mediated by structural and institutional factors, and are not uniformly experienced by those affected.

However, the ontological and epistemological assumptions underlying these principles and their deployment remain undertheorized. For example, the notion that people are integral to the ecosystem and therefore that environmental management decisions should take human health issues into account raises complex questions about people–environment relations that are often theorized from simplistic neo-Malthusian perspectives. Such theorizing sometimes leads to the attribution of environmental health problems to individual level factors, such as people's reckless behavior, and "inappropriate" land use practices, and relieves of accountability the role of power relations, societal, and institutional factors in producing these environmental health risks. Also, there tends to be an uncritical adoption of *a priori* notions of environmental and health problems without evaluating how these problems came to be constructed as such, by whom and for whom. Concepts such as "community," "participation," "gender relations," "ecosystem degradation," and "human health" are sometimes constructed to suit the needs of the namers and should be "deconstructed" to reveal their social and historical locations. The failure to adopt a critical stance to analyzing environmental health problems risks producing interventions that will not only fail to address the underlying causes of ecological and public health problems, but also augment existing social inequities and vulnerabilities that we set out to address.

Examining issues from such perspectives draws on conceptual tools in poststructuralism, postmodernism, and critical theory. All three theories are critical of positivism and provide new ways to theorize and study contemporary society (Agger, 1991). As critiques of positivism, they interrogate taken-for-granted assumptions about the ways in which science is written, read, and interpreted. They reject positivist notions of the possibility to represent the world without presuppositions, arguing that such representations are both politically imbued and philosophically impossible, and every knowledge must be contextualized by its historical and cultural nature (*ibid.*). While it is often difficult to cleanly separate the three overarching theories of poststructuralism, postmodernism, and critical theory, they differ with respect to their origins and areas of emphasis. I will discuss the central tenets of the theories and subsequently illustrate their application to the key concepts of ecohealth.

CENTRAL TENETS OF CRITICAL THEORETICAL PERSPECTIVES

Critical theory emerged from the German Frankfurt School, and while comprised of different strands of theory, tends to be heavily influenced by the work of Habermas (Agger, 1991). Critical theory functions effectively as a critique of ideology and relations of social dependencies. Critical theory encourages a systematic critique of ideologies through critical self-reflection to unravel an understanding of the social and institutional constraints of a problem, and then use this knowledge to take the necessary action toward change. This concept of critique, reflection, and action underlies the philosophical assumptions of participatory action research (PAR) (Freire, 1971). This perspective is informed by Paulo Freire's concept of praxis of how action and reflection reinforce each other, leading to critical consciousness and transformative power (Freire, 1972). Critical theorists also believe that knowledge is steeped in historical contexts and is shaped by everyday life experiences (Habermas, 1971). The work of phenomenologists has also legitimated experience as the basis of knowing. People's feelings, emotions, and experiences collected through narratives and group processes serve as the basis for new knowledge (Tandon, 1996). For example, in rural Northern Ghana, Dakubo (2004) used a PAR approach to involve community members and a transdisciplinary research team in a cyclical process of collective reflection and analysis of the environmental health conditions in the community. Through this exercise, community members were able to critically evaluate the underlying factors influencing community health and devised strategies to respond to them, leading to the development of a healthy community. Critical theorists argue that by conforming to dominant ideologies, and social and economic rules, positivism is not only a flawed philosophy, but also an uncritical political theory, that exempts the role of power structures in shaping problems. This perspective is particularly important when examining how power relationships shape people's interactions with the environment or access to health resources.

Poststructuralism, on the one hand, is seen as a theory of knowledge and language, and is dominated by the work of French theorists such as Derrida.

Poststructuralists take the critique of science further by illustrating how the writings of mainstream ideologies (science) are not neutral but rather an attempt to use language and text to constitute reality and perpetuate certain discourses and subject positions. According to Derrida, every text should be seen as a contested terrain, in the sense that it conceals conflicts within it, and such conflicts can only be revealed through contextualization. Poststructuralists are therefore averse to the use of definitions and categories. According to Derrida (1976), every definition "deconstructs" itself, further unraveling as we probe into its foundational assumptions and meanings. Feminists scholars have written against such categories as they serve to label groups in particular ways or construct people as deviating from the norm, based on which they can then diagnose and treat them. The goal of poststructuralist analysis is to try and deconstruct any hidden assumptions embedded in scientific writings and to lay bare their social, historical, and political connotations. This exercise is intended to politicize and democratize science and allow readers to be informed of the assumptions of the texts they read and to contest these if necessary.

Postmodernism, which is closely affiliated with poststructuralism, is a theory of society, culture, and history, and is influenced by the writings of Foucault, Barthes, Lyotard, and Baudrillard (Agger, 1991). One key tenet of postmodernism is the rejection of totalizing discourses of history and society, what Lyotard calls *grand narratives*. These grand narratives try to explain the world in terms of patterned relationships by overgeneralizing and extrapolating local conditions to global scenarios. According to Lyotard we cannot tell large stories about the world but only small stories that bring to bear the heterogeneous "subject" positions of individuals and other population groups. Postmodernists therefore rejects the notion of a universal science and the associated totalizing claims, instead relying on particular modes of knowledge to be defined by the multiplicity of people's subject positions. Postmodernists also examine the social world from the multiple perspectives of social groups and individuals filtered through race, age, gender, and class. They prefer to interpret knowledge based on people's everyday experiences and not from a central universal source. Foucault writes about the close association between knowledge and power, and the need to trace knowledge to the discourses/practices that frame the knowledge from within them (Foucault, 1980).

What then are the implications of critical theory for the conceptual and methodological tools used in ecohealth? Critical theory forces researchers to be self-reflective of the taken-for-granted assumptions that are embedded in the concepts and methodologies they use. As an emerging field of study, ecohealth stands to benefit from engaging with such critical theoretical perspectives, so as to become a more theoretically informed field of study. Next, I will illustrate how we might begin to apply these theoretical perspectives in examining people–environment interactions, and in working with communities and transdisciplinary research teams. I make use of analytical frameworks such as political ecology, feminist perspectives, and the sociology of science to support the discussion.

SOCIETY, ENVIRONMENT AND HEALTH: BEYOND LINEAR LINKAGES

Central to the ecohealth approach is the notion that human beings are integral to the ecosystem, and not apart from it. Hence, human health and well-being must be central to ecosystem management decisions. Ecohealth also provides a framework from which to examine the interrelationships between people and the various dimensions of the environment (social, cultural, economic, physical), and to evaluate the mechanisms through which both ecosystems and human health are impacted through ecosystem transformations.

Yet, the complexities surrounding people–environment interactions, the underlying forces that shape these interactions, and the coping and resistance strategies employed by people receive little theorizing. Similarly, issues surrounding the emergence of an environmental problem, how it came to be labeled as a problem, and whether or not the affected stakeholders share the same views are issues that should be critically analysed. It is still common to find neo-Malthusian conceptions of environmental problems, simply constructed to serve certain research and institutional agendas. More often than not, many environment and health problems in the developing world continue to be attributed to people's reckless behaviours and "inappropriate" land-use practices, without situating them in broader macro–micro power structures and how these shape people's land-use practices, health behaviours, and access to health and environmental resources (Mayer, 1996; Turshen, 1984). What is often not evident is how powerful actors such as state governments, institutions, and multinational companies materially control weaker actors' access to environmental resources through protected areas zoning or how weaker actors' actions are discursively constructed through language as harmful, without evaluating these within the context of who has the power to name, define, prioritize, and objectively intervene.

The narrow views of how human–environment relationships were conceptualized and the deterministic, neo-Malthusian models sometimes used to explain environmental change and their lack of attention to power, gave rise to political ecology (Blaikie, 2008; Robbins, 2004). Informed by political economy and critical theory, political ecology aims to understand societal and ecological processes as completely intertwined, with political and economic factors at multiple levels shaping and being shaped by the environment. Political ecology has two broad strands; the first strand is concerned with understanding human–environment relations from a political–economic context, such as Marxist debates of unequal power relationships in a global capitalist political economy, and how this influences access to, use, and control of resources; and how social categories such as gender, class, and ethnicity mediate access to environmental resources (Bryant & Bailey, 1997). The second strand, the critical component, is concerned about the scientific constructions of environmental problems, and the totalizing and grand narratives sometimes used by such framing. Some examples include the exaggerated framing of land degradation and desertification from a global perspective and how these are designed to suit the global agendas of mega institutions (Binns, 1997). Critical political ecology is concerned about the concepts used in people–environment interactions, particularly how environmental issues are defined, categorized, named, explained, and solved through positivist notions of objective reality.

Political ecologists see the environment as a "politicized" space in which environmental problems are simultaneously political–economic problems and should be examined from that perspective (Bryant & Bailey, 1997; Forsyth, 2003). Central to political ecology is the role of power relations among various actors, and how such power relations shape access to, control, and use of environmental resources, as well as the distribution or displacement of environmental costs. Hence, rather than focus on the activities of land users, attention should be drawn to the state, corporate giants, multinational corporations, and how their activities influence the way people interact with the biophysical environment. Political ecology, informed by poststructuralist theorizing, argues that the selective identification, prioritization, and representation of environmental problems is not a neutral process, but acts that either reinforce or challenge existing social and economic arrangements (Peet & Watts, 1996).

For example, an exploration of the underlying causes of environmental health problems and health disparities in rural Northern Ghana pointed to continuous marginalization by the government, and the austerity measures that were implemented by the World Bank and the International Monetary

Fund (IMF). As a result of the IMF/World Bank structural adjustment policies (SAP) subsidies were removed from basic commodities such as food, agricultural products, and fuel, resulting in increased rural poverty, household food insecurity, and increased dependence on natural resources. The rural poor had no choice but to encroach on marginal lands and fragile ecosystems, and to use intensive modes of farming practices that were deemed unsustainable (Alubo, 1990; Kanji, Kanji, & Manji, 1991; Kessler & Van Dorp, 1998). Similarly colonial land-use policies restricted harvesting of traditional food and hunting from so-called forest and game reserves, thereby forcing locals to resort to poaching and "illegal" harvesting of traditional food. Such practices, unfortunately, tend to be labeled by "experts" as "inappropriate" land-use practices that contribute to ecosystem degradation. While some of these practices may lead to ecosystem degradation, it is not sufficient to blame the land users and their land-use practices without situating them in their broader sociopolitical, historical, and economic contexts. Political ecology therefore cautions against the adoption of preexisting notions of environmental problems and their associated causes without sufficient contextualization (Blaikie, 1995; Bryant & Bailey, 1997).

Extending this thinking to the field of public health, political ecology seeks to illustrate that most environmental health problems are rooted in unequal power structures that shape people's interaction with the environment, and consequently expose them to varying environmental and occupational health risks (Kalipeni & Oppong, 1998; Mayer, 1996, 2000). The onset of environmental conditions that produce ill health are seen as the result of a complex interplay of socio-ecological and political factors, shaped by uneven power relations in society. Hence, the analysis of the determinants of health must be situated within political, economic, and ecological contexts, and broadened to include power dynamics and social organization.

Application of political ecology to health also seeks to challenge the simplistic explanations that focus on individual level factors, such as people's behaviours and belief systems as the cause of poor health. These factors alone are not sufficient to explain the persistence of poor health outcomes in certain population groups, nor are they sufficient to explain the health disparities that characterize rural–urban, Northern–Southern communities in Ghanaian communities. Just like environmental problems, poor health problems must be situated within the purview of the broader socioeconomic and political factors that constrain people's access to health-enhancing resources.

Environment–people relations in ecohealth-based studies must therefore always be contextualized and examined through a power lens, and how differential power structures influence people's access to environmental costs and assets. The failure to critically analyse these issues risk a wrongful diagnosis of environmental health problems and hence inappropriate interventions.

COMMUNITY: BEYOND A RESEARCH SITE

Most ecohealth-based research takes place in community settings. More often than not, communities tend to be portrayed as coherent spatial units whose membership is given, they are seen as having shared norms and values and sometimes work toward achieving communal goals (Jewkes & Murcott, 1996). For example, the Alma-Ata Declaration views community as a place-based setting of people with commonly shared social, economic, political, and cultural needs, problems, and characteristics (World Health Organization, 1978). Communities may also be seen as social systems, whereby the focus is on social interaction, social institutions, and social control (Haglund, Weisbrod, & Bracht, 1990). Such conceptions of community align with modernist and critical social theories' notions of shared social realities whereby people's needs and aspirations can be nurtured to achieve collective emancipatory goals (Boutain, 1999). Such notions assume that people hold a single social, political, and cultural view about which common emancipatory truths/goals can be achieved (*ibid.*). The adoption of a common view of problems tends to result in generalizing knowledge and solutions, which may or may not be suitable to all community members.

Alternative theorizing of community, especially by poststructuralist, postmodernist, and feminist theorists urges researchers to see communities as contested sites embedded with social exclusions, inequalities, and social divisions (Crow & Maclean, 2006). Poststructuralists' caution against adopting static or preexisting notions of community. They argue that communities do not exist *a priori*. They are constructed to suit our needs as researchers and academics, and mostly associated with the problem to be investigated (e.g., mining town). Similarly, there are no preexisting identities in communities, we construct them to suit our research agendas, or construct them as deviating from the "norm," and in need of professional expertise to bring them back to conformity. Also, Davies (1994) draws attention to the subjective and the constantly shifting, fragmented, multifaceted, and contradictory nature of people's experiences. Attention to multiple identities and subjective experiences means that there is not just one truth, but multiple truths which must be acknowledged. Hence, people's knowledge and experiences should always be contextualized within the social and institutional structures in which they emerge.

Based on this, then, communities should be seen as sites of struggle, where there is continuous negotiation of needs, problems, and priorities. They are also guided by informal rules and norms of who belongs or does not belong, who can and cannot speak, what can be said publicly about the community and what cannot be, and by whom? For example, during a focus group discussion in the ecohealth project in Ghana, the facilitator asked a question on chieftaincy issues, and a young man attempted to answer but was admonished by an elder "not to say a word." This is because the young man was born in Southern Ghana, and later migrated to his home community in the North; as such he was considered an "outsider." Besides, he was too young to comment on chieftaincy issues. Chieftaincy issues in this region are regarded as important and sacred and can only be spoken to by elders. The young man, in turn, argued that as part of the focus group, he had all rights to answer the question. He was later admonished by his peers for challenging the "received" wisdom of a community elder. In this instance, place of birth and age have implications for how one is defined as a community member, and what can and cannot be commented on. There are also implications for the use participatory approaches that aim to give all voices a chance to speak candidly about their problems, while being oblivious about the unspoken sanctions that are part of the community fabric. It also points to the power dynamics at play in focus group settings, as the views we collect may just represent those in powerful positions and who are sanctioned to speak.

Communities must therefore be viewed from a critical perspective, not just as the research sites we perceived before embarking on the project. The uncritical adoption of community risks masking these conflicting differences, views, needs, and issues that exist in the community and could result in the development of interventions that end up augmenting existing inequities. As cautioned by Cameron and Gibson (2005), communities and community members should always be seen as being constituted according to the needs of various discourses. Their membership should be seen as embedded in a variety of social and power structures, with varying and conflicting needs, views, and experiences. As such, environment and health problems should always be contextualized based on people's respective circumstances, identities, and positions.

PARTICIPATION: BEYOND SIMPLE INVOLVEMENT

Ecohealth-based studies encourage the use of participatory approaches to engage community members in investigating and responding to specific environmental health problems. Participatory approaches are used as a means to generate learning among community members about how existing environmental conditions impact human health outcomes, and to take the necessary actions to promote health. Participatory approaches are also used to operationalize the concept of transdisciplinarity, whereby disciplinary perspectives are integrated with local or traditional views. A variety of participatory techniques have been used for ecohealth studies, including PAR. For example, Dakubo (2004) used a PAR approach to conduct an ecohealth research project in rural Northern Ghana. PAR usually involves the beneficiaries of a study as co-researchers in investigating the problem at hand (Fals-Borda & Rahman, 1991; MacIntyre, 2008). As coresearchers, community members participate in all phases of the research process, including identification of the problem, development of the research agenda, data collection and analysis, and implementation of emerging solutions.

The philosophical underpinnings of PAR are aligned with "postmodern tradition that embraces a dialectic of shifting understandings" whereby "objectivity is impossible" and "multiple or shared realities exist" (Kelly, 2005, p. 66). PAR challenges the notion of seeking knowledge through mainstream research; it instead suggests that knowledge grounded in everyday life experiences can be developed through active participation in research. Yet these are not explicitly explained in most ecohealth studies. Why do people engage in the cyclical process of critiquing ideologies, understanding the circumstances that shape their life, and taking the necessary action to change these? Such actions lead to informed knowledge, liberation, and emancipation. The theory linking ideology, reflection, action, and emancipation have been well-articulated by Habermas and Freire. For example, with respect to environmental health issues, Stevens and Hall (1992) articulate that the process of critiquing unhealthy environments potentiates the process of liberation as critique serves to enlighten people about the position they occupy and the interests they serve, thus raising their level of consciousness. Informed by critique, people enter into dialogue with each other and reflect critically on the factors shaping environmental conditions and how these impact their health. This reflection allows them to design and implement context-specific actions aimed at addressing environmental issues.

Despite the usefulness of participatory approaches, their implementation can be fraught with a number of challenges, including difficulty in fully integrating community members as coresearchers, achieving full participation, capturing "accurate" responses through group processes, and contesting power dynamics. For example, Wallerstein (1999, p. 50) outlines power conflicts encountered in research on New Mexico's Healthier Communities Initiatives and concluded that handling such conflicts requires "a painful self-reflective process." Feminist criticism of PAR have focused on issues such as: which participants are involved in a PAR project and whose ideas are reflected in the results; how categories such as "poor," "oppressed," and "marginalized" are conceived, with gender often being ignored as a factor of oppression; and the masculine conceptions of knowledge and gender dynamics in the research process (Maguire, 1987, 1996; Rose, 2001).

These challenges are sometimes augmented as a result of varying factors, including competing interests and the lack of confidence in the knowledge and research skills of local actors. For example, in the Ghanaian case study, members of the research team were mainly front-line extension workers who mostly used top-down approaches when working with community members. As "experts" in their respective disciplines, they spoke from a privileged perspective and made decisions about what was right and wrong. In such circumstances, it was difficult for community members to see themselves as colleagues or consider their views as equally important. Equal partnership tends to be illusive by virtue of differences such as education, knowledge, gender, age, socioeconomic status, and place of residence. Yet, there have been some success stories reported internationally with the use of PAR to integrate traditional views (Henry et al., 2002; Pyett, 2002). For example, Tsey, Patterson, and Witeside (2004) reported using PAR in a remote Aboriginal Australian community to work with a men's self-help group to plan and implement activities.

Another challenge posed by participatory approaches is the use of focus group discussions to gather people's views about a particular issue. The composition of focus groups, local power dynamics, the extent to which the views gathered are truly representative of the issue under investigation, are all challenges that ecohealth researchers have to deal with. Yet as Maguire (1996) points out, it is not just about the techniques used in PAR, it is broader than that. It is an approach that is firmly situated within the human enquiry paradigm that challenges the positivist research paradigm with its objective and apolitical statements, methods, and dehumanizing assumptions. PAR, in contrast, is an emancipatory approach that allows people to internalize their situations and act to change them. For example in the Ghanaian study, focus groups were gender-disaggregated because the realization that women were reluctant to speak candidly about their health concerns in the presence of men, for fear of being called "the lazy wife." Similarly, men were reluctant to speak about their health concerns before women for fear of being perceived as a "weak" male figure. However, as was seen in the men's focus group discussions, factors such as age and where one is born became important factors, as younger men could not challenge the "received" wisdoms of elderly men, nor could people born outside the community speak about chieftaincy issues (Dakubo, 2004, 2010).

In the women's group religion, dress code and socioeconomic status played key roles in determining who dominated group discussions. For example, in the women's group, they decided to hold focus group discussions on Sundays after church. In a community where going to church means wearing one's best clothes, other women did not feel comfortable participating in these discussions as they did not fit in, nor did they think their views would be taken seriously. These issues have implications for the data we collect through group processes, and whether or not these views are representative of elite voices. As Mosse (1994) cautions, data from group processes obscure power relations and other group dynamics and tend to reflect the voices of those culturally sanctioned to speak in public. Also, information gathered through participatory approaches is usually of mixed opinions, including combinations of fact and value, consensus and difference, openness and sensitivity, as well as public and private accounts. Care must therefore be taken when interpreting data from group process, and being careful not to assume such information as representative of the group or community. Going over the information with the group and conducting follow up in-depth interviews may help clarify some issues.

Participatory approaches also seek to provide opportunities for the vulnerable, weak, and marginalized to speak candidly about their problems. However, given that many research projects are initiated by outside researchers, who sometimes gain entry into the community through local power structures and gatekeepers, identifying such population groups can be a problem. Besides, if these gatekeepers are part of the research team and are responsible for identifying study participants, it is unlikely that they will choose the marginalized. At the same time, working against local power structures could augment existing inequalities when the research project is completed. Hence, while participatory approaches are encouraged in ecohealth research, their use and the information gathered must be evaluated and contextualized in light of the audience, group composition, power dynamics, and topics under discussion. Their transformative and emancipatory benefits should be evaluated against those of positivist approaches.

TRANSDISCIPLINARITY: BEYOND INTEGRATION OF DISCIPLINARY EXPERTISE

A central pillar of ecohealth-based research is transdisciplinarity. Ecohealth research calls for the integration of expertise from the social, natural, and health sciences when investigating environmental health problems (Rosenfield, 1992). Such integration can be achieved through the establishment of a transdisciplinary research team, which, in turn, collaborates with community members to integrate local knowledge systems in investigating and responding to the problem at hand. The integration of scientific and traditional knowledge systems allows for a comprehensive understanding of the problem as well as the emergence of new ideas that otherwise would not have emerged from uni-disciplinary investigations.

Such knowledge integration is particularly important for ecohealth research, since many rural and Indigenous peoples tend to have close ties with the biophysical environment, which provides them with sophisticated knowledge of the structure and functions of the ecosystem. This information, together with historical accounts of ecosystem patterns, and the complex ways in which they interact with the biophysical environment all serve as useful information for external researchers and the research project.

However, despite these benefits, putting together a transdisciplinary research team can be fraught with challenges, including difficulties integrating local knowledge systems with Western scientific knowledge, dealing with competing interests among group members, and addressing with the underlying political and power relations of participants and other stakeholders (Agrawal, 1995; Nadasdy, 1999). In rural and remote settings, the challenges may be compounded by limited access to expertise from the natural, health and social sciences. Whereas in industrialized countries there is ready access to professionals such as entomologists, medical doctors, veterinary doctors, ecologists, anthropologists, and toxicologists, it can be difficult finding such expertise in many communities in the global South. Not only are such experts scarce, they also are not accustomed to working together in a transdisciplinary fashion. In such settings, it takes some amount of creativity to establish a functional transdisciplinary research team. For example, in the Ghanaian case study, a team of front-line workers from various government departments, including forestry, health, environment, and adult education, were brought together to work as a transdisciplinary research team. Prior to the study, workshops were organized with the research team to introduce them to a variety of data gathering techniques such as the strategic planning process and focus group discussions. It was also an opportunity for them to get to know each other, acquaint themselves with other department's roles in the community, and to map out a research agenda for the study (Dakubo, 2004, 2010).

Once the transdisciplinary team is in place, the next challenge is to foster an environment where all members can work together collegially and explore ways to integrate and accept local knowledge systems as valid ways of knowing. Even though local people are involved as coresearchers, their views are sometimes tokenistic. As noted by Sillitoe (1998, p. 226), the implication of considering indigenous and scientific perspectives side by side does not mean scientists have to revise their working suppositions on objectivity, positivism, and reductionism to accommodate other views. Instead, Sillitoe considers it an opportunity to compare local views and explanations with scientific views. Others have criticized this noncritical view of Sillitoe's, arguing that he ignores the political context in which indigenous knowledge research takes place (Posey, 1998; Stirrat, 1998). Berkes and Henley (1997) argue for pluralism, pointing to the recognition of indigenous knowledges as a legitimate way of knowing, as opposed to so-called value-free scientific knowledges.

The privileging of scientific views over local views, compounded with other factors such as different disciplinary expertise and ranks of members in the research team, can result in power struggles, with some members dominating the research process, controlling ideas, subduing other opinions and deciding what is "right" or "wrong." Such power struggles can be challenging for young researchers with little experience. For example, drawing on my experience in Ghana, and as a young, female doctoral candidate studying abroad, I was cognizant of how my varying identities might pose challenges as the principal research investigator. Then, as a second-time researcher in the community, and now a clan member, I was not sure how I would be received and perceived – was I an "insider" or "outsider?" However, as there are no fixed positions, I negotiated the various identities throughout the research process. On one hand, striving to be an "insider," a "sister," and a "daughter," when that accorded me the opportunity to gain in-depth information about an issue, or be accepted as part of the community. On the other hand, I strove to be an "outsider," a "visitor" when it allowed people to cooperate with me and to provide me with the necessary support, as is mostly given to foreigners (*ibid.*). Naples (1996) has also documented experiences with negotiating "insider–outsider" relations in an ethnographic study of two towns in Iowa, and concludes that positions of "outsiderness" and "insiderness" are ever shifting and permeable social locations, and not static or fixed as often perceived.

Another challenge in transdiciplinary research teams is the extent to which local knowledge and expertise are integrated with scientific perspectives, and accepted or rejected as valid ways of knowing. Foucault's (1997) concept of "subjugated knowledges" is particularly important wherein historical knowledges that were present within functional and systematic ensembles have been suppressed or silenced by dominant forms of knowledge; and other types of knowledges disqualified as nonconceptual, naïve, inferior, primitive, or scientifically non-vigorous. However, "hybrid" research that integrates local and traditional views has been useful in understanding contemporary social and ecological processes in an historical context (Batterbury, Forsyth, & Thomson, 1997; Rocheleau, 1995). Others suggests moving beyond the analytical separation of local and scientific knowledges systems to a concentration on the complex processes of epistemic negotiation in different settings (Agrawal, 1995; Wynne, 2004/1996).

The debates surrounding the value and validity of local knowledges and the extent to which they can be integrated with scientific knowledge systems have merits for ecohealth-based research. Given that ecohealth sits at the interface of the natural, social, and health sciences, certain topics may be very technical in scope and my require dominance of scientific perspectives, such as understanding the pathways through which mercury bioaccumulates move through the food chain, and affects the nervous system. In other instances, it may be important to depend heavily on local knowledge systems to understand the historical context of ecosystem transformation. Hence the concept of "hybrid" research is appealing, while being cognizant that no knowledge is value-free. Local knowledges should also be interrogated for their accuracy, completeness, and currency.

GENDER AND SOCIAL EQUITY: BEYOND SPECIFIC IDENTITIES

Ecohealth-based research emphasizes the importance of gender relations and social equity in the research agenda. This is based on the recognition that people, by virtue of societal roles, responsibilities, and rights, interact differently with the biophysical environment, and through such interactions become exposed to different health risks. The extent to which people are able to cope with or respond to these health risks is further determined by the social, political, economic, and institutional factors in which they are embedded. Those with limited resources, coping skills, or embedded in power relations that constrain their access to and use of health and environmental resources tend to bear a disproportionate portion of the disease burden. Hence, the use of a social equity lens allows us to understand how axes of difference, such as gender, ethnicity, race, and socioeconomic status all affect people–environment relations and consequently influence their health outcomes.

Within the spheres of political ecology, gender has been an important theme for researchers. Feminist political ecologists see gender as a critical factor in shaping access to, and control over, resources, environmental decisions, and technologies (Rocheleau, Thomas-Slayter, & Wangari, 1996). They focus on how local realities and their constantly changing roles and identities shape their access to and use of environmental resources (Jackson, 1998). Feminist political ecologists explore environmental change through the lens of gender, with a focus on identifying "the constraints and opportunities that shape gendered land-use behaviour [...to assure] a more accurate assessment of environmental change at the scale where decisions are made" (Steinmann, 1998, p. 81). They also believe that scientific discourse is "gendered" because gender inequities in science can deny women the knowledge necessary to address environmental problems, and have set out to deconstruct the "myth of value-free objectivity and universality in science" (Rocheleau et al., 1996). Also of importance is how gender roles, rules, and norms influence men and women's interactions with the biophysical space, and shape gender-patterned knowledges and experiences.

While the adoption of a gender and social equity lens is important in ecohealth research, its deployment is sometimes essentialist in scope. Critical scholars caution against overly focusing on specific identities as though these were fixed identities with commonly shared experiences (Flax, 1990; Frug, 1992). For example, the focus on women and women's health outcomes may lead to a false assumption of commonly shared experiences, while undermining the complexity of factors that endow women with unique and constantly shifting experiences. Critical scholars caution that the search for a universal women's experience runs the risk of smoothing over differences, conflicts, and contradictions that are inherent in people's everyday lives. It also risks interpreting women's experiences out of context, and creating a sense of homogeneity, coherence, and timelessness

(Abu-Lughod, 1993). Such oversimplification of women's health experiences, particularly in the global South, has led to the development of women-focused programs that undermine the ability to understand women's health as influenced by other axes of difference and institutional structures (Whiteford & Manderson, 2000). A critical approach to gender in ecohealth research, therefore, cautions against the adoption of preexisting identities and experiences of gender. Instead paying attention to individual experiences, and as contextualizing these as embedded in place, socio-economic status, ethnicity, and as shaped by other socioecological and political factors is considered important.

CONCLUSION

This chapter examines how ecohealth, as an emerging field of study, can benefit from critical theoretical perspectives so as to become a more theoretically informed subdiscipline. In particular, the chapter evaluates how key concepts such as people–environment relations, transdisciplinarity, community, participation, gender and social equity are deployed in traditional ecohealth literature, and how these practices can be examined from a critical perspective. The challenge is that by adopting uncritical notions of these concepts, we run the risk of oversimplifying the inherent challenges, conflicts, and experiences that characterize people–environment relations. In sum the application of critical theoretical perspectives unravels the hidden assumptions embedded in these concepts, allows for an informed understanding of the causal basis of environmental health problems, and leads to the development of interventions that are both liberatory and transformative.

REFERENCES

Abu-Lughod, L. (1993). Introduction. In L. Abu-Lughod (Ed.), *Writing women's world. Bedouin stories* (pp. 1–42). Berkeley, CA: University of California Press.

Agger, B. (1991). Critical theory, poststructuralism, postmodernism: Their sociological relevance. *Annual Review of Sociology*, *17*, 105–131.

Agrawal, A. (1995). Dismantling the divide between indigenous and scientific knowledge. *Development and Change*, *26*, 413–439.

Alubo, S. O. (1990). Debt crisis, health and health services in Africa. *Social Science and Medicine*, *31*, 639–648.

Batterbury, S., Forsyth, T., & Thomson, K. (1997). Environmental transformations in developing countries: Research and policy trends. *Geographical Journal*, *163*, 2.

Berkes, F. & Henley, T. (1997). Co-management and traditional knowledge: Threat or opportunity? *Policy Options*, 29–31.

Binns, T. (1997). Is desertification a myth? In A. Goudie (Ed.), *The human impact reader: Readings and case studies* (pp. 400–407). Oxford: Blackwell.

Blaikie, P. (2008). Epilogue: Towards a future for political ecology that works. *Geoforum*, *39*, 765–772.

Blaikie, P. M. (1995). Understanding environmental issues. In S. Morse & M. Stocking (Eds.), *People and environment*. London, UK: University College London Press.

Boutain, D. M. (1999). Critical nursing scholarship: Exploring critical social theory with African American studies. *Advances in Nursing Science*, *21*(4), 37–47.

Bryant, R. L., & Bailey, S. (1997). *Third world political ecology*. London: Routledge.

Cameron, J., & Gibson, K. (2005). Participatory action research in a poststructuralist vein. *Geoforum*, *36*, 315–331.

Crow, G., & Maclean, C. (2006). Community. In G. Payne (Ed.), *Social divisions* (2nd ed). Basingstoke: Palgrave.

Dakubo, C. (2004). Ecosystem approach to community health planning in Ghana. *EcoHealth*, *1*, 50–59.

Dakubo, C. (2010). *Ecosystems and human health: A critical approach to ecohealth research and practice*. New York, NY: Springer.

Davies, B. (1994). *Poststructuralist theory, gender, and teaching*. Melbourne, Australia: Deakin University.

Derrida, J. (1976). *Of grammarology*. Baltimore, MD: John Hopkins University.

Fals-Borda, O., & Rahman, M. A. (1991). *Action and knowledge: Breaking the monopoly with participatory action research*. New York, NY: Apex Press; Intermediate Technology Publications.

Flax, J. (1990). *Thinking fragments: Psychoanalysis, feminism, and postmodernism in the contemporary west*. Oxford: University of California Press.

Forget, G., & Lebel, J. (2001). An ecosystem approach to human health. *International Journal of Occupational Environmental Health*, *7*(Suppl. 2), S3–38.

Forsyth, T. (2003). *Critical political ecology: The politics of environmental science*. New York, NY: Routledge.

Foucault, M. (1980). *Power/knowledge*. New York, NY: Pantheon.

Foucault, M. (1997). *Society must be defended: Lectures at the college de France, 1975-76* (D. Macey, Trans.). London: Allen Lane.

Freire, P. (1971). *Pedagogy of the oppressed*. New York, NY: Continuum. Translated by Myra Bergman Ramos.

Freire, P. (1972). *Pedagogy of the oppressed*. Harmondsworth: Penguin.

Frug, M. J. (1992). *Postmodern legal feminism*. London: Routledge.

Habermas, J. (1971). *Knowledge and Human Interests*. Boston, MA: Beacon.

Haglund B., Weisbrod, R., & Bracht, N. (1990). Assessing the community: Its services, needs, leadership and readiness. In N. Bracht (Ed.), *Health promotion at the community level*. Newbury Park, CA: Sage.

Henry, J., Dunbar, T., Arnott, A., Scrimgeour, M., Matthews, S., Murakami-Gold, L., & Chamberlain, A. (2002). *Indigenous research reform agenda: Rethinking research methodologies*. Casuarina, NT: Cooperative Research Centre for Aboriginal Health.

Jackson, S. (1998). Feminist social theory. In S. Jackson & J. Jones (Eds.), *Contemporary feminist theories* (pp. 12–33). Edinburgh: Edinburgh University Press.

Jewkes, R., & Murcott, A. (1996). Meanings of community. *Social Science and Medicine*, *43*, 555–563.

Kalipeni, E., & Oppong, J. (1998). The refugee crisis in Africa and implications for health and disease: a political ecology approach. *Social Science and Medicine*, *46*, 1637–1653.

Kanji, N., Kanji, N., & Manji. (1991). From development to sustained crisis: Structural adjustment, equity and health. *Social Science and Medicine*, *33*, 985–993.

Kelly, P. J. (2005). Practical suggestions for community interventions using participatory action research. *Public Health Nursing*, *22*(1), 65–73.

Kessler, J. J., & Van Dorp, M. (1998). Structural adjustment and the environment: The need for an analytical methodology. *Ecological Economics*, *27*, 267–281.

MacIntyre, A. (2008). *Participatory action research*. Thousand Oaks, CA: Sage Publications.

Maguire, P. (1987). *Doing participatory research: A feminist approach*. Massachusetts, MA: Centre for International Education, University of Massachusetts.

Maguire, P. (1996). Proposing a more feminist participatory research: Knowing and being embraced openly. In K. de Koning & M. Martin (Eds.), *Participatory research in health: Issues and experiences* (pp. 27–39). London: Zed Books.

Mayer, J. D. (1996). The political ecology of disease as one new focus for medical geography. *Progress in Human Geography*, *20*(4), 441–456.

Mayer, J. D. (2000). Geography, ecology and emerging infectious diseases. *Social science and Medicine*, *50*, 937–952.

Mosse, D. (1994). Authority, gender and knowledge: Theoretical reflections on the practice of participatory rural appraisal. *Development and Change*, *25*(2), 497–526.

Nadasdy, P. (1999). The politics of TEK: Power and the "Integration" of knowledge. *Arctic Anthropology*, *36*(1–2), 1–18.

Naples, A. N. (1996). A feminist revisiting of the insider/outsider debate: The "outsider phenomenon" in rural Iowa. *Qualitative Sociology*, *19*(1), 83–105.

Peet, R., & Watts, M. (1996). Liberation ecology: Development, sustainability and environment in an age of market triumphalism. In R. Peet & M. Watts (Eds.), *Liberation ecologies: Environment, development, and social movements* (pp. 1–45). London: Routledge.

Posey, D. A. (1998). 'Comment on the development of indigenous knowledge: A new applied anthropology' by Paul Sillitoe. *Current Anthropology*, *39*(2), 241–242.

Pyett, P. (2002). Working together to reduce health inequalities: Reflections on a collaborative participatory approach to health research. *Australian N Z Journal of Public Health*, *26*, 332–336.

Robbins, P. (2004). *Political ecology: A critical introduction*. Malden, MA: Blackwell.

Rocheleau, D. (1995). Maps, numbers, text and context: Mixing methods in feminist political ecology. *Professional Geographer*, *47*, 458–466.

Rocheleau, D., Thomas-Slayter, B., & Wangari, E. (Eds.). (1996). *Feminist political ecology: Global issues and local experiences*. London: Routledge.

Rose, D. (2001). *Revisiting feminist research methodologies*. Status of Women Canada, Research Division.

Rosenfield, P. L. (1992). The potential of transdisciplinary research for sustaining and extending linkages between the health and social sciences. *Social Science and Medicine*, *35*(11), 1343–1357.

Sillitoe, P. (1998). The development of indigenous knowledge: A new applied anthropology. *Current Anthropology*, *39*, 223–252.

Steinmann, S. H. (1998). Gender, pastoralism and intensification: Changing environmental resource use in Morocco. In J. Albert et al. (Eds.), *Transformation of Middle Eastern natural environments: Legacies and lessons*. Yale F&ES Bulletin No.103, pp. 81–107.

Stevens, P. E., & Hall, J. M. (1992). Applying critical theories to nursing in communities. *Public Health Nursing*, *9*(1), 2–9.

Stirrat, R. L. (1998). Comment on 'The development of indigenous knowledge: A new applied anthropology' by Paul Sillitoe. *Current Anthropology*, *39*, 242–243.

Tandon, R. (1996). The historical roots and contemporary tendencies in participatory research: Implications for health care. In K. De Koning & M. Martin (Eds.), *Participatory research in health: Issues and experiences* (pp. 19–26). London: Zed Books.

Tsey, K., Patterson, D., Witeside, M., et al. (2004). A micro analysis of a participatory action research process with rural Aboriginal men's health group. *Australian Journal of Primary Health*, *10*, 64–71.

Turshen, M. (1984). *The political ecology of disease in Tanzania*. New Brunswick, NJ: Rutgers University Press.

Wallerstein, N. (1999). Power between evaluator and community: Research relationships within New Mexico's healthier communities. *Social Science and Medicine*, *49*, 50.

Whiteford, L. M., & Manderson, L. (2000). *Global health policy, local realities. The fallacy of the level playing ground*. London: Lynne Rienner Publishers.

World Health Organization. (1978). *Primary health care: Report of the international conference on primary health care*. Alma Ata, U.S.S.R., 6–12 September, Geneva.

Wynne, B. (2004/1996). May the sheep safely graze? A reflexive view of the expert-lay knowledge divide. In S. Lash, B. Szerszynski, & B. Wynn (Eds.), *Risk, environment and modernity: Towards a new ecology*. Sage. (Brazilian translation of a book previously published in 1996.)

EXPLORING ABORIGINAL PEOPLE'S CONNECTION TO COUNTRY TO STRENGTHEN HUMAN–NATURE THEORETICAL PERSPECTIVES☆

Jonathan Yotti Kingsley, Mardie Townsend and Claire Henderson-Wilson

ABSTRACT

Purpose – *Aboriginal people across Australia have diverse practices, beliefs and knowledges based on thousands of generations of managing and protecting their lands (Country). The intimate relationship Aboriginal*

☆In this chapter, 'Aboriginal' refers to Aboriginal and/or Torres Strait Islander people, as this is the preferred terminology of the peak body of Aboriginal community health (NACCHO, 2012). Aboriginal and Torres Strait Islander people are members/descendants of Aboriginal cultures of Australia or the Torres Strait Islands, through identification and acceptance by the community. The Australian Human Rights Commission (2013) notes that Aboriginal peoples are the first inhabitants of Australia and are diverse in geography, language and tradition. 'Indigenous' refers to Traditional Custodian groups in the international context in accordance with international law to represent over 350 million people (Stephens, Porter, Nettleton, & Willis, 2006). However, it must be acknowledged that Traditional Custodian groups worldwide have their own unique practices, beliefs and knowledge systems.

Ecological Health: Society, Ecology and Health
Advances in Medical Sociology, Volume 15, 45–64

ISSN: 1057-6290/doi:10.1108/S1057-6290(2013)0000015006

people have with their Country is explored in this chapter because such knowledge is important for building insight into the relationship between social and ecological systems. Often in research Aboriginal views have been marginalised from discussions focused on their lands to the detriment of ecosystems and human health. This chapter aims to understand if such marginalisation is evident in Western human–nature relationship discourses.

Approach – *This chapter provides a critical literature review which examines whether Aboriginal people's diverse understanding of their ecosystems have been incorporated into human–nature theories using the biophilia hypothesis as a starting point. Other concepts explored include solastalgia, topophilia and place.*

Findings – *Critiques of these terminologies in the context of Aboriginal people's connection to Country are limited but such incorporation is viewed in the chapter as a possible mechanism for better understanding human's connection to nature. The review identified that Aboriginal people's relationship to Country seems to be underrepresented in the human–nature theory literature.*

Value – *This chapter emphasises that the integration of Aboriginal perspectives into research, ecological management and policy can provide better insight into the interrelationships between social and ecological systems.*

Keywords: Aboriginal people; human-nature relationship; biophilia; topophilia, solastalgia; place

INTRODUCTION

Reviewing Aboriginal people's connection to the land offers new ways of researching ecosystem-based approaches to health as it allows for a more in-depth cultural exploration of the nexus between social and ecological systems. A challenge to Ecohealth research is how to appropriately incorporate diverse Aboriginal knowledges, which are often situated outside of current dominant scientific contexts. This is an important investigation as Aboriginal ecological perspectives have often been marginalised or rejected from discussion and debate, and have instead been framed as the 'other' (as compared with 'scientific' knowledge), thus restricting Aboriginal people's ecological practices and opportunities to share and shape their knowledges. Foucault (1982) explores such power relations, viewing them as ingrained in the social nexus, 'conducted' as a form he terms 'agonism' (permanent provocation/taunting often rather than direct confrontation) and linked to social practices, including domination. Yet, Aboriginal ecological perspectives are based on some 70,000 years of experience in sustaining the Australian ecosystem, practices which continue to evolve and cultivate contemporary insights into the relationship between people and place. Clearly, there is value in reviewing such diverse knowledge both theoretically and practically.

Using the biophilia hypothesis as the starting point, this chapter provides an overview of nuanced human–nature theories and compares them to the diverse ecological knowledges that contextualise Aboriginal people's connection to their Country. To explain the human–nature relationship, biophilia is used as it has to a degree cut across different academic disciplines. Biophilia suggests that attachment to the natural environment is innate, while other discourses examined in this chapter – topophilia, solastalgia and place attachment – frame attachment as built or damaged over the lifespan. By examining these human–nature theories through the philosophical lens of a population still strongly connected to nature, the hope is to understand and suggest ways in which these knowledges can be better incorporated in ecology and public health research and practice. This chapter does not aim to develop new research nor be an opinion piece but rather to serve as a review of literature and commentary. It draws mainly on Aboriginal people's perspectives but also refers to Indigenous connections to traditional lands to fill knowledge gaps through a critical literature review.

BACKGROUND

> There is a power in country that radiates to all those who live there and becomes incorporated in humans by virtue of their spiritual makeup. (Taylor, 2012, p. 21)

Aboriginal people have a deep spiritual connection to their lands known as Country (Weir, 2012). This connection governs social, economic and cultural structures, as evidenced by Aboriginal Law being based on the relationship between traditional knowledge and environmental interaction (Johnston, Jacups, Vickery, & Bowman, 2007; Strang, 2005). The knowledge and interaction with Country consists of every living and non-living element, whether that is land, fire, water, minerals or animals (Burgess & Morrison, 2007). Country is dynamic, having multiple meanings for Aboriginal individuals and communities (Nursey-Bray & Hill, 2010; Weir, 2012).

For Aboriginal people their Country represents a critical element of their health and wellbeing, with the management of their lands ensuring ecological sustainability. Improved understanding of this relationship can offer new insights for public health research and practice by better incorporating humankind's relationship with nature.

Aboriginal people's cultural obligation to manage land is often referred to as 'caring for Country'. This practice is evidenced by Aboriginal customary behaviour, direct education through exposure to cultural practices and belief systems expressed through ceremonies, song, dance, story-telling, visiting sacred sites and taboos (Tonkinson, 2011). Often this knowledge builds over an Aboriginal person's lifetime, based on reciprocity with and respect for the natural world through collective actions, with Elders being the major knowledge-holders. Caring for Country is rooted in hunting, harvesting bush food and traditional medical practices (Nursey-Bray & Hill, 2010). In Australia, landscape burning is an example of Aboriginal people managing ecosystems in a state of equilibrium through traditional knowledge and practice. Johnston and colleagues (2007) explain that Aboriginal Northern Territory communities deliberately burn the landscape to hunt animals, clear the land, assist travel and promote plant growth – a process embedded in spiritual, social and cultural systems refined over thousands of years.

Such relationships are often unfamiliar to Western environmental managers, who may not have intimate connections to places based on responsibility, identity and stewardship of the land, river and sea (Nursey-Bray & Hill, 2010). This does not mean that non-Indigenous people cannot have spiritual connections with the Australian landscape, but the relationships Aboriginal people have is based on a considerable longer period of time. Therefore, it is not surprising that evidence highlights the health benefits of involving Aboriginal people in caring for Country projects (Burgess, Berry, Gunthorpe, & Bailie, 2008; Campbell, Burgess, Garnett, & Wakerman, 2011; Richmond & Ross, 2009). However, Aboriginal people are often excluded from discussions and research, in respect to their Country, and Country is often viewed as no longer relevant unless financially viable (Palsson, 1995; Strang, 2005; Weir, 2012). Shiva (1995) highlights that the marginalisation of local knowledge is a necessary method of Western monocultural thinking/action (agriculture based on singular/intensive crop production), which consequently homogenises the environment. In Australia, exclusion from policy and action in environmental management has led to Aboriginal people feeling restricted in their access to Country and political rights (Nursey-Bray & Hill, 2010). Non-Indigenous academics and bureaucrats often dismiss Aboriginal paradigms by generalising and codifying such topics rather than viewing this diverse knowledge as relevant (Muller, 2012). There is also evidence of the exploitation of Aboriginal experiences and beliefs without involving the appropriate knowledge-holders or gaining their consent.

Internationally, there have been increased attempts by forestry managers, scholars and policy-makers to gain better understanding of Indigenous people's connections to their lands (e.g. Karjala & Dewhurst, 2003). This is because the connection Indigenous people have with their land is based on community survival, guardianship of biological resources and livelihood, leading to reciprocal relationships with nature (Gadgil & Rao, 1995; Nettleton et al., 2007). For Indigenous people this connection is a basic human right (Heinamaki, 2009), acknowledged in the United Nations Declaration on the Rights of Indigenous Peoples (2007), which recognised the importance of Indigenous lands, knowledges and practices. Continuing these practices will build environmental sustainability, as traditional ecological knowledge has the ability to inspire, generate creativity and improve designs to enhance ecological management (e.g. Martin, Roy, Diemont, & Ferguson, 2010). As Thomas Berry emphasised, 'Indigenous knowledge holders therefore offer a way of creating a new ecological spirituality for society' (Tucker, 2013).

Nonetheless, researchers should never romanticise connections Aboriginal people have with their Country, nor assume they remain as they were in the past (Panelli & Tipa, 2009). Popular discourse often either paint a romantic notion of Aboriginal people's connections with land as being innate and harmonious, or the opposite, as an unscientific relationship of the past which does not allow this population to grow, revitalise and reinvent an Indigeneity that reflects today's realities. Internationally, this theme is evident in narratives such as those of Noble Savages and Ecological Indians, where stereotypes are applied. Unpacking such stereotypes can allow Aboriginal knowledge-holders and their colleagues' ability to share their ecological philosophies and concepts, in an attempt to develop new kinds of cultural, theoretical, legal and conservation discourses.

BIOPHILIA HYPOTHESIS

When Wilson (1984) popularised the biophilia hypothesis, he was essentially providing the first Western framework for the analysis of the 'taken-for-granted' knowledge that humans have an inherent connection with nature.

Although biophilia's broad concept of 'love of life or living systems' had been around since 1964, Wilson's contribution was to propose that this human affinity with and attraction to nature is genetically based (Chapman, 2002; Simaika & Samways, 2010). Wilson (1993) emphasised that for 99% of our existence, humans have been hunters and gatherers. Hence, humans have a preference for natural environments based on the evolution of the brain as well as nuanced emotional responses to the environment conjuring both negative and positive emotions (Gullone, 2000; Kahn, Sverson, & Ruckert, 2009).

Wilson and colleagues justified these negative responses through the complementary biophobia hypothesis. Biophobia refers to the fear of nature and natural processes (e.g. the fear of snakes, spiders or being in nature) and offers insight into human's embrace of, and dependency upon, urban spaces and technology (Kahn & Kellert, 2002). A more recent yet similar theoretical construct is that of urbanophilia. Félonneau (2004) focuses on human's attraction (urbanophilia) or aversion (urbanophobia) to cities rather than focusing on relationships with nature. Since Wilson's hypothesis, there has been a rise in urbanisation, with evidence indicating that this disconnection from nature leads to negative health outcomes (Maller et al., 2008; Pretty, 2007), possibly due to humans' inherent need to connect with nature. Kellert (1993, p. 21) highlighted the daunting task of proving ideas such as inherent connection; nature as part of our 'species evolutionary heritage'; 'human competitive advantage and genetic fitness'; philosophies of nature's meaning; and human ethics of conservation which seems intuitive within Aboriginal cultures.

Difficulties in explaining biophilia come down to whether the concept is based on human biology and genetics or affinity, cognitive, culture or learned experiences (Bird, 2007; Kahn, 1999). In *The biophilia hypothesis* (1993), biophilia is mediated through spirituality (Kellert, 1993); visual stimulation and engagement (Ulrich, 1993); aesthetic pleasure (Heerwagen & Orians, 1993); cultural engagement (Nubhan & Antoine, 1993); religion/tradition (Sagan and Margulis, 1993); environmental activism (Orr, 1993); and language symbolism, myths and environmental philosophy. Kellert (1996) noted that biophilia has nine distinctive tendencies (Table 1).

A concept that recognises humankind's previous experiences with nature, reflected in biophilia, is that people prefer savannah-like environments as a consequence of their origins in the East African savannahs two million years ago (Gullone, 2000). Diamond (1993) highlights that biophilia fails to address the notion of evolution, as this focus on savannah-like habitats ignores the fact that humans evolved from such environments thousands of

Table 1. Biophilia Tendencies.

Tendencies	Description
Utilitarian value	Material/physical benefit gained from nature in terms of consumption, protection and security
Naturalistic value	Satisfaction associated with contact with nature through feelings of joy, fascination and wonder
Ecological–scientific value	Knowledge and recognition of material uses of nature through direct exploitation. The human willingness to learn and study about nature believing nature can be understood by science
Aesthetic value	The pleasure of visually viewing nature
Symbolic value	Nature as a form of symbols, through language and communication of ideas, religion and stories that build cultural identity
Humanistic value	Emotional attachment to elements of nature, expressed in feelings of love
Moralistic value	Ethics and norms around the treatment of the non-human world
Dominionistic value	Feelings of domination/mastery of nature
Negativistic value	Feelings of fear of certain forms of nature, associated with biophobia

generations ago. Further, because of the complexity of biophilia, this theory is rarely tested (Simaika & Samways, 2010) with minimal recent studies (e.g. of studies include Delavari-Edalat and Abdi, 2009, 2010). Simaika and Samway reiterate 'Indeed, biophilia has not been shown to be innate, and it should not be assumed heritable until demonstrated to be so. It is therefore beyond scientific reason … [to] assume biophilia is innate' (2010, p. 903).

As evolution progresses, it seems reasonable that humans might adapt and become more urban beings, but advocates for the biophilia hypothesis view connection to nature as embedded in human biology (Nisbet, Zelenski, & Murphy, 2011). In Australia, Aboriginal people living in urban areas maintain a connection to Country in indirect rather than direct ways. This indirect relationship has been precipitated through a history of poor government policy (e.g. forced removal from Country, increased urbanisation and poor land management practices) causing ecological disasters. For example, current native fish numbers in waterways like the Murray–Darling Basin are around 10% of the levels that existed prior to colonisation (Ladson & Finlayson, 2004). Evidence indicates that even when Aboriginal people are physically removed or damage occurs to Country this connection remains (Kingsley, Townsend, Phillips, & Aldous, 2009). It is important to note, however, that not all Aboriginal people who live on their Country or in urban regions maintain or want to retain this intrinsic relationship.

In *The biophilia hypothesis*, a section titled 'Culture' is dedicated to Indigenous and local people's connection to nature. By lumping Indigenous and local knowledges together, as is common throughout biophilia literature, Indigenous people are seen as just another group, rather than a culture with valuable knowledge to share. Nonetheless, in the 'Culture' section Nelson (1993, p. 224) acknowledged that biophilia is 'a deep, pervasive, ubiquitous, all embracing affinity to life that lies at the very core of traditional hunting-fishing-gathering cultures ... [Indigenous Interior Alaskan people] manifest biophilia in virtually every dimension of their existence'.

Nelson (1993) reflected that Indigenous people respected the land, with ancestry leading them in the right direction by offering codes and beliefs that engendered respect of nature. Direct experiences Indigenous people have of their lands can provide insight into how to incorporate biophilia and sustainable ecological relationships into current practice (Cajete, 1999). Deb and Malhotra (2001), however, note that this biophilic symbolic recognition of nature (Table 1) was clearly evident in Indigenous populations but not commonly valued in Western culture. In Australia, biophilia for Aboriginal people has been loosely applied by Rose (2004), Kingsley et al. (2009) and Babaian and Twigg (2011) to reflect connectedness to place, belonging, stories of creation, sustenance of life, identity, sacred geography and ancestry. Aboriginal people are interconnected with nature, unable to separate themselves from their Country because of its ancestral significance (Strang, 2005), demonstrating a biophilic connection. McKnight (1999, p. 176), an anthropologist who spent 40 years studying Aboriginal people, highlighted that for this population 'nature is humanized. Frilled lizards, lightning, moon ... act like human beings, and like human beings, they speak language'.

This connection with Country is explained as transcending generations, tying Aboriginal people to their ancestry based on some 70,000 years of history (Muller, 2012; Rasmussen et al., 2011). This connects Aboriginal people back to place, and can take many forms beyond being just human or animal, as objects within the world take on meaning and connection to the past, present and the future (Kingsley, Townsend, Henderson-Wilson, & Bolam, 2013; McKnight, 1999; Muller, 2012). Preserving these biophilic elements is fundamental to Aboriginal people's identities and the maintenance of their traditional knowledge generates an intimate understanding of Australia's harsh climate and how to live sustainably within it (Packer et al., 2012). Perhaps an increased recognition of theories such as biophilia could strengthen acceptance of Aboriginal people's connection to nature and lay a framework for understanding that relationship to place and health intertwine. Such recognition might result in a fundamental shift in attitudes called for by Cajete (1999, p. 190):

> Implicit ... is the importance of moving beyond idealization and patronization of Indigenous knowledge, something that inadvertently can lead to marginalization of the most profound epistemology regarding the interaction of human beings and nature. Indigenous people must be supported in their collective attempts to restore their traditions while also revitalizing themselves in ways they feel are appropriate in contemporary society.

Biophilia could be a medium to explain Aboriginal ecological knowledge, as it provides researchers with a tool to indicate how humans connect with nature and a means to test empirical questions around whether this connection is innate, cultural or learned. In Australia, Aboriginal people's deep connection to nature underpins their genealogical, social and cultural responsibility to protect and pass onto appropriate others knowledge of their interminable relationship with their ecosystems. This intrinsic connection remains true for both Aboriginal people living remotely and in urban environments, and may offer methods of improving both ecosystem management and urban development/design.

BEYOND THE BIOPHILIA HYPOTHESIS

Beyond the biophilia hypothesis, academics have tried to identify how humans relate with nature. Three complementary concepts are explored in this section: solastalgia, topophilia and place attachment.

Solastalgia

Across Australia, recent droughts, environmental degradation and dryland salinity have been associated with poor mental health (Speldewinde, Cook, Davies, & Weinsteins, 2011; Stain et al., 2011). Solastalgia refers to the grief and negative distress associated with changes to a local environment for an individual or community who has an attachment with that locality (Albrecht et al., 2007). Albrecht (2005) refers to the pain, distress, suffering and psychological consequences experienced by Aboriginal people and also residents of Australia's Hunter Valley, with coal mining development of the land they call home. These psychological consequences are described as manifesting in emotional pain, including loss of control, identity and isolation (Albrecht, 2005; Connor, Albrecht, Higginbotham, Freeman, & Smith, 2004). Kingsley and colleagues (2009) question whether the loss of identity described is complete, noting that since colonisation Aboriginal people have viewed environmental destruction, and although hampering biophilic attachments and causing considerable pain, can still remain attached to Country.

By focusing on the negative psychological consequences of environmental destruction, solastalgia in some ways (but for different reasons) relates to the distancing of people from the natural environment, which relates to biophobia. Orr (1993, p. 419) highlighted that 'biophobia sets in motion a vicious cycle … caus[ing] people to act in a fashion that undermines the integrity, beauty and harmony of nature'. Soliphilia is the antidote to solastalgia, defined as the love, responsibility and unified effort to protect our home and planet (Albrecht, 2011). Interestingly, Albrecht (2005, p. 55) draws a link between solastalgia and biophilia, noting that 'the innate desire to be connected to life and living things, what … Wilson calls 'biophilia' … is, in part, an innate desire to overcome solastalgia by finding an earthly 'home' in connection with living things'.

While solastalgia is a relatively new concept, Aboriginal people where considered by Albrecht as an example of distress felt as a result of a home being destroyed. However, like all cultures, Aboriginal culture is evolving and these connections remain critical no matter the colonial history as they do for non-Indigenous people who may also have attachments to their local environments. Since colonisation, Aboriginal people have experienced genocide, forced assimilation and removal from their Country, hampering their capacity to manage the ecosystems on which they, and all Australians, depend. As Country is central to the existence, education and governance of Aboriginal people, a notion often denied by non-Indigenous people, such removal is an attack on their belief systems and has been linked to poor social, health and ecological outcomes (Richmond & Ross, 2009). Such experiences seem evident for non-Indigenous people living in rural/regional areas where health outcomes are poor, possibly due to the degradation of land.

Despite literature identifying the impact of the destruction of Country on health, wellbeing and identity, there are few studies that focus on solastalgia in relation to Aboriginal people. Some examples include that of McNamara and Westoby (2011), who interviewed Torres Strait Islander women about the solastalgic impact of climate change, noting it caused a sense of sadness, fear, distress and lack of identity. Another example focused on Aboriginal and non-Indigenous artists collaborating to interpret the concepts of solastalgia, describing the sadness of local environmental changes (Kirker, 2012). This phenomena reiterates the UN Declaration of Rights of Indigenous Peoples, which emphasises the importance of Indigenous people having the ability to control the management of their lands and negotiate what happens on them in a more sustainable/equitable way. This would allow Aboriginal people to strengthen their capacity to participate in a global market to improve health and to ensure self-determination over their own futures.

Another concept similar to solastalgia is ecological identity. Ecological identity refers to how individuals perceive themselves within natural settings. Perceptions that individuals hold about their local area and ecological systems have implications for their health, whereby 'damage of the planet is seen as damage to self' and to biophilic continuity (Nisbet et al., 2011, p. 304). Connor and colleagues (2004, p. 50) suggest that ecological identity/solastalgia may be fundamental to understanding place attachment and may 'mediate relationships between ecosystem distress and human health'. Reviewing Aboriginal people's solastalgic experiences may provide insights into this integrated health topic.

Topophilia

Topophilia refers to the 'love of a place' (O'Hare, 2007) and is often used to explain the importance that spaces and structures hold in everyday life (Ruan & Hogben, 2007). Topophilia not only focuses on the natural environment, emphasised in biophilia and solastalgia, but also on urban settings and how to make physical structures more accommodating. This approach is significant in light of increased urbanisation, as people need to understand how to make urban spaces health promoting, attractive and green (Ogunseitan, 2005). Topophilia focuses on the affective relationships people have within the settings they find themselves in, which in turn builds their identity (Kohane, 2007). Ogunseitan (2005, p. 143) described topophilia as 'a vivid and personal experience, but research is scarce on the determinants of individual preferences and on the potential health benefits derived from such experience'.

The relationship individuals have with urban spaces refers to places they interact within and the social/cultural characteristics of the physical environment (Murphy, 2007). As with biophilia, topophilia too has a phobia, described as the negative experience of a place 'overcome by a sense of insecurity and dislocation' (Jenner, 2007, p. 115). Topophobia is referred to as the lack of freedom, mobility and security within a place (Ruan & Hogben, 2007). Tuan (2007) notes that for some people, dread and fear of nature (evident in biophobia) causes people to want to distance themselves from it through structures that have been built, such as skyscrapers. However, humans enjoy having nature in their grasp in a controlled manner, such as courtyards, tourist beach resorts and holiday homes (Como, 2007; O'Hare, 2007). Contradicting this, West-Newman (2008, p. 166) identifies the topophilic attachment New Zealand Maoris have to beaches: it 'flows in our veins: the chemical composition of our blood is a reminder of our remote ancestry in the primordial oceans'. This seems to be the only study in topophilia focusing on Indigenous knowledge and biophilia.

Topophilia has relevance to Aboriginal people's relationships to places. However, few (if any) studies use the lens of topophilia theory to focus on Aboriginal peoples' relationships to place. Embodiment theory, which focuses on understanding knowing, being and acting, could potentially identify what experiences/relationships to place look and feel like for Aboriginal people as embodied in their daily practices and rituals. It has been recognised that place-making for Aboriginal people must incorporate the experience of place-based ancestry, Aboriginal identity and colonisation (Faintin, 2003). Faintin (2003) highlighted that Aboriginal cultural production is usually interpreted from the perspective of non-Indigenous people. For topophilia to capture Aboriginal people's experiences of place it should be understood through an Indigenous worldview.

Place Attachment

'Place' is a complex term. White, Virden, and Riper (2008) acknowledge that human–nature experiences in place are categorised as place identity, place dependency, place attachment, place bonding or sense of place. Fried (2000) and Cresswell (2004) highlight that place is often viewed as static rather than fluid, dynamic and always changing. Identity and place in Aboriginal cultures are intertwined with ancestry, and although fluid and dynamic, are often linked to a locality (Burgess & Morrison, 2007). People everywhere are involved in 'place-making', which can be progressive, declining or frozen in time. In the context of urban Aboriginal Australians that have weathered a history of cultural assimilation, connection to Country is constantly evolving, as is building sense of self, cultural identity and place (Kingsley et al., 2013; Souza & Rymarz, 2007). However, Aboriginal perspectives have been difficult to integrate into the Australian context, especially in school settings (Harrison & Greenfield, 2011). Sutherland and Swayze (2012, p. 87) note that the context of place varies considerably in Aboriginal communities and 'requires distinct methods of facilitating "cultural-border crossing"'. Therefore, Indigenous place literature needs to be considered from a place-based contextual perspective (Willox et al., 2012). White et al. (2008, p. 649) point out, Aboriginal people 'may see a place as part of the self and simultaneously as a resource for satisfying goals of explicitly felt behaviours'.

Fried (2000) highlighted that functional place attachment is hampered by physical destruction, but identifies that the social aspect of a community is as critical. Accordingly Aboriginal people's disconnection from land is associated with powerlessness, not only for the individual, but also in terms of social structures (Burgess & Morrison, 2007). However, Aboriginal people live in a global world and not all Aboriginal people feel disempowered when being off Country, and should be allowed to orbit between the two worlds they find themselves in order to engage in cultural practices and their own self-determination (Cape York Partnership, 2013). Although place is embodied by physical characteristics, these attachments are affected by 'product of risks and opportunities, the nature of the social organization attached to the locale, its political, social and economic relationships with other places' (Frumkin, 2003, p. 1451). Attachment to place can be cultivated through different forms and may extend beyond one's home to a wider locality where a sense of belonging is developed (Fried, 2000). Aboriginal people's sense of place has symbolic meaning seen in place-based rituals, knowledge, values and ceremony (Lowan, 2009), which is evidence of the biophilic symbolic tendency.

'Sense of place' refers to this psychological attachment to a place (Creswell, 2004). Massey (2005) develops a notion of 'progressive' sense of place that reconstructs the way we view place, combining bodies, objects and flows to explain the multiple meanings of place not frozen but immersed in change. This resonates with Aboriginal understandings where nature is often the embodiment of oneself and therefore are always re-creating their sense of its meaning. Massey's sense of place refers to a place as process defined by the outside; a site of multiple identities and histories defined by its interactions. It focuses on many elements such as landmarks, movement, orientation and body configurations within the places people move between. Expressing a similar idea, Gesler and Kearns (2002) highlight that place attachment is influenced if one is considered an 'insider', 'outsider' or 'somewhere in between'.

Spartz and Shaw (2011, p. 346) point out that 'personal connection[s] to land tend to come through in stories told about meaningful places and convey something deeper than basic attitude'. Larsen and Johnson (2012) acknowledged that Indigenous geography and place-based theories are caught 'in between worlds'. With increased globalisation there has been a loss of local cultures and 'homogenised global spaces' leading to a loss of sense of place (Creswell, 2004). This may be represented through rainforests turning into massive treeless cattle stations that have no biophilic connection. Perhaps it is represented by Indigenous people's connection to nature being viewed as the 'other' rather than integrated into research appropriately. Their connection to Country then posits Aboriginal people as the other–easy to erase as a people, as is the Country they care for, rather than as the rich and evolving sustainable cultures that they are and intend to be in the future.

CONCLUSION

This chapter uses the biophilia hypothesis as a catalyst to consolidate and explore theoretical concepts investigating the human–nature relationship to illustrate the connection Aboriginal people have with their Country. Currently, the application of the biophilia hypothesis in the research arena seems limited, perhaps due to this theory being contested. When reviewing four complementary human–nature theories, it is evident that Aboriginal perspectives and insights vary in the degree to which they are integrated into these approaches. It is the belief of the authors that by incorporating Aboriginal understandings of Country into health and ecological literature, we will better understand humans' relationship with nature.

Theories such as the biophilia hypothesis have gained popularity in academic circles in the last three decades as a way to explain the human–nature relationship. However, a central premise of modernity is its foundation in reductionism, duality and linearity, which is unable to cope with or accommodate the diversity and life-producing capacity of nature. This premise has supported the urbanisation movement across the globe. There may be reasons for this, like access to amenities, social contact, employment and impacts of climate change, but there also seem to be underlying conflicts with human–nature relationships. Overall, although there are ways of having incidental contact with nature in urban spaces, there seems to be a prevading disconnect from nature in these spaces.

There may be innate, learned, social and ecological reasons for destroying or keeping some forms of nature, but to ensure a consistent and justifiable response academics, policy-makers and communities must thinking about long-term approaches and question what nature means. As Gesler and Kearns (2002, p. 75) note, 'names are often used to classify, to put the often jumbled items or concepts we encounter into some kind of order'. However, perhaps by doing this (by calling something 'biophilia' or 'solastalgia') people are ignoring the complexity of the issues being addressed. This can create homogenisation and difficulty in describing the diversity of views, approaches and ways of life in this space. Advocates in the Ecohealth field need to be strategic in the way they promote such approaches so it is compelling enough for populations to want to protect plants, animals, diverse ancient environments, and to engage in land management that is health promoting, ecologically sound and socially just. One strategy would be to integrate Aboriginal knowledge into contemporary ecology management by adopting and adapting existing theoretical frameworks to support a clearer understanding of the links between connections to nature and human health.

Biophilia and theories focused on human–nature relationships, perhaps as a consequence of their complexity or because of the perception of being 'unscientific', have struggled to gain traction in political agendas, often overshadowed by other environmental issues. It is the authors' belief that the human–nature relationship is crucial to debates on issues such as climate change, public health and sustainability, because if people have no relationship with the natural environment, they will no longer see the point of protecting the planet and in turn the health of the human species as a whole. Better integration of Aboriginal perspectives of connection to nature into such theories will correspondingly enhance understandings of Aboriginal health and provide new insights in the interrelationship between society and ecology.

ACKNOWLEDGEMENT

Thank you Dr. Steve Trudgill and Dr. Cecily Maller for assisting in this chapter's development.

REFERENCES

Albrecht, G. (2005). 'Solastalgia': A new concept in health and identity. *Philosophy, Activism and Nature*, *3*, 41–55.

Albrecht, G. (2011, October 29). *Environment change, distress & human emotion solastalgia [TEDxSydney]*. Retrieved from http://www.youtube.com/watch?v=-GUGW8rOpLY. Sydney: TEDxTalks.

Albrecht, G., Sartore, G. M., Connor, L., Higginbotham, N., Freeman, S., Kelly, B., ... Pollard, G. (2007). Solastalgia: The distress caused by environmental change. *Australasian Psychiatry*, *15*, S95–S98.

Australian Human Rights Commission. (2013, April 18). *Questions and answers about aboriginal & torres strait islander peoples.* Retrieved from http://www.humanrights.gov.au/publications/questions-and-answers-about-aboriginal-torres-strait-islander-peoples. Sydney: AHMRC.

Babaian, C., & Twigg, P. (2011). The power of plants: Introducing ethnobotony and biophilia into your biology class. *The American Biology Teacher*, *73*(4), 217–221.

Bird, W. (2007). *Natural thinking: Investigating the links between the natural environment, biodiversity and mental health.* Sandy: Royal Society for the Protection of Birds.

Burgess, C. P., Berry, H. L., Gunthorpe, W., & Bailie, R. S. (2008). Development and preliminary validation of the 'Caring for Country' questionnaire: Measurement of an indigenous australian health determinant. *International Journal for Equity in Health*, *26*(7), 1–14.

Burgess, P., & Morrison, J. (2007). Country. In B. Carson, T. Dunbar, R. Chenhall & R. Bailie (Eds.), *Social determinants of indigenous health* (pp. 177–202). Crows Nest: Allen & Unwin.

Cajete, G. (1999). Reclaiming biophilia: Lessons from indigenous peoples. In G. A. Smith & D. R. Williams (Eds.), *Ecological education in action: On weaving education, culture and the environment* (pp. 189–206). Albuny: SUNY Press.

Campbell, D., Burgess, C. P., Garnett, S. T., & Wakerman, J. (2011). Potential primary health care savings for chronic disease care associated with Australian aboriginal involvement in land management. *Health Policy*, *99*, 83–89.

Cape York Partnership. (2013, April 23). *Orbiting between two worlds.* Retrieved from http://www.cyp.org.au/about/history/orbiting-between-two-worlds. Cairns: Cape York Partnership.

Chapman, R. J. (2002). Exploiting the human need for nature for successful protected area management. *The George Write Forum*, *19*(3), 52–56.

Como, A. (2007). The voyage and the house: Bernard rudofsky's search for place. In X. Ruan & P. Hogben (Eds.), *Topophilia and topophobia: Reflections on twentieth-century human habitat* (pp. 150–169). London: Routledge.

Connor, L., Albrecht, G., Higginbotham, N., Freeman, S., & Smith, W. (2004). Environmental change and human health in upper hunter communities of New South Wales, Australia. *EcoHealth*, *1*, 47–58.

Cresswell, T. (2004). *Place: A short introduction.* Malden: Blackwell.

Deb, D., & Malhotra, K. C. (2001). Conservation ethos in local traditions: The West Bengal heritage. *Society & Natural Resources: An International Journal*, *14*(8), 711–724.

Delavari-Edalat, F., & Abdi, M. R. (2009). Decision support system for monitoring environmental-human interactions. *Environmental Monitoring and Assessment*, *153*(1–4), 9–26.

Delavari-Edalat, F., & Abdi, M. R. (2010). Human–environment interactions base on biophilia values in an Urban context: Case study. *Journal of Urban Planning and Development*, *136*(2), 162–168.

Diamond, J. (1993). New guineans and their natural world. In S. R. Kellert & E. Wilson (Eds.), *The biophilia hypothesis* (pp. 251–274). Washington: Shearwater.

Faintin, S. (2003). Aboriginal identities in architecture. *Architecture Australia, September/October*, 84–87.

Félonneau, M. L. (2004). Love and loathing of the city: Urbanophilia and urbanophobia, topological identity and perceived incivilities. *Environmental Psychology*, *24*(1), 43–52.

Foucault, M. (1982). The subject of power. *Critical Inquiry*, *8*(4), 777–795.

Fried, M. (2000). Continuities and discontinuities of place. *Journal of Environmental Psychology*, *20*, 193–205.

Frumkin, H. (2003). Healthy places: Exploring the evidence. *American Journal of Public Health*, *93*(9), 1451–1455.

Gadgil, M., & Rao, P. R. S. (1995). Designing incentives to conserve India's biodiversity. In S. Hanna & M. Munasinghe (Eds.), *Property rights in a social and ecological context* (pp. 53–62). Washington: The Beijer International Institute of Ecological Economics and The World Bank.

Gesler, W. M., & Kearns, R. A. (2002). *Culture/place/health*. London: Routledge.

Gullone, E. (2000). The biophilia hypothesis and life in the 21st century: Increasing mental health or increasing pathology? *Journal of Happiness Studies*, *1*, 293–321.

Harrison, N., & Greenfield, M. (2011). Relationship to place: Positioning aboriginal knowledge and perspectives in classroom pedagogies. *Critical Studies in Education*, *52*(1), 65–76.

Heerwagen, J. H., & Orians, G. H. (1993). Humans, habitats, and aesthetics. In S. R. Kellert & E. Wilson (Eds.), *The biophilia hypothesis* (pp. 138–172). Washington: Shearwater.

Heinamaki, L. (2009). Rethinking the status of indigenous peoples in international environmental decision-making: Pondering the role of arctic indigenous peoples and the challenge of climate change. Climate Governance in the Arctic. *Environment & Policy*, *50*, 207–262.

Jenner, R. (2007). Potential places, places of potentiality: Levitation and suspension in modern Italian architecture. In X. Ruan & P. Hogben (Eds.), *Topophilia and topophobia: Reflections of twentieth-century human habitat* (pp. 114–132). London: Routledge.

Johnston, F. H., Jacups, S. P., Vickery, A. J., & Bowman, D. (2007). Ecohealth and aboriginal testimony of the nexus between human health and place. *EcoHealth*, *4*(4), 489–499.

Kahn, P. H. (1999). *The human relationship with nature: Development and culture*. Cambridge, MA: MIT Press.

Kahn, P. H., & Kellert, S. R. (2002). *Children and nature: Psychological, sociocultural, and evolutionary investigation*. Cambridge, MA: MIT Press.

Kahn, P. H., Sverson, R. L., & Ruckert, J. H. (2009). The human relation with nature and technological nature. *Current Directions in Psychological Science*, *18*(1), 37–42.

Karjala, M. K., & Dewhurst, S. M. (2003). Including aboriginal issues in forest planning: A case study in central interior british Columbia, Canada. *Landscape and Urban Planning*, *64*, 1–17.

Kellert, S. R. (1993). Introduction. In S. R. Kellert & E. Wilson (Eds.), *The biophilia hypothesis* (pp. 20–27). Washington, DC: Shearwater.

Kellert, S. R. (1996). *The value of life: Biological diversity of human society*. Washington, DC: Island Press.

Kingsley, J., Townsend, M., Henderson-Wilson, C., & Bolam, B. (2013). Developing an exploratory framework linking Australian aboriginal peoples' connection to country and concepts of wellbeing. *International Journal of Environmental Research and Public Health*, *10*(2), 678–698.

Kingsley, J., Townsend, M., Phillips, R., & Aldous, D. (2009). "If the land is healthy... it makes the people healthy": The relationship between country and health for the yorta yorta nation, boonwurrung and bangerang tribes. *Health and Place*, *15*(1), 291–298.

Kirker, A. (2012). Perilous paths life in your hands: Art from solastalgia and janet laurence: After Eden. *Art Monthly Australia*, *251*, 41–43.

Kohane, P. (2007). Agreement and decorum: Conversations within the architecture of louise kahn. In X. Ruan & P. Hogben (Eds.), *Topophilia and topophobia: Reflections on twentieth-century human habitat* (pp. 62–91). London: Routledge.

Ladson, T., & Finlayson, B. (2004). Specifying the environment's right to water: Lessons from victoria. *Dialogue*, *23*(3), 19–28.

Larsen, S. C., & Johnson, J. T. (2012). In between worlds: Place, experience, and research in indigenous geography. *Journal of Cultural Geography*, *29*(1), 1–13.

Lowan, G. (2009). Exploring place from an aboriginal perspective: Consideration for outdoor and environmental education. *Canadian Journal of Environmental Education*, *14*, 42–58.

Maller, C., Townsend, M., St Leger, L., Henderson-Wilson, C., Pryor, A., & Prosser, L. (2008). *Healthy parks, healthy people: The health benefits of contact with nature in a park context: A review of relevant literature*. Burwood: Deakin University.

Martin, J. F., Roy, E. D., Diemont, S. A. W., & Ferguson, B. G. (2010). Traditional ecological knowledge (TEK): Ideas, inspiration, and designs for ecological engineering. *Ecological Engineering*, *36*, 839–849.

Massey, D. B. (2005). *For space*. London: Sage Publications.

McKnight, D. (1999). *People, countries, and the rainbow serpent: System of classification among the lardil of mornington Island*. New York, NY: Oxford University Press.

McNamara, K. E., & Westoby, R. (2011). Solastalgia and the gendered nature of climate change: An example from Erub Island, Torres Strait. *EcoHealth*, *8*(2), 233–236.

Muller, S. (2012). 'Two ways': Bringing indigenous and non-indigenous knowledge together. In J. K. Weir (Ed.), *Country, native title and ecology* (pp. 59–80). Canberra: Australian National University E Press.

Murphy, P. (2007). Economy and affect: People–place relationship and the metropolis. In X. Ruan & P. Hogben (Eds.), *Topophilia and topophobia: Reflections on twentieth-century human habitat* (pp. 202–220). London: Routledge.

National Aboriginal Community Controlled Health Organisation. (2012, December 26th). *Definitions*. Retrieved from: http://www.naccho.org.au/aboriginal-health/definitions/. Canberra: NACCHO.

Nelson, R. (1993). Searching for the lost arrow: Physical and spiritual ecology in the hunters world. In S. R. Kellert & E. Wilson (Eds.), *The biophilia hypothesis* (pp. 201–328). Washington, DC: Shearwater.

Nettleton, C., Stephens, C., Bristow, F., Claro, S., Hart, T., McCausland, C., & Mijlof, I. (2007). Utz wachil: Findings from an international study of indigenous perspective on health and environment. *EcoHealth*, *4*, 461–471.

Nisbet, E. K., Zelenski, J. M., & Murphy, S. A. (2011). Happiness is in our nature: Exploring nature relatedness as a contributor to subjective well-being. *Journal of Happiness*, *12*, 303–322.

Nubhan, G. P., & Antoine, S. (1993). The loss of floral and faunal story: The extinction of experience. In S. R. Kellert & E. Wilson (Eds.), *The biophilia hypothesis* (pp. 229–250). Washington, DC: Shearwater.

Nursey-Bray, M., & Hill, R. (2010). Australian indigenous peoples and biodiversity. *Social Alternatives*, *29*(3), 13–19.

Ogunseitan, O. A. (2005). Topophilia and the quality of life. *Environmental Health Perspectives*, *113*(2), 143–148.

O'Hare, D. (2007). Not another waikiki? Mobilizing topophilia and topophobia in coastel resorts areas. In X. Ruan & P. Hogben (Eds.), *Topophilia and topophobia: Reflections on twentieth-century human habitat* (pp. 185–201). London: Routledge.

Orr, D. W. (1993). Love it or lose it: The coming biophilia revolution. In S. R. Kellert & E. Wilson (Eds.), *The biophilia hypothesis* (pp. 415–440). Washington, DC: Shearwater.

Packer, J., Brouwer, N., Harrington, N., Gaikwad, J., Heron, R., Elders, Y. C., … Jamie, J. (2012). An ethnobotanical study of medicinal plants used by the Yaegl Aboriginal community in northern New South Wales, Australia. *Journal of Ethnopharmacology*, *139*, 244–255.

Palsson, G. (1995). Learning by fishing: Practical science and scientific practice. In S. Hanna & M. Munasinghe (Eds.), *Property rights in a social and ecological context* (pp. 85–98). Washington, DC: The Beijer International Institute of Ecological Economics and The World Bank.

Panelli, R., & Tipa, G. (2009). Beyond foodscapes: Considering geographies of indigenous well-being. *Health & Place*, *15*(2), 455–465.

Pretty, J. (2007). *The earth only endures: On reconnecting with nature and our place in it*. London: Earthscan.

Rasmussen, M., et al. (2011). An aboriginal Australian genome reveals separate human dispersal into Asia. *Science*, *334*, 94–98.

Richmond, C. A. M., & Ross, N. A. (2009). The determinants of first nation and inuit health: A critical population health approach. *Health & Place*, *15*, 403–411.

Rose, D. (2004). Fresh water rights and biophilia: Indigenous Australian perspectives. *Dialogue*, *23*(3), 35–43.

Ruan, X., & Hogben, P. (2007). Architectural enclosure: A prologue to topophilia and topophobia. In X. Ruan & P. Hogben (Eds.), *Topophilia and topophobia: Reflections on twentieth-century human habitat* (pp. 1–12). London: Routledge.

Sagan, D., & Margulis, L. (1993). God, gaia, and biophilia. In S. R. Kellert & E. Wilson (Eds.), *The biophilia hypothesis* (pp. 345–364). Washington, DC: Shearwater.

Shiva, V. (1995). *Monocultures of the mind: Perspectives on biodiversity and biotechnology*. Malaysia: Third World Network.

Simaika, J. P., & Samways, M. J. (2010). Biophilia as a universal ethic for conserving biodiversity. *Conservation Biology*, *24*(3), 903–906.

Souza, M., & Rymarz, R. (2007). The role of cultural and spiritual expressions in affirming a sense of self, place, and purpose among young Urban, indigenous Australians. *International Journal of Children's Spirituality*, *12*(3), 277–288.

Spartz, J. T., & Shaw, B. R. (2011). Place meanings surrounding an Urban natural area: A qualitative inquiry. *Journal of Environmental Psychology*, *31*, 344–352.

Speldewinde, P. C., Cook, A., Davies, P., & Weinsteins, P. (2011). The hidden health burden of environmental degradation: Disease comorbidities and dryland salinity. *EcoHealth*, *8*, 82–92.

Stain, H. J., Kelly, B., Carr, V. J., Lewin, T. J., Fitzgerald, M., & Fragar, L. (2011). The psychological impact of chronic environmental adversity: Responding to prolonged drought. *Social Science & Medicine*, *73*, 1593–1599.

Stephens, C., Porter, J., Nettleton, C., & Willis, R. (2006). Disappearing, displaced, and undervalued: A call to action for indigenous health worldwide. *The Lancet*, *367*, 2019–2228.

Strang, V. (2005). Knowing me, knowing you: Aboriginal and European concepts of nature as self and other. *Worldviews*, *9*(1), 25–56.

Sutherland, D., & Swayze, N. (2012). The importance of place in indigenous science education. *Cultural Studies of Science Education*, *7*, 83–92.

Taylor, L. (2012). Connections of spirit: Kuninjku attachments to country. In J. Weir (Ed.), *Country, native title and ecology* (pp. 21–46). Canberra: Australian National University.

Tonkinson, R. (2011). Landscapre, tranformations, and immutability in an aboriginal Australian culture. In P. Muesburger (Ed.), *Culture memories, knowledge and space* (pp. 329–345). Berlin: Springer Science and Business Media.

Tuan, Y. (2007). Time, space, and architecture: Some philosophical musing. In X. Ruan & P. Hogben (Eds.), *Topophilia and topophobia: Reflections on twentieth-century human habitat* (pp. 22–30). London: Routledge.

Tucker, M. E. (2013, April 19th). *Thomas berry "The universe is a communion of subjects, not a collection of objects"*. Retrieved from: http://www.thomasberry.org/Biography/tucker-bio.html

Ulrich, R. S. (1993). Biophilia, biophobia, and the natural landscapes. In S. R. Kellert & E. Wilson (Eds.), *The biophilia hypothesis* (pp. 73–137). Washington, DC: Shearwater.

United Nations. (2007). *The UN declaration on the rights of indigenous peoples*. New York, NY: UN.

Weir, J. (2012). Country, native title and ecology. In J. Weir (Ed.), *Country, native title and ecology* (pp. 1–20). Canberra: Australian National University.

West-Newman, C. L. (2008). Beach crisis: Law and love of place. *Space and Culture*, *11*(2), 160–175.

White, D. D., Virden, R. J., & Riper, C. J. V. (2008). Effects of place identity, place dependency, and experience-use history on perceptions of recreation impacts in a natural setting. *Environmental Management*, *42*(4), 647–657.

Willox, A. C., Harper, S. L., Ford, J. D., Landman, K., Houle, K., Edge, V. L., & Rigolet Inuit Community Government. (2012). 'From this place and of this place': Climate change, sense of place, and health in nunatsiavut, Canada. *Social Science & Medicine*, *75*, 538–547.

Wilson, E. O. (1984). *Biophilia*. Cambridge, MA: Harvard University Press.

Wilson, E. O. (1993). Biophilia and the conservation Ethic. In S. R. Kellert & E. Wilson (Eds.), *The biophilia hypothesis* (pp. 31–41). Washington, DC: Shearwater.

PART II
SOCIAL-ECOLOGICAL HEALTH INJUSTICES AND PATHS TO JUSTICE

TIBETAN PROTEST SELF-IMMOLATION IN CHINA: REFLECTIONS ON ECOLOGY, HEALTH AND POLITICS

Colin D. Butler

ABSTRACT

Purpose – *This chapter explores the protest self-immolations since 2009 of over 100 Tibetans in China. It investigates whether these events have ecological as well as social causes and may thus be relevant to the emerging discipline of 'EcoHealth'.*

Method – *Targeted literature review and reflective analysis, presented as a narrative.*

Findings – *Chinese citizens identifying as Tibetan have experienced substantial ethnically based discrimination for over 60 years, manifest as attempted cultural destruction, pervasive disrespect and linguistic suppression. Tibetans, now a minority in much of their former territory, have witnessed and at times been forced to participate in ecological destruction, much of it led by Chinese settlers, endorsed by occupying authorities. Tibetans have for decades protested against the Chinese they regard as invaders and occupiers, but Tibetan acts of protest*

Ecological Health: Society, Ecology and Health
Advances in Medical Sociology, Volume 15, 67–89

ISSN: 1057-6290/doi:10.1108/S1057-6290(2013)0000015007

self-immolation are a recent response. Academic analysis has been scarce, particularly by Chinese scholars. Until now, EcoHealth practitioners have also denied any relevance, as if in a waltz led by the Chinese government.

Practical and social implications – *Attempts to identify rational causes for Tibetan self-immolation conflict with themes of liberation and fairness central to Communist Chinese ideology. Most Chinese analysis of Tibetan self-immolation is superficial, nationalistic and unsympathetic. Also disturbing is the reaction to these issues shown by the International Association of Ecology and Health. It is suggested that this illustrates a failure to translate rhetoric of 'speaking truth to power' to reality, a retreat from idealism common to many social movements.*

Originality and value – *Increasing human demand on a limited biosphere necessitates a deepened understanding of eco-social factors. Practitioners concerned with sustaining our civilisation are encouraged to explore the integrated dimensions revealed by this case study.*

Keywords: China; EcoHealth; protest immolation; social medicine; suicide; Tibet

"Science is that magic word that turns value-laden knowledge into value-free knowledge"

Vicente Navarro, 1980

INTRODUCTION: ECOHEALTH AND PLANETARY OVERLOAD

EcoHealth is a recently coined term for a fairly recently identified field of study. Its innovation is to link 'eco' with 'health'. The Greek word '*ecos*' is widely translated as 'home' and is linked to two important disciplines, each relevant to EcoHealth. These are economy (management of the human system) and ecology (study of the living home of which humans are a part). It is also linked to 'ecological economics'. This term, apparently tautological ('eco eco'), tries to draw attention to the idea that the management of 'home' should also include its wider, non-human components. After all, living organisms, including humans, continually consume and transform other matter, living and inert. Ecological economists (Daly, 1996) have long argued that mainstream, dominant economics are pre-occupied with a small subset of the wider economy, that which can be easily monetised. There is a tacit assumption that the much larger *ecos* – the biosphere in which we evolved, and which we continually transform, is so large and abundant that we need not worry about its depletion and degradation.

In recent decades there has been growing multi-disciplinary appreciation that the impact of humanity upon the Earth system, including its ecosystems, is on a scale sufficient to harm the whole human enterprise (Butler, 2008). For example, Lord Martin Rees, President of the Royal Society, UK, published *Our Final Century* (Rees, 2003). Hundreds of other articles, books and reports have in recent years similarly warned that a massive change in human affairs is required if civilisation is to survive (Ehrlich & Ehrlich, 2013; Oreskes & Conway, 2013). While a few prominent social scientists have disputed these broad conclusions (Demeny, 1988; Johnson, 1999) an increasing number of prominent social scientists appear to be awakening to this unpleasant prospect (Graham, 2013; Urry, 2010).

From 1970 to the mid-1990s a gradually increasing number of health scientists have also linked global environmental 'overload' with the diminishing prospect of sustainable health (Butler, 1994; King, 1990; McMichael, 1993; Ward & Dubos, 1973). In this context, the sub-discipline of 'ecosystem health' emerged in the 1990s, surviving, by that name, for about a decade (Rapport, 1995). This term, confusing to some, tried to position human health and well-being within its broader ecological and environmental context. But many ecologists were critical, suggesting that to apply the term 'healthy' to ecosystems was anthropocentric, judgemental and arbitrary. However, to pragmatic health workers, such comments had little relevance, since neither the journal nor its governing society was primarily about ecosystems. Furthermore, the idea that (human) health is desirable and can be recognised is central to health workers.

The society and journal folded in the early 2000s, in part fed by semantic controversies. Soon after, in 2004, many of the concepts of Ecosystem Health were re-born in a new journal, *EcoHealth* (Parkes et al., 2005). Two years later the International Association of Ecology and Health (IAEH) was founded (Wilcox & Daszak, 2006). It governs the journal. Confusingly, however, an NGO called 'Wildlife Trust' a few years later renamed itself as the 'EcoHealth Alliance'.

ECOHEALTH AND ONE HEALTH

There has been an even more recent recognition that some of the ideas of EcoHealth and Ecosystem Health intersect with an older school, called 'One

Health'. This term, probably coined about 40 years ago, itself evolved from 'One Medicine', the insight, more than 100 years old, of the similarities and connections between veterinary and human medicine (Zinsstag, Schelling, Waltner-Toews, & Tanner, 2011). One Heath practitioners strive to strengthen the cross-sectoral cooperation between the human and veterinary communities (Mackenzie & Jeggo, 2011).

This understanding, widely attributed to two of the leading medical (sic) figures of the late 19th century – Rudolf Virchow and William Osler – is based on two key insights. The first is that humans are animals, at least physiologically. The second was that humans and animals sometimes share infections, including parasites, which can cross the species barrier. Some of these organisms are called zoonoses, a term credited to Virchow. Originally it was thought that zoonoses originated in non-human animals and passed onto humans. There is more recent recognition that some infections have originated in the human animal, then passed onto animals. Tuberculosis is one such example (Smith, Hewinson, Kremer, Brosch, & Gordon, 2009).

Importantly, however, One Health has little history of interest in the social determinants of health, perhaps because this has little perceived relevance to veterinary medicine, at least historically. More recently, however, awareness of social dimensions to non-human animal behaviour has expanded greatly (de Waal, 2009).

In the last decade, there has been a revival of One Health, and an attempt to incorporate an awareness of local and regional ecological and environmental change, as reflected in the phrase 'One World, One Health' (http://www.oneworldonehealth.org/sept2004/owoh_sept04.html).[1] Until recently, both One Health and EcoHealth developed in parallel for several years, largely mutually unaware of each other, and then unsure whether to be rivals, friends or partners.[2]

Both disciplines have narrow and wide interpretations. In each case the larger perspective includes an attempt to understand the importance of the global ecological and environmental milieu. However, to date, One Health has made little effort to encourage its practitioners to grapple with the social causes and consequences of global change, though some One Health leaders work intensively to try to alleviate infectious diseases of poverty, itself a consequence and cause of poverty (Schelling, Wyss, Bechir, Moto, & Zinsstag, 2005). Their lack of apparent interest in this area may partly reflect veterinary public health training which is likely to have far less exposure to these ideas than human public health courses.

On the other hand, EcoHealth (as espoused by the IAEH and the journal) claims that social factors are important. This is best supported by its frequent use of the term 'transdisciplinarity'. While this term has several meanings, it generally includes principles of equity, though perhaps not social justice. For example, the prominent human ecologist Lawrence proposes that an 'interdisciplinary' holistic approach that combines 'biological, cultural, economic, political, psychological and social factors' forms an 'ecological' perspective which is in turn a foundation for 'transdisciplinary' research (Lawrence, 2004). Some founders of EcoHealth define it as the (transcendant?) integration of 'knowledge across and beyond academic disciplines' (Parkes et al., 2005). Transdisciplinary research is also sometimes used to stress the equivalence in importance and respect, if not power, between researcher and researchee, similar to participatory action research. This is a field in which the hypotheses of the community being studied are given substantial weight, and where the research findings are carefully communicated. In contrast, the EcoHealth Alliance is mainly concerned with the interaction between newly recognised and emerging infectious diseases and ecological change (Butler, 2012; Robbins, 2012).

ECOHEALTH AND SOCIAL MEDICINE

There is an under-appreciation, both within EcoHealth and One Health, of the relevance and importance of the work of Virchow, the outstanding German medical scientist of the 19th century, still remembered for his contribution to pathology, social medicine and, as mentioned, 'One Health'. However, these achievements are not equally recognised by their three intellectual descendants. The European Society of Pathology still publishes *Virchows Archiv* (http://www.springer.com/medicine/pathology/journal/428). However, in this chapter, I primarily focus on Virchow's other two main contributions.

In 1848 Virchow helped establish a weekly publication called *die Medizinische Reform* (Medical Reform). In the brief window of time and liberal thought then crossing Europe, this journal promoted the cause of what is still widely called 'social medicine'. This publication is reported to have had headlines such as 'medicine is a social science' and 'the physician is the natural lawyer of the poor' (Brown & Fee, 2006). Social science, at that time very young, was then as now, focussed on efforts to reduce inequality. Virchow is said to have been inspired by his countryman, Friedrich Engels, who had recently published (in German) '*The conditions of the working class in England*' (Engels, 1845 (reprinted 1958)), a foundational text of social medicine. Engels described the dreadful poverty and squalor in the slums of the industrialising city of Manchester, which he linked to bad health.

It is as if Virchow has three lineages, none of which integrate even two of his major contributions (including social medicine and health promotion, neither of which show convincing understanding of ecological issues) (Butler & Friel, 2006). On the other hand, EcoHealth seems close to blending at least two. But while one foundational paper relevant to EcoHealth states 'pioneers such as Snow, Koch and Budd knew that social factors had a strong influence on infections' (Parkes et al., 2005), this paper does not mention Virchow, though his younger colleague and rival,[3] Robert Koch, is named. This would seem hair-splitting were it not that Virchow is so central to infectious diseases and One Health. There is at least one exception, however:

> nowhere in the recommendations do the ... authors discuss altering the economic and political causes of disease emergence. ... one would think that health practitioners should be making strong health representations to organisations ... on how to prevent the mess in the first place. This is one of the many instances where we can see that the ideological lenses through which health and disease are studied constrain the opportunities to find solutions. This, if nothing else, should raise a warning flag that those who study disease are not necessarily well-equipped to promote health and that new modes of thought which can incorporate multiple perspectives are required. (Waltner-Toews, 2004) (p. 12)

Perhaps this exception (in a textbook) proves the rule.

ECOHEALTH AND TIBET

With this background, I now hope the reader is in a position to assess whether what follows is a contribution to contemporary social medicine and also to EcoHealth. I next document and reflect upon the reaction to a paper I submitted to the 4th biannual conference of the IAEH, held in Kunming, Yunnan, China, in October 2012. This was shortly before the appointment of the new Chinese head of state, Paramount Leader Xi Jinping, and thus an unusually sensitive time. My rejected abstract proposed a link between public self-immolation – suicide and protest by burning – and 'eco-social' distress. It was scored as 0/5 – not warranting presentation even as a poster. As a co-editor of the journal (since 2010) and as an invited keynote speaker at the 2nd annual conference (held in Mexico in 2008) this score was unusual.

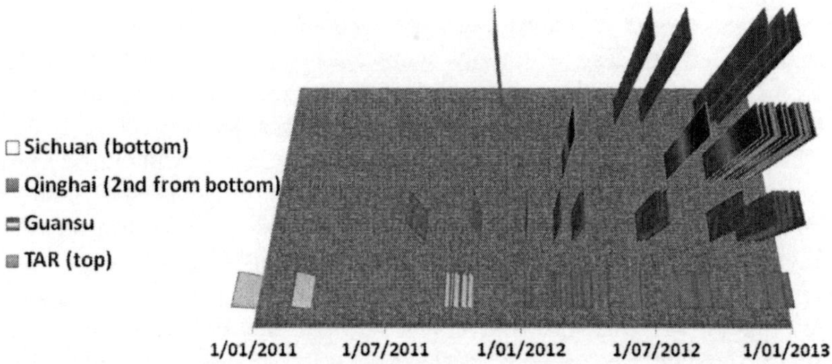

Fig. 1. The Location and Date for 93 Known Protest Immolations in Tibet, Occurring Between March 2011 and December 9, 2012. Most were Initially in Sichuan. In Late 2012 the Focus Shifted to the North, to Guansu and Qinghai Provinces. A Few have also Recently Occurred in the Tibetan Autonomous Region (TAR), where the Population Remains More Ethnically Homogenous (Data from http://www.freetibet.org/news-media/na/full-list-self-immolations-tibet).

I was motivated to submit this paper because of the rate of protest self-immolation by Tibetans in China. At the time I submitted the abstract about 50 people had died this way, an extraordinarily high rate considering there are only about 6 million ethnic Tibetans. In the following 10 months (as I write) this number has more than doubled. A few Tibetans living in exile in India and Nepal have also self-immolated in sympathy.

In India, suicide by fire is an ancient custom, but there is no Tibetan equivalent, perhaps because of the scarcity of firewood, and the consequent rarity of cremation. Most cases of self-immolation in China were originally in Sichuan, but more recently, their focus has shifted to the north, to Guansu and Qinghai (see Fig. 1), the Chinese provinces with significant Tibetan minorities. A list of Tibetan protest-immolations is maintained at http://www.freetibet.org/news-media/na/full-list-self-immolations-tibet.

PROTEST SELF-IMMOLATION

Protest self-immolation has a long history. Some, unlike in China to date, have had significant political impacts, most notably in Vietnam in 1963 and Tunisia in 2010. When Buddhist monk Thích Quáng Ðuc, aged 67, set himself on fire on a busy Saigon street to protest Buddhist persecution by the South Vietnamese government, then led by the Roman Catholic Ngô Đình Diêm, it became a potent symbol of distress in America's ally. Malcolm Browne (who had been alerted) filmed Đuc's burning, surrounded by his students, and was awarded the Pulitzer Prize (Biggs, 2008). It didn't end the Vietnam War, but within six months, Diêm's government had fallen to an army coup, despite strong US support for it. Other monks, nuns and lay Buddhists have since self-immolated in Vietnam, according to Thich Quáng Đô, leader of the banned Unified Buddhist Church of Vietnam, however, proof of this is hard to find.[4]

In January 1968, 20-year-old Czech student Jan Palach sef-immolated in Wenceslas Square to protest the Soviet-led invasion of his country that crushed the 'Prague Spring', preceding liberation from Russia by over two decades. Palach's sacrifice is still commemorated on each anniversary. But Palach's life was not threatened by Soviet occupation. No one starved in egalitarian Czechoslovakia at that time, yet, much of its population was unhappy. Thousands risked Soviet wrath by supporting the dissident Czech leader, Alexander Dubček. What would drive this resistance? It seems plausible to consider the possibility of social ecology, a group consciousness, which collectively yearned for something higher on Maslow's hierarchy of needs than food or water, including self-determination and freedom.

Uzzell traces the idea that some suicides can have an altruistic motivation to the pioneering sociologist Durkheim (Uzzell, 2012). Uzzell describes how the self-immolation of the impoverished, publically humiliated[5] Tunisian street vendor Mohamed Bouazizi acted as a catalyst for the Arab Spring, eventually triggering the overthrow of four Middle Eastern dictators (Uzzell, 2012). Many acts of protest self-immolation are imitated, a form of contagious social movement. Bouuaziz's act is attributed as motivating over 100 other self-immolations protesting social and economic conditions in the Middle East, including Algeria and Saudi Arabia.

The Oxford sociologist Michael Biggs has tabulated over 500 acts of protest self-immolation occurring between 1963 and 2002,[6] with the highest rate among Kurds.[7] But the highest number occurred in 1990 in India, when in only 10 weeks over 200 people self-immolated. These were mostly students protesting at affirmative action aimed at redressing entrenched Indian inequality, thus threatening to reduce access to university and civil service positions to some from more privileged castes (Biggs, 2008). However, in India, self-death by fire is mostly *not* political protest, but the third most common method of private suicide, after poisoning (often with pesticides) and hanging (Vijaykumar, 2007). Recently, young Bulgarians have been committing self-immolation to protest deepening poverty and inequality in one of the most deprived parts of Europe.

Outside India, self-immolation retains its capacity to shock and horrify. But few political self-immolations have achieved their desired reforming effect, including, to date, those by Tibetans. Although Tibetan suicides have generated considerable publicity in some countries, they have scarcely been reported in China (Carrico, 2012) and seem to have had negligible influence. Some Chinese spokespeople attribute the cause of increasing incidents of self-immolation among Tibetans to appeals from the Dalai Lama, claiming that this Nobel Peace Laureate and spiritual leader is orchestrating self-immolation. The Tibetan government in exile, based in Dharamsala, India, denies this.

I next develop arguments that Tibetan protest-immolation is related to ecology and health (hence EcoHealth). To do that, more description is needed of the political, social and ecological circumstances of Tibetans in China.

THE CHINESE IN TIBET

For centuries, the relationship between China and Tibet has been fractious. Although Tibet and China signed a peace treaty as equals in the 9th century, for hundreds of years China, the much larger and more populous neighbour, has claimed dominance over geographically large but sparsely populated Tibet. But until the second half of the 20th century, this dominance was mostly theoretical. In 1910 a Chinese army briefly occupied Tibet's capital, Lhasa, but was soon forced to retreat, allowing Tibet's political (and religious) leader the 13th Dalai Lama to return from brief exile in India. Britain had also invaded Tibet in 1903; in 1914 it signed a treaty with Tibet, which gained British India a large swathe of fertile land in what is today the north eastern Indian state of Arunachal Pradesh.[8] The Chinese rejected this treaty, but could do no more than protest. For several decades, Tibet then operated as an effectively independent country (Goldstein, 1989).

In 1949, the People's Liberation Army of the new Communist Chinese regime were able to achieve what their nationalist opponents[9] could not. The main stated Chinese justification for this violent attack was that Tibet was under-developed, inequitable and that its people needed 'liberation'. Typical of this genre is the book *Lhasa the Open City* by the British–Chinese author Han Suyin who praised the invasion for freeing the Tibetans from their feudal past (Suyin, 1979).

Prior to 1949, Tibet was indeed feudal, beset with political intrigue, (Goldstein, 1989) xenophobia, monasticism and unique and complex religious practices. Some reform was underway, but more was desirable. But it was free of self-immolation as a means of protest, and claims that the Chinese have delivered net benefits to the ordinary Tibetans since 1949 that would not otherwise have occurred cannot be substantiated, and in any case do not justify the great suffering subsequently inflicted by China upon Tibetan people and their culture.

The high altitude of Tibet is unpleasant for most ethnic non-Tibetans, who lack physiological adaptation to the thin air (Beall et al., 2001). If the Chinese did not invade for altruism then what was their motivation? Mao Tse Tung is reported as stating: 'Although Tibet does not have a big population, its international [strategic] position is extremely important. Therefore we must occupy it and transform it into a people's democratic Tibet' (Jian, 2006). Another clue lies in a Chinese name for Tibet; Xizang, 'Land of the Western Treasure' (Avedon, 1986). While this treasure could be spiritual, it is chiefly material. Tibet holds immense material and strategic value to China, including its fresh water, vast area and minerals. Its location on the roof of the world gives it power to dam and divert many of Asia's major rivers, for both electrical power and irrigation. The most plausible reason that China strives to control Tibet's physical, ecological and security resources is in order to enhance the prosperity and long-term security of the Chinese people, who are predominantly of Han ethnicity. The well-being of Tibetans is subsidiary.

THE PLUNDERING OF TIBETAN ECOLOGY

A century ago Tibet was rich in biodiversity, with immense herds of antelope and other species. It is difficult to document the scale of biodiversity change in Tibet since then, but it appears enormous. British explorer Cecil Rawling described, while exploring parts of the Tibetan plateau between 1903 and 1905, a place he called 'Antelope Plain' in which herds of 15,000–20,000 Tibetan antelope (*Pantholops hodgsonii*) congregated, visible at one time (Rawling, 1905).

By 2000, Tibet's once densely forested south-eastern region had been almost completely converted to agricultural land (Cui & Graf, 2009), with consequent loss of habitat for numerous animals. Several Tibetan species are now classed as endangered or vulnerable, including the snow leopard (*Uncia uncia*) and the Marco Polo sheep (argali) (*Ovis ammon polii*).

The antelope population is reported to have declined by 90% between 1900 and the early 1990s, to about 70,000 due to its deliberate hunting, particularly to provide shahtoosh, a very warm and light form of wool (International Fund for Animal Welfare, 2001).

The claim by the exiled Tibetan community that Buddhism has long restricted the practice of hunting has recently been challenged. Yeh (2012) cites Huber (1997) as noting that none of the 14th Dalai Lama's seven books published before 1985, nor interviews that he gave from his arrival in India in 1959 through the mid-1980s, made any reference to the relationship between Tibetan Buddhism and ecology (Huber, 1997:107e108). However, in opposition to this, Rawling (1905) mentions that while 'indiscriminate' killing of antelope in 'great numbers' was done by men from 'Ladakh, Rundor and Rudok' he also noted that herds of yak and antelope had no fear of man, and were in fact curious. Describing a similar time and phenomenon, the Swedish explorer Sven Hedin wrote: 'No one sends a bullet after a kiang (*Equus kiang*) within sight of the mountain of the Gods ... the animals know that the holy lake and its shores are a sanctuary' (1913). Hedin was commenting on an area in Western Tibet near Mt Kailash, sacred to Buddhists and Hindus.

The existence of a sanctuary implies that other parts of Tibet were not sanctuaries, but it is also evident that pious Buddhists traditionally practised a degree of self-restraint and wildlife protection. Tibet has numerous sacred mountains and lakes, the wildlife of which is likely to have experienced at least some degree of protection. Tibetan Buddhists were not, however, generally vegetarian. The consumption of products from domesticated animals, especially of yak butter, was vital. The population of Tibetans, in comparison to its vast area was tiny. This, more than any non-violent practice, may have protected its herds and flocks. However, several Tibetan customs also limited population growth, principally polyandry and monasticism.

In 1958, soon after the Chinese occupation of Tibet, and 'continuing for two decades', blue sheep (*Pseudois nayaur*) were killed in 'commercial quantities' in Qinghai, for export to Germany. In his book 'Wildlife of the Tibetan Steppe' the renowned conservationist and biologist Schaller provides numerous references suggestive of a drastic decline in wildlife populations in the vast rangelands of northern Tibet (1998). He notes:

> The beauty of these steppes and peaks will persist, but without wildlife they will be empty and the Tibetans will have lost part of their natural and cultural heritage. My vision for tomorrow is the past when humans, livestock, and wild animals lived in the vast steppes of the Chang Tang in ecological harmony. (Schaller, 1998) (cited by Miller, 2009)

While subsistence hunting of grazing mammals and the deliberate killing of wolves in Tibet is probably ancient, there is still a view that contemporary Tibetans are comparatively less likely to kill wildlife than many other people (Yeh, 2012). Yeh also describes how in 2006, in response to an appeal by the Dalai Lama, Tibetans in China set fire to more than a million dollars' worth of otter, leopard and tiger pelts, imported (illegally) from India. The Chinese interpreted these fires as acts of loyalty to the Dalai Lama and threatening to their territorial sovereignty. Some salaried employees were then forced to wear endangered animal pelts, including tiger-skin, thus violating Chinese national law (Yeh, 2012).

Tibetans may be more culturally sensitive to the loss of wildlife than some other human populations – they may have comparatively high *biophilia* (a deep appreciation for other species).[10] In Sichuan, due to population expansion and deforestation, there is increasing contact and conflict between bears and people. Though killing bears is illegal, poaching has occurred, but comparatively less is undertaken by Tibetans (Liu et al., 2011). In summary, the Chinese occupation of Tibet has been accompanied by important adverse ecological effects, and it is plausible that these effects are depressing the spirit of at least some Tibetans. This hypothesis should be of interest to an association concerned with ecology and health.

LAND SICKNESS, SOLASTALGIA AND BIOPHILIA

The American ecologist Leopold wrote extensively of what he called 'land sickness' (1949). This is most commonly interpreted as harm to biodiversity and ecosystems arising from the intrusion of humans upon non-human species, whether plants, animals, or fungi. But could land sickness also affect human well-being? Clearly, Leopold was himself moved by the harm to nature which he witnessed, as have many other leading conservationists, from John Muir to Thoreau and Bob Brown, until recently the leader of the Australian Green Party.

The philosopher Albrecht has developed a concept related to *biophilia* called *solastalgia*, the pain experienced by some people when landscapes are altered in unwanted ways, such as if an open coal pit replaces pasture, or a wetland is converted to a car park (Albrecht et al., 2007). Indeed, such thoughts have appeared in popular Western culture for decades, such as in the song 'Big Yellow Taxi' by Joni Mitchell.[11] Some people have deep traditions of 'sacred' groves, and sites (Ramakrishnan, Saxena, & Chandrashekara, 1998), illustrative of what the Millennium Ecosystem Assessment called 'cultural' ecosystem services (Butler et al., 2003).

It is not implausible, though speculative, that the psychological health of large numbers of people, perhaps if especially biophilic, will be harmed by their witness of significant landscape changes, such as deforestation and the hunting of charismatic and endangered species. In some cases, such threatened species may not be harmed directly, such as the transmissible facial tumour disease afflicting Tasmanian devils.

Many oppressed people are denied environmental and ecological 'entitlement'. Such examples are particularly clear concerning the historical displacement of indigenous peoples. More recent examples include Australian Aboriginals, tribal people in Ecuador, and the San in Botswana (Good, 1992). In some of these groups, rates of suicide and other forms of self-harm are high, though protest-immolation is rare or unknown. Collectively, there is often a deep and shared sense of loss. In summary, it is impossible to exclude ecological factors, including mourning for the loss of wild spaces from the mix of causal determinants for self-immolation. We can be confident that such suffering and loss is not purely social, but more properly described as 'eco-social'.

TIBETAN CULTURAL OPPRESSION BY CHINA

The scale of cultural oppression of Tibetan people is bitterly contested. For decades, the Chinese have imposed severe restrictions on reporting and travel within occupied Tibet. Such restrictions themselves are the most convincing evidence that China has something to hide, perhaps something shameful. The investigative film maker and journalist Vanya Kewley at great risk undertook a clandestine trip for six weeks to Tibet in the late 1980s, funded by UK Channel 4. This resulted in a film and a book (Kewley, 1990), both of which support claims made by the exiled Tibetan community of over one million premature Tibetan deaths since the 1949 invasion. While some of these deaths were part of the great Mao-made famine from 1959 to 1961 there is sufficient other evidence, including mass uprisings, imprisonment for 're-education' and torture[12] to be confident that there is much unhappiness among Tibetans due to their restricted freedom and the ongoing discrimination they face.

There is also consensus that numerous monasteries in Tibet were looted and destroyed. Countless religious artefacts were either sold to collectors or melted and sold for their metallic value (Sakya, 1999). Monks and nuns were forced to copulate in public (Abel, 2012). From the Chinese communist perspective such acts were justified by a value system that despised religion, but for Tibetans this abuse exemplifies disrespect and attempted cultural genocide. The plight of Tibetans continues to generate sympathy among supporters in several Western countries and India, where the Tibetan government-in-exile has been based since 1959 (Avedon, 1986).

SELF-IMMOLATION OF TIBETANS IN CHINA

The Tibetologist Barnett (2009) documents an increase in tension and protests during the 2008 Beijing Olympics, soon before the first act of Tibetan self-immolation in 2009. For the first time, protestors in China used information technology, especially cell phones and the internet. The rate of protest-immolation increased sharply in September 2011. They have mainly been restricted to the east of the central and western Tibet, now called the Tibetan Autonomous Region. These are mostly areas where ethnic Tibetans are now a minority, due to the resettlement of high numbers of Chinese, mainly from the lowlands. At least 20% of these self-immolators are monks, and about 10% have been female, including several nuns. The age of 56 people (of 93 at the time of the analysis) has been reported. The average of these is 24.5.

SELF-IMMOLATION, HEALTH AND POLITICS

Two issues are relevant to the putative relationship between ecology, health and Tibetan self-immolation. The first is whether political actions such as this have a health dimension. I argue they do, using the concepts integral to social medicine. Suicide is a particularly tragic form of death, with flow-on effects, both political and health-affecting throughout the community. Some people attempting self-immolation have survived due to the use of fire extinguishers by police. Survivors suffer from disfiguring burns. The families of Tibetan self-immolators (some of whom had young children) are likely to be traumatised, as may the wider community of Tibetans and supporters.

The funerals of protest-immolators are reported to have recently attracted thousands of people, even of people normally considered unremarkable, such as nomads. This illustrates the power of modern communication technology (even where state media is censored). It also suggests large-scale social suffering.

Lack of access and direct research makes the scale of suffering impossible to measure, but the starting position should assume harm. The limited evidence available, such as final letters and statements made by people about to die, suggests political solidarity; some hint at deep regret, perhaps sadness, on behalf of the Tibetan people (see box).

Tenzin Wangmo, Ngaba, 17 October 2011, died

Tenzin Wangmo, a 20-year-old nun, set herself on fire around 1 pm outside Dechen Chokorling nunnery (also known as Mame nunnery), three kilometres from Ngaba Town. She died at the scene.

Tenzin called for religious freedom in Tibet and the return of the Dalai Lama. She walked about for up to eight minutes before she collapsed and died. She reportedly said to fellow nuns that morning that she 'will have to do something of great importance'.

Source: With permission from http://www.freetibet.org/news-media/na/full-list-self-immolations-tibet#.UHVkCNmt-So

However, some analysts (including assessors for the IAEH meeting in Kunming) quarantine protest-immolation as entirely 'political' and irrelevant to health. They thus do not conceptualise health with the social medicine lens advanced by Virchow, Engels and more recently the Commission on the Social Determinants of Health (Vincent Navarro, 2008). Some with this apparent view hold influential positions within the IAEH.

Political actions such as protesting a hospital closure have an explicit health dimension, as do protests against polluted air or water (common in China, though news is often suppressed) (Bradsher, 2012). The Occupy protests can also be considered to have an implicit health dimension, because protestors lobby for the wealthy to pay higher taxes, in turn to increase funding for public goods including health care (Sachs, 2012). It is not necessary for this argument that such protestors understand or explicitly claim that adverse social conditions, including inequality, have adverse health consequences.

Tibetan protest-immolation in China is clearly related to a loss of well-being, from which health cannot be fully disentangled. Protestors and their supporters are trying to improve their social conditions, and that would improve their health. They may not fully articulate this, but if so, health workers who can see these links have a role to act as amplifiers and interpreters. The IAEH has missed an important opportunity to do so.

I have not interviewed any people who subsequently died by self-immolation, and it is impossible to independently verify the accounts on the internet, compiled by Tibetan supporters. As mentioned, most self-immolations have occurred in a fairly small part of eastern Tibet, and it is probably not coincidental that most have been reported in areas with the greatest contact with Han Chinese. Eastern Tibetans – especially the Khampas – also have a reputation as warriors, a sub-national identity which may be further emboldening them. Perhaps Tibetans in western Tibet have a sufficiently strong culture – a sufficiently resilient human ecology – to not resort to this desperate means. But the same might have been said for Sichuan in 2009, before the current epidemic started.

REFLECTIONS

I have been aware since the 1960s of the Chinese communist invasion of Tibet. I have long been involved with efforts to publicise the mainstream exiled Tibetan community perspective, mainly through the NGO Benevolent Organisation for Development, Health and Insight (BODHI) which I co-founded in 1989 (www.bodhi.net.au). I have also been involved with practical efforts to enhance the education and living standards of people in Tibet, especially of rural people. So, I am not neutral (but nor are Sinophilic observers).

Because of my expectation and previous experience of Chinese censorship, my Kunming abstract avoided mention of Tibet or China. My most direct previous experience of Chinese censorship occurred not in China, but in Hanoi, Vietnam, in 2008. At that time, I was chair of an academic meeting on the environment, a parallel session at the fifth United Nations Day of Vesak meeting, held at the national convention centre. This was an unusual event in Communist Vietnam, bringing together Buddhists from about 90 countries. It attracted about 4,000 people, mostly Vietnamese. In a briefing held before these sessions, a senior Vietnamese monk instructed chairs to immediately suppress any discussion arising from the floor that was critical of China, especially if about Tibet.

I witnessed other evidence of how apprehension of China shadowed that meeting, in the three organising conferences that I attended. For example, despite rhetoric about the contribution that Buddhism could make to improve global social conditions there was no mention of the most honoured living Buddhist, the Dalai Lama. Tibetans were absent from both the planning and the final meeting. Representatives of the Indian dalit Buddhist community, which now number over 10 million, were also missing.

In the run-up to the Kunming meeting it was clear that mention of self-immolation in China would guarantee rejection, which I thought in any case likely. If it had been accepted, I planned to discuss the general principles that I perceive as linking eco-social distress with self-immolation, to describe self-immolation beyond China and to conclude with a brief discussion of the recent and current situation of the Tibetans, a few hundred kilometres from the meeting's venue.

THE RESPONSE OF THE IAEH

Concern with vulnerable peoples and populations is frequently asserted as central to EcoHealth philosophy. The issue of vulnerable people (which must surely include Tibetans) is a regular theme at IAEH biennual meetings, and Kunming was no exception. Images of peaceful Tibetan scenes (e.g. monasteries), suggestive of a happy, respected ethnic minority, were central to promoting the Kunming meeting. I have now concluded that such assertions are purely figurative.

Many reform movements, especially after their initial establishment phase, appear to come to value pragmatism over idealism. Examples are the World Health Organisation (WHO) and the Health Promotion Movement. WHO in 1978 called for 'Health for All by the Year 2000'; but as 2000 approached its cry became a whisper. The Health Promotion movement, at the Ottawa Charter in 1986, asserted the importance of the environment to health (and challenges to power) – but quickly diluted these messages (Butler & Friel, 2006). The liberal arm of the EcoHealth spectrum appears more open to the ideas that link politics, health and ecology described in this chapter, but the conservative pole is dominant.

The journal *EcoHealth* publishes many papers relevant to aspects of psychological loss, solastalgia and land sickness, but none have concerned minorities in China. If the 2012 IAEH meeting had been held in a free society then my abstract may have been accepted. Although no explanation was provided for its rejection, the senior academic in charge of the Kunming conference explicitly denied any censorship. But nor did he provide any academic reason. Fish are not necessarily aware they swim in water.

A very senior IAEH figure advised me that my abstract lacked scientific credibility, but suggested that this could be enhanced by multi-disciplinary studies, preferably including psychiatrists. This is an excellent suggestion, were it only possible. Both Russia and China have a well-documented history of psychiatric abuse for political reasons (Parry & Cui, 2010). Bonnie writes, 'in the 1980s, China also established a system of maximum-security forensic hospitals (Ankang), modeled after the Soviet "special hospitals", for confining offenders who present a "social danger". A substantial proportion of cases referred for forensic assessment involved "political" offenses, ... most of these offenders were diagnosed as schizophrenic, found non-responsible, and committed to the Ankang facilities' (Bonnie, 2002).

I invite the reader to reflect. What is apolitical? Some could argue that this means to accept the prevailing political environment without challenge or question. In Australia, that might mean to abstain from voting and to show no interest in parliamentary affairs. Presumably, one could get on with things that really matter, such as science. If I was transported to Germany in November 1938, to participate in a conference, would it be "apolitical" to ignore *Kristallnacht* – the night of broken windows – that had just occurred? What if I was in Rwanda during the 1994 genocide?

In such cases, would 'apolitical' ignorance be very different to participation in atrocity? Does not silent assent to injustice, justified as 'scientific' and thus acceptably 'apolitical', allow the perpetuation of injustice? If an official were to dismiss Virchow's report on typhus in Silesia as simply 'political', would that be partisan or neutral? I suggest that to ignore matters because to do otherwise is to risk being called political is itself a deeply political act (Butler, 2003).

A neutral scientific response could call for an experiment. Occupied Tibet could be divided into parts with existing policies and others with genuine autonomy, opportunity and greater ecological protection. Protest-immolation rates could, then, be monitored to test the hypothesis that a lower rate would occur in the intervention area. By such means, evidence could accumulate in support of the hypothesis that self-immolation has an eco-social causation. But this is only a thought experiment; other forms of evidence are actually needed.

One might argue that genuine Tibetan autonomy would discriminate against the lowlanders. Alternatively, authorities might learn that means other than ruthless suppression and repression of the Tibetans bring benefits. But if less harm to Tibetans caused more harm to the Han, then this would be evidence of a zero-sum game, a context in which winners co-exist with losers. This would be consistent with theories of limits to growth and limited human carrying capacity, advanced by Donella Meadows and other theorists (Butler, 2004).

As we look forward on a fixed planet to a future with ever-diminishing resources per person, better ways for humans to co-operate are vital (Butler, 2012). Sixty years of Chinese indoctrination have not generated sufficient Tibetan well-being. Minorities can rarely be made extinct, but they can be marginalised and silenced; their problems trivialised and suppressed. So far, such actions have kept the Chinese majority supportive and ignorant of Chinese polices in Tibet. The Dalai Lama is widely regarded within China as an evil pariah; state-run Chinese media has carried exhortations for the Dalai Lama to self-immolate (Carrico, 2012).

But such actions, over the longer term, are neither just nor sustainable. Even if Tibetan protest immolation abates (which seems likely as the current generation gradually realises its lack of effectiveness) the underlying pain of the Tibetans will remain. This ongoing suffering appears now as a tolerable and a minor irritant to an indifferent majority, but it is a psychic splinter that must be removed if Chinese civilisation is to truly flourish.

The IAEH, when forced in a corner, bowed to the powerful East wind. However, if practitioners of EcoHealth are to occupy more than a tiny niche in the mosaic of health then they need to display far more courage. The forces which are currently steering the world towards its precipice are immense. Our time is one of a slow emergency, as we drift towards four degrees of warming and to continuing biodiversity loss. As individuals age, most become more conservative. Not only does energy wane, but as retirement and death approaches, most elderly, who also disproportionately occupy positions of authority and power, take fewer risks, including to genuinely challenge power.

The issues of ecology and health are far broader than of novel infectious diseases, the human burden of which is trivial and likely to remain so (Butler, 2012). Practitioners of public health, including EcoHealth, especially those who are more junior, need to collaborate and co-operate.

But to have any genuine effect on the juggernaut of linked ecological loss and political oppression they probably also need to be active in non-academic organisations. They should be prepared that such activities will arouse suspicion, especially if they work in conservative milieus, such as China and the United States. But they should also recall the life and work of Rudolf Virchow, and the ongoing relevance of social medicine.

NOTES

1. This is a trademark of the Wildlife Conservation Society.
2. In 2010 the proceedings of the inaugural One Health congress, held in Melbourne Australia, were published in *EcoHealth*, the journal for the Association of Ecology and Health. In 2012, *EcoHealth* selected 'One Health' as one of its three keywords.
3. Koch (1843–1910) lived long enough to be awarded the Nobel Prize (in 1905). In later life Virchow, by then the doyen of German science, tried to suppress Koch's growing recognition (Feldberg, 1995; Strick, 1998). Strick also argues that Koch was politically conservative, with little if any sympathy for social medicine.
4. Biggs reports 92 cases in Vietnam.
5. Allegedly slapped by a female official in front of the busy market crowd on the day he died.
6. And up 2,500 undocumented cases.
7. 14 per million 'urban population'.
8. This forms the basis of ongoing Chinese claims over this Indian state.
9. Who also claimed Tibet as part of China.
10. Love of the bio, or life. A term coined by EO Wilson (Wilson, 1984).
11. 'They paved paradise, put up a parking lot.'
12. In 1997 the International Commission of Jurists reported 25%–33% of over 600 known political prisoners in Tibet at that time were nuns. It found torture in detention to be widespread in Tibet, and that 'women, particularly nuns, appear to be subjected to some of the harshest, and gender-specific, torture, including rape using electric cattle-prods and ill treatment of breasts' (http://www.tibetjustice.org/reports/un/unint5.pdf).

ACKNOWLEDGEMENTS

I thank Dr Maya K. Gislason, Dr Delia Grace, Dr Annie Carroll, Susan Woldenberg Butler, Professor Colin Soskolne and several anonymous correspondents from the EcoHealth community for their encouragement, comments and, in some cases, criticism.

REFERENCES

Abel, S. (2012). Self-Immolation (Chinese-Tibet Relations). In A. Su, S. Abel, A. Bahta, W. Barnard, M. P. Diaz, J. Hamrick, C. Jones, K. McDougal, C. Orellana, L. Reilly, M. Tarantino & J. Tibbals (Eds.), *Field guide to socially engaged Buddhism* (pp. 47–52). Seattle: Seattle University.

Albrecht, G., Sartore, G.-M., Connor, L., Higginbotham, N., Freeman, S., Kelly, B., & ... Pollard, G. (2007). Solastalgia: The distress caused by environmental change. *Australasian Psychiatry*, *15*(1), S95–S98. doi:10.1080/10398560701701288

Avedon, J. (1986). *In exile from the land of snows: The definitive account of the Dalai Lama and Tibet since the Chinese conquest*. New York, NY: Vintage Books.

Barnett, R. (2009). The Tibet protests of spring 2008: Conflict between the Nation and the State. *China Perspectives*, *3*, 6–23. Available at http://chinaperspectives.revues.org/4836

Beall, C. M., Laskowski, D., Strohl, K. P., Soria, R., Villena, M., Vargas, E., ... Erzurum, S. C. (2001). Pulmonary nitric oxide in mountain dwellers. *Nature*, *414*, 411–412.

Biggs, M. (2008). Dying for a cause – alone? *Contexts*, *7*(1), 22–27.

Bonnie, R. J. (2002). Political abuse of psychiatry in the Soviet Union and in China: Complexities and controversies. *Journal of the American Academy of Psychiatry and Law*, *30*, 136–144.

Bradsher, K. (2012). Bolder protests against pollution win project's defeat in China. *New York Times*. Retrieved from http://www.nytimes.com/2012/07/05/world/asia/chinese-officials-cancel-plant-project-amid-protests.html?_r=0

Brown, T. M., & Fee, E. (2006). Rudolf Carl Virchow: medical scientist, social reformer, role model. *American Journal of Public Health*, *96*(12), 2104–2105.

Butler, C. D. (1994). Overpopulation, overconsumption, and economics. *The Lancet*, *343*, 582–584.

Butler, C. D. (2003). How political should a general medical journal be? Politics, health and justice are intertwined. *BMJ*, *326*, 821.

Butler, C. D. (2004). Human carrying capacity and human health. *PLoS Medicine*, *1*(3), 192–194.

Butler, C. D. (2008). Environmental change, injustice and sustainability. *Journal of Bioethical Inquiry*, *5*(1), 11–19.

Butler, C. D. (2012). Infectious disease emergence and global change: Thinking systemically in a shrinking world. *Infectious Diseases of Poverty*, *1*, 5.

Butler, C. D., Chambers, R., Chopra, K., Dasgupta, P., Duraiappah, A., Kumar, P., ... Wen-Yuan, N. (2003). Ecosystems and human well-being. In Millennium Ecosystem Assessment (Ed.), *Ecosystems and human well-being. A framework for assessment* (pp. 71–84). Washington: Island Press.

Butler, C. D., & Friel, S. (2006). Time to regenerate: Ecosystems and health promotion. *PLoS Medicine*, *3*(10), e394.

Carrico, K. (2012). *Chinese state media representations – Cultural anthropology, hot spot forum, cultural anthropology*. Retrieved from http://culanth.org/?q=node/531

Cui, X., & Graf, H-F. (2009). Recent land cover changes on the Tibetan Plateau: A review. *Climatic Change*, *94*, 47–61. doi:10.1175/ei223.1

Daly, H. (1996). *Beyond growth*. Boston, MA: Beacon Press.

de Waal, F. (2009). *The age of empathy*. New York, NY: Harmony Books.

Demeny, P. (1988). Demography and the limits to growth. *Population and Development Review*, *14*(Supplement), 213–244.

Engels, F. (1845) (reprinted 1958). *The conditions of the working class in England* (W. O. Henderson & W. H. Chaloner, Trans.). Stanford: Stanford University Press.

Ehrlich, P. R., & Ehrlich, A. H. (2013). *Can a collapse of global civilization be avoided?* Proceedings of the Royal Society B, 280, 20122845. Retrieved from http://dx.doi.org/10.1098/rspb.2012.2845

Feldberg, G. D. (1995). *Disease and class: Tuberculosis and the shaping of modern North American society*. New Brunswick, NJ: Rutgers University Press.

Goldstein, M. C. (1989). *A history of modern Tibet, 1913–1951. The demise of the lamaist state*. Berkeley, CA: University of California Press.

Good, K. (1992). Interpreting the exceptionality of Botswana. *Journal of Modern African Studies*, *30*(1), 69–95.

Graham, H. (2013). Ensuring the health of future populations. *BMJ*, *345*, e7573.

Hedin, S. (1913). *Trans-Himalaya: Discoveries and adventures in Tibet* (Vol. III). London: MacMillan.

Huber, T. (1997). Green Tibetans: A brief social history. In F. Korom (Ed.), *Tibetan culture in the diaspora* (pp. 103–119). Wien: Verlag der Osterreichischen Akademie der Wissenschaften.

International Fund for Animal Welfare. (2001). Wrap up the trade. An international campaign to save the endangered Tibetan antelope. Yarmouth Port, MA, USA.

Jian, C. (2006). The Tibetan rebellion of 1959 and China's changing relations with India and the Soviet Union. *Journal of Cold War Studies*, *8*(3), 54–101.

Johnson, D. G. (1999). The growth of demand will limit output growth for food over the next quarter century. *Proceedings of the National Academy of Sciences*, *96*(11), 5915–5920.

Kewley, V. (1990). *Tibet. Behind the ice curtain*. London: Collins.

King, M. (1990). Health is a sustainable state. *The Lancet*, *336*, 664–667.

Lawrence, R. (2004). Housing and health: from interdisciplinary principles to transdisciplinary research and practice. *Futures*, *36*(4), 487–502.

Leopold, A. (1949). *A sand country almanac. With essays on conservation from Round river*. Oxford: Oxford University Press.

Liu, F., McShea, W. J., Garshelis, D. L., Zhu, X., Wang, D., & Shao, L. (2011). Human-wildlife conflicts influence attitudes but not necessarily behaviors: Factors driving the poaching of bears in China. *Biological Conservation*, *144*(1), 538–547. doi:10.1016/j.biocon.2010.10.009

Mackenzie, J. S., & Jeggo, M. H. (2011). 1st international one health congress. *EcoHealth*, 7(Supplement), S1–S2.

McMichael, A. J. (1993). *Planetary overload. Global environmental change and the health of the human species*. Cambridge: Cambridge University Press.

Miller, D. J. (2009). New views of the Tibetan plateau and Himalaya for conserving biodiversity. *Mountain Forum Bulletin*, 56–57.

Navarro, V. (1980). Work, ideology, and science: The case of medicine. *Social Science & Medicine. Part C: Medical Economics*, *14*(3), 191–205. doi:10.1016/s0160-7995(80)80003-7

Navarro, V. (2008). Politics and health: A neglected area of research. *European Journal of Public Health*, *18*(4), 354–356.

Oreskes, N., & Conway, E. M. (2013). The collapse of western civilization: A view from the future. *Daedalus*, *142*(1), 40–56.

Parkes, M. W., Bienen, L., Breilh, J., Hsu, L.-N., McDonald, M., Patz, J. A., & ... Yassi, A. (2005). All hands on deck: Transdisciplinary approaches to emerging infectious disease. *EcoHealth*, *2*, 258–272.

Parry, J., & Cui, W. (2010). China's psychiatric hospitals collude with officials to stifle dissent, say civil rights' groups. *BMJ*, *340*, c3371.

Ramakrishnan, P. S., Saxena, K. G., & Chandrashekara, U. M. (1998). *Conserving the sacred: For biodiversity management*. New Delhi: UNESCO, Oxford, IBH.

Rapport, D. J. (1995). Ecosystem health: Exploring the territory. *Ecosystem Health*, *1*(1), 5–13.

Rawling, C. G. (1905). *The great plateau, being an account of exploration in central Tibet, 1903, and of the Gartok expedition, 1904–1905*. London: Edward Arnold.

Rees, M. (2003). *Our final century*. London: William Heinemann.

Robbins, J. (2012). *The ecology of disease*. New York Times.

Sachs, J. (2012). *The price of civilization*. London: Vintage.

Sakya, T. (1999). *The dragon in the land of snows. A history of modern Tibet since 1947*. New York, NY: Columbia University Press.

Schaller, G. B. (1998). *Wildlife of the Tibetan steppe*. Chicago, IL: University of Chicago Press.

Schelling, E., Wyss, K., Bechir, M., Moto, D. D., & Zinsstag, J. (2005). Synergy between public health and veterinary services to deliver human and animal health interventions in rural low income settings. *BMJ*, *331*, 1264–1267.

Smith, N. H., Hewinson, R. G., Kremer, K., Brosch, R., & Gordon, S. V. (2009). Myths and misconceptions: the origin and evolution of Mycobacterium tuberculosis. *Nature Reviews Microbiology*, *7*, 537–544.

Strick, J. (1998). Preface. In T. D. Brock (Ed.), *Robert Koch: A life in medicine and bacteriology* (pp. 374). Berlin: Springer.

Suyin, H. (1979). *Lhasa, the open city a journey to Tibet*. St. Albans, UK: Panther Books.

Urry, J. (2010). Consuming the planet to excess. *Theory, Culture & Society*, *27*(2–3), 191–212.

Uzzell, J. (2012). Biopolitics of the self-immolation of Mohamed Bouazizi. *e-International Relations*. Retrieved from http://www.e-ir.info/2012/11/07/biopolitics-of-the-self-immolation-of-mohamed-bouazizi/

Vijaykumar, L. (2007). Suicide and its prevention: The urgent need in India. *Indian Journal Psychiatry*, *49*(2), 81–84.

Waltner-Toews, D. (2004). *Ecosystem sustainability and health*. Cambridge: Cambridge University Press.

Ward, B., & Dubos, R. (1973). *Only one Earth: The care and maintenance of a small planet*. Harmondsworth: Penguin.

Wilcox, B. A., & Daszak, P. (2006). Launching the international ecohealth association. *EcoHealth*, *3*, 125–126.

Wilson, E. O. (1984). *Biophilia*. Cambridge, MA: Harvard University Press.

Yeh, E. T. (2012). Transnational environmentalism and entanglements of sovereignty: The tiger campaign across the Himalayas. *Political Geography*, *31*(7), 408–418. doi:10.1016/j.polgeo.2012.06.003

Zinsstag, J., Schelling, E., Waltner-Toews, D., & Tanner, M. (2011). From "one medicine" to "one health" and systemic approaches to health and well-being. *Preventive Veterinary Medicine*, *101*, 148–156.

ECOHEALTH THROUGH AN ABILITY STUDIES AND DISABILITY STUDIES LENS

Gregor Wolbring

ABSTRACT

Purpose – *The goal of this chapter is to cultivate interest in the societal dynamic of ability expectations and ableism, a dynamic first thematized by the disabled people rights movement but which is also broadly applicable to the study of the relationship between humans, animals, and environments. Another aim of this chapter is to think about disabled people within ecosystem approaches to health through the ableism framework and to show that insights gained from disability studies are applicable to a broader study of health within contexts of environmental degradation. Building from this approach, the reader is invited to consider the utility of the conceptual framework of eco-ability "expectations" and eco-ableism as a way to understand health within coupled social-ecological systems.*

Methodology/approach – *This chapter uses an ability expectation and ableism lens and a disability studies and ability studies approach to analyze the relationship between humans, animals, and environments.*

Ecological Health: Society, Ecology and Health
Advances in Medical Sociology, Volume 15, 91–107

ISSN: 1057-6290/doi:10.1108/S1057-6290(2013)0000015008

Findings – *Certain ability expectations and ableism are responsible for (a) the invisibility of disabled people in ecological health discourses; (b) the standoff between anthropocentric and biocentric/ecocentric approaches to health; and (c) the application of scientific and technological advancements to address problems arising out of current relationships between humans, animals, and environments.*

Originality/value of chapter – *The reader is introduced to the concepts of ableism and eco-ableism, which have not yet been used in EcoHealth discourses and flags the need for further engagement with disability issues within the field.*

Keywords: Ableism; eco-ableism; disabled people; ecology of health; ability studies; disability studies

INTRODUCTION

The linkages between humans, animals, and nature within and outside the health[1] arena have been discussed for some time as has the importance of developing an integrated approach to health research and practice (e.g., at the United Nations Conference on Environment and Development (UNCED), Rio de Janeiro). Two recent notable developments emerging from these health discourses are the "One Health" framework, which focuses on the impact of the human-animal relationships on animal and human health (Monath, Kahn, & Kaplan, 2010; Zinsstag et al., 2009, 2012), and the EcoHealth field (Butler & Weinstein, 2011; Wilcox et al., 2004), which engages in active debates around important environment and health issues facing the world today (Charron, 2012). A defining mandate of the EcoHealth field is that any intervention must improve the health and well-being of people, animals, and ecosystems. An important feature of the application of an ecological perspective to health research is that it "emphasises both individual and contextual systems and the interdependent relations between the two" (McLaren & Hawe, 2005, p. 6). McLaren and Hawe paraphrasing Kickbusch (1989) also observe that when taken up within public health

> key features of an ecological public health include understanding health as a pattern of relations rather than as a quantitative outcome (that is, viewing health as a process nested in contexts rather than as a static attribute of individuals), and a new type of

> dialogue between the natural and social sciences that enables the explanation of both physical and social processes contributing to health. (2005, p. 9)

Given its integrative capacities, ecologically informed health research is being seen as a way forward. For example, recent work on the post-2015 Sustainable Development Goals (which has direct implications for ecosystem health) is acknowledging that strategies to eliminate poverty will be more effective if they highlight the full gamut of linkages between sustainable development, global environmental change, health and well-being (Langlois, Campbell, Prieur-Richard, Karesh, & Daszak, 2012). Still improving the health and well-being of people, animals, and ecosystems is a multilevel problem, making it difficult to come up with a unified vision for producing "health for all." Supporting the development of integrated visions through effective medical, environmental, and social health governance strategies and coherent health laws capable of working at the intersection of human, animal, and the environmental health is also a challenge, in part because there are specific populations who continue to be marginalized and disenfranchised.

One population that continues to slip through the cracks are disabled people, not only in terms of their health needs but also in terms of their involvement in forming new public and global health agendas. Involving disabled people and drawing on insights from disability studies, such as cultural concepts of ability expectations (want stage), ableism (perceiving a given ability as essential; need stage), and disablism (negatively impacting people seen as missing "essential abilities") (Wolbring, 2012b), are good places to begin redressing these inequities. Using disability studies and ability studies lenses in a foresight manner, this chapter focuses on the specific contributions that the social theories and experiences of disabled people can make to ecosystem-based approaches to health. Issues faced by disabled people need to become more visible as this is a population disproportionately impacted by environmental and social degradations. In addition, disabled people's insights into disabling practices are worth contemplating as are some of the conceptual tools this community uses. The cultural dynamics of ability expectations and ableism in particular (Wolbring, 2012b), this chapter argues, are useful concepts when analyzing issues, such as governance strategies and the formation of laws intended to improve the well-being for voiceless populations, including animals and places, such as ecosystems.

This chapter starts by looking at the EcoHealth topic through a disability studies lens and in particular investigates the portrayal and involvement of disabled people within the EcoHealth literature. In its second section, the concept of ability expectations and its impact on the relationship between humans, animals, and the environment is introduced as is eco-ableism as an ecologically focused analytical lens that can be used in inter-, trans-, and intradisciplinarily ways to generate policies and advance the relationship between humans, animals, and their environments. The third section looks at where the EcoHealth field could move and asks some questions I believe the EcoHealth field has to address. Overall, it is hoped that this chapter enables the reader to become more aware of both existing and emerging challenges and to realize the potential utility of an ability expectation and ableism lens for studying health as a social-ecological phenomenon.

ECOHEALTH THROUGH A DISABILITY STUDIES LENS

Disability studies is an interdisciplinary academic discipline that brings together insights from the natural sciences, social sciences, and humanities to address the societal treatment disabled people encounter due to not meeting the species-typical, body-linked ability expectations of the so-called nondisabled. Disability studies champions the social model of disability that frames problems disabled people experience as being the result of barriers that originate within the physical, social, and cultural environment over the medical model of disability which locates the problem the person has as originating within the body. For example, a "disabling" environment is when people exhibit a form of ableism that expects certain body-linked abilities and considers anything existing outside of these expectations to be an impairment (Ayim, 1997; Campbell, 2008; Carlson, 2001; Imrie, 1996; Loja, Costa, Hughes, & Menezes, 2013; Wolbring, 2008a, 2008b, 2012b). How we portray disabled people, how we construct the meaning of disability and the visibility of disabled people, and how we depict the issues they contend with are key areas that disability studies scholars research. These are, I argue, also important arenas of investigation to pursue for those working on health issues from a social-ecological vantage point. According to Kone et al., "the complexity of today's environmental health problems calls for multilevel, intersectorial interventions, involving the coordinated efforts of researchers, policy-makers, nongovernmental organizations, and affected communities" (Kone et al., 2012, p. 413). However, one of the largest communities of people impacted by environmental health drivers globally, namely disabled people, are for the most part absent from these initiatives. This trend also appears to be true within intersectoral work being conducted under the rubric of EcoHealth.

One way to get a sense of the ways in which disability is being defined and addressed within EcoHealth discourses is by conducting a discourse analysis of material gathered through keyword searches. On December 20, 2012, for example, a keyword search using Google Scholar generated 35 hits with the keyword combination "ecological health" and "disabled people" and 102 hits with the keyword combination "ecological health" and "people with disabilities." In contrast, results with the keyword combination "ecological health" and "women" generated 35 times more hits. On May 15, 2013, a keyword search of the journal *EcoHealth* via the Springer Web site using the terms "disabled people," "people with disability," and "people with disabilities" generated $n = 0$ hits, while a search for the term "disability" returned $n = 23$ hits. A further analysis of how the term disability was used shows that overall the trend has been to adopt a medical and not a social understanding of disability; furthermore, when used, the term 'disability' is often linked to the phrase "disability adjusted life years" (see for example: Choudhury et al., 2013; Dakubo, 2004; Ebi, Woodruff, von Hildebrand, & Corvalan, 2007; Khan, Phillips, Fernando, Fowles, & Lea, 2007; Narrod, Zinsstag, & Tiongco, 2012; Thomson, Connor, Ward, & Molyneux, 2004; Tserendorj et al., 2011). Additional ways in which the term 'disability' is used include references to: impairment caused by pesticides and issues of compensation, including eligibility for disability pensions (Yanggen, Cole, Crissman, & Sherwood, 2004); the disability burden created when impairment in children is caused by exposure to contaminants and the related considerations of the "economic costs of childhood diseases and disabilities" (Davies, 2006); the need to address issues of learning disabilities by controlling the use of and exposure to contaminants (Wernham, 2007); the prevalence of mental disorders and their significance as a source of disability (Jalaludin & Garden, 2011); and, finally, the role of zoonoses in causing permanent disability (Narrod et al., 2012). The overarching observation that can be drawn from this search is that there exists a concern about disability within the literature albeit only in the medical sense; therefore, few studies have used critical social theory and disability studies tools to give a new kind of breadth and depth to existing research so that the concept of disability is given dimension and complexity beyond the medical framework.

A similar conclusion can be drawn from an analysis of one key EcoHealth publication of the International Development Research Centre (IDRC), an organization in building the field of ecosystem approaches to health (or EcoHealth research), which has the mandate "to connect ideas of environmental and social determinants of health with those of ecology and systems thinking in an action-research framework applied mostly within a context of social and economic development" (Charron, 2012, p. 6). For example, in a recent IDRC-affiliated publication, terms such as disabled people or people with disabilities returned $n = 0$ hits and the term disability/ies were mentioned only thrice and all three where within a medical framework. One mention highlighted that "certain pollutants can damage the neurological system with irreversible effects. Therefore, early detection of subtle damage is important to prevent long-term disability" (Charron, 2012, p. 84). A second described that "[Cystic echinococcosis] CE is found worldwide and its total burden is estimated to be about one million disability adjusted life years (DALYs) per year with approximately 200,000 new diagnosed cases reported annually" (Charron, 2012, p. 191). A final example related to the observation that "human exposure to chemicals and metals has been linked with biological alterations, subtle adverse health effects, and eventually, illness, major disabilities, and sometimes death" (Charron, 2012, p. 84). This observation means that there is work to be done. According to the IDRC, ecosystem approaches to health (or EcoHealth research) have six principles: systems thinking; transdisciplinary research; participation; sustainability; gender and social equity; and knowledge to action (Charron, 2012). Given that conflicting values are common to EcoHealth problems (Cleland & Wyborn, 2010), I submit that both disabled people and a disability studies lens could be better integrated into work that covers the nexus between the social and the ecological system.

While these hit count results and the quotes highlighted here are just one indicator of the lack of visibility of disabled people in EcoHealth discourses, overall these findings suggest, at least as a preliminary gauge, that disabled people are relatively invisible within the newly emerging field of EcoHealth and that the medical model of disability prevails. These absences exist despite the fact that (a) the term "disability" plays a central role in the discourse around temporary to long-term disability and DALY lost, caused by the environment; (b) the vulnerability of people with disabilities is often due to a decrease in environmental health; and (c) discrimination against disabled people is often linked to how a given social entity models human-nature relationships, for example, how a society deals with environmental disasters (Wolbring, 2009, 2011b, 2011d; Wolbring, Leopatra, & Yumakulov, 2012). This medical bias, when working on issues of disability, has important implications including that there is a lack of visibility of disabled people in work addressing health at the interface of social and environmental determinants of health. When investigating this phenomenon from another angle – looking at the links between disability and common social-ecological health issues – the trend persists. Disabled people are largely absent from the water and sanitation (Noga & Wolbring, 2012; Wolbring et al., 2012), climate change (Wolbring, 2009), the social determinants of health (Wolbring, 2011c), and sustainable development discourses, including from work on the Millennium Development Goals (8th Disabled People's International (DPI) World Assembly, 2011; Wolbring & Burke, 2013). These silences in the literature inspire the question: What kinds of approaches to disability should social-ecological approaches to health research be considering?

Shifting the focus and taking the lead from disability studies scholars, another area of relevance to the study of health at the humans-animals-environments interface is the way in which ability and disability are imagined in relation to states of health. Specifically, disabled people are engaging in debates around what disablement really is and how health and biological "deviance" coexist (Blume, 2010; de Clerck, 2010; Glickman & Carey, 1993; Lane, 1995; Preston, 1995). As the rich literature contained within the disability studies suggests, the biological and ecological sciences should do more than looking at bodily deficiencies when it looks at "disability." Such a shift would require moving beyond a medicalized understanding of disability just as health sciences should move beyond a default use of the medical model of disability. In addition, the EcoHealth field could be strengthened by further responding to, and incorporating, knowledge derived from the realities of living with a body ability difference caused by a variety of circumstances. In general for the field of EcoHealth, it is my contention that the low visibility of disability studies scholars and disabled people and the one-sided medical understanding of disability are a loss for EcoHealth discourses. A gain for EcoHealth discourses would be to incorporate insights from disability studies, thereby gaining access to a wealth of theoretical insights about overlapping concepts such as vulnerability and interdependence. Interdependence is, in fact, an apt concept to use to elaborate upon this point.

Within disability studies, interdependence, and relatedly embodying the ability for interconnectedness, is a principle that leads "to action motivated by compassionate understanding and appreciation rather than competition; the experience of feeling with all beings now and into future generations" (Besthorn & McMillen, 2002, p. 226). For example, the late disability studies scholar Eric Leipoldt proposed that "the disability perspective of interdependence is a practical guide from the margins to making new choices that may lead to a just and sustainable world – one that reduces the distance between each other and our environment" (Leipoldt, 2006, p. 21). As the section that follows argues, another potential contribution that can be made to the study of health at the society-ecology interface is the concept of ableism, which was coined by the disabled people rights movement and disability studies scholars. Specifically, the next section in this chapter looks at the EcoHealth field through the lens of ability studies and the concept of eco-ableism and underscores that mutual learning can occur between the discipline of disability studies, ability studies (Wolbring, 2008c), and the EcoHealth field.

ABILITY STUDIES: INVESTIGATING ECO-ABILITY EXPECTATION AND ECO-ABLEISM AND THEIR IMPACT ON ECOSYSTEM APPROACHES

Ability studies investigates (a) the abilities that are seen as essential in a given context, (b) the dynamic of how an ability expectation consensus is reached, if it is reached, and (c) the impact of ability expectations. The disabled people rights movement was the first to look at the cultural dynamics of framing and reproducing ability expectations and in doing so coined the term ableism as a cultural concept (Wolbring, 2012b). Although ableism was developed to make visible disablism (the lack of support and active disablement by the ones who see themselves as able) (Miller, Parker, & Gillinson, 2004), the yearning for an ability (want stage), the internalization of the idea that a specific ability is necessary (need stage) (ableism) (Wolbring, 2012a) and the notion of disablement linked to ableism are all concepts that have broader social applicability. For example, individuals, households, communities, groups, sectors, regions, and countries make numerous decisions daily based on ability expectations (Wolbring, 2008b, 2012b). Some perceive the ability to protect one's privacy, to buy a house, to be competitive, productive, or efficient, or the ability to consume to be essential. Others do not hold these values and rather give significance to the ability to live in a harmonious equitable society. Ability expectations and ableism lead to an ability-based and ability-justified understanding of, among other things, human's relationships with other humans, other species such as animals, and other entities such as nature (Wolbring, 2011a).

Within the disability arena, the term "ableism" is used in a negative way – to problematize a form of discrimination. However, ableism does not have to be negative, for example, a society could decide that the harmonious interaction of humans, animals, and the environment is an *essential ability* of a functioning society, and this could be perceived as a positive normative expectation (Wolbring, 2008b). The abilities one favors and the ableist frameworks one subscribes to contour not only how people approach human-human, human-animal, and human-nature relationships (Wolbring, 2011a) but also inform how people define ecological problems and imagine solutions to these problems. Players involved in the shaping of ecological discourses exhibit ability expectations (eco-ability expectations) and forms of ableism (eco-ableism) that influence how they define ecological problems and solutions to the problem and, therefore, whom they invite to the table as stakeholders and knowledge producers (Wolbring, 2011d). According to Naess (1973), the central objective of the "shallow ecology movement" is to increase the ability to enhance the health and affluence of people in developed countries, while the "deep ecology movement" has its focus on building the ability to promote "biospheric egalitarianism, the view that all living things are alike in having value in their own right, independent of their usefulness to others" (Stanford Encyclopedia of Philosophy, 2008). At root, social discourses and practices are at work in the spheres of giving, taking, and acting on perceptions of worth.

One example of how different ability expectations contour ideas about how humans relate to nature (Wolbring, 2011a) is evident in the basic presumptions held within anthropocentric and biocentric worldviews. The anthropocentric school of thought is human-centered in its interpretation of the relationship between humans and nature (Stanford Encyclopedia of Philosophy, 2008; Verhagen, 2008) and thus humans view animals and nature as tools for fulfilling their needs and wishes. This outlook is disabling for nature (Wolbring, 2011a) and animals as both are valued and cared for only when considered capable of decreasing negative environmental impacts or increasing states of health and well-being for humans. The eco- or biocentric school of thought is, in contrast, centered on valuing the natural world and within this framework humans are expected to live in harmony with the needs of the biosphere (Wolbring, 2011a). In addition, as much as the concept of ableism originated from the disabled people rights movement, what forms of ableism are exhibited in other discourses such as the human-nature relationship also impacts disabled people. For example, an anthropocentric view is based on one entity dominating another entity, which one could say is reflected in how disabled people are treated by the ones who feel entitlement based on having certain body-related abilities. Living in harmony with nature as suggested in the bio/ecocentric approach might also require that humans live in harmony with one another, which might be better suited for disabled people as it emphasizes interconnectivity and respect for the "other." However, as little work has been done in this arena, this outcome would depend on how a bio/ecocentric view would be operationalized.

Anthropocentric and biocentric views not only differ in how humans are perceived as causing the ill health of nature and what to do about it, but also in how to use nature to improve the health of humans. Anthropocentric environmental protection fights pollution, resource depletion, and now climate change with the goal of preserving a particular human way of living, which today is still often driven by favoring the ability to consume, the ability to outperform others, and the ability to generate a high Gross Domestic Product. For today's energy and climate discourse, these three ability expectations precondition people to look for more "eco-friendly" energy sources as long as these sources do not challenge the status quo. While sources of eco-friendly energy may reduce the deleterious impacts of consumption on nature, the motivation for doing so has little to do with providing for nature's needs and much to do with a growing realization that the "old" ways of treating nature threatens existing hegemonic structures and practices. Another way in which the status quo is maintained is through a turn toward techno-tools such as geoengineering and modifying humans. In principle, geoengineering (Abashian, 1998; Boyd, 2009; Corner & Pidgeon, 2010; Gardiner, 2011; Schneider, 2008) can be pursued either to address the needs of natural systems (biocentric approach) or to fix nature to suit the need of humans (anthropocentric approach); so far the anthropocentric approach prevails. Any geoengineering approach has to have a global impact if it is to fulfill the positive expectation of alleviating climate change problems. But that means it can also have negative global impacts such as exhibiting geotoxic problems that will disproportionately impact humans that are already marginalized as well as animals and could ultimately redefine the relationship between humans and nature.

Another venue for intervention can be imposed on human bodies. In adaptation scenarios, for example, human survival is sometimes premised upon the ability to modify the human body to cope with harsh or disruptive climates. To give one example: within the article "Human Engineering and Climate Change" (Liao, Sandberg, & Roache, 2012), the authors propose human engineering that has the end goal of changing bodily abilities as an adaptation strategy as it would enable people to withstand the impacts of climate change. The authors propose that human engineering is a potentially necessary alternative to geoengineering because they believe that efforts to change the ability expectations of humans (e.g., modifying aspirations toward consumerism through educational programs) are not working and geoengineering might be too dangerous. These same frameworks are also informing the treatment of animals such as those impacted by scientific laboratory experimentation. For example, some actors in the enhancement field are selling the idea of biological, especially cognitive, enhancements of animals to ameliorate the impacts of their maltreatment (Chan, 2009), thereby applying an enhancement form of ableism to animals (Wolbring, 2008b). The ground for accepting the idea of modifying biological structures (such as humans, animals, and the environment) to give them beyond biological structure-typical abilities is directly linked to other body-linked ability expectations discourses such as the ones related to disabled people and the ability expectation judgments they face and has direct implications for what it means to be healthy (Wolbring, 2005) and whose future and what types of health (human, animal, or the environment) we are trying to secure.

WHAT DO THESE PERSPECTIVES MEAN FOR THE FUTURE?

So where are we heading? In relation to ability expectations, some concrete points of departure have been highlighted. At the scale of the social, however, there might be a way to bring problems of equity and justice together by linking issues of health both for disabled people and the Earth in social movements akin to those that forged links between justice for women and the Earth. Some think that a shift in ability expectations at the scale of societies is already occurring based on instances in which anthropocentric frameworks are seen to be being dismantled. For example, according to Verhagen, "evidence of an emerging biocracy in the modern Western world is legislation about endangered species and the representation of other life forms during political assemblies when persons or organizations become spokespersons and keepers of rivers, forests etc." (2008, p. 14). In another example, Ecuador may be construed as the first country to have become a legal biocracy or ecocracy by giving a new set of rights to nature. However, close attention needs to be paid to the underlying assumptions being reproduced in these forward-looking efforts. For example, how biocentric is the Ecuadorian approach? Articles 71–74 of its new constitution describe the relationship of humans to nature. Article 71 can be interpreted as giving rights to the "entity" nature. Article 72 talks about nature's right to an integral restoration and Article 73 talks about what actions against nature are prohibited. Yet, Article 74 retains anthropocentric reasoning: "Persons, people, communities and nationalities will have the right to benefit from the environment and form natural wealth that will allow wellbeing" (Republic of Ecuador, 2008). Examples of other social movements coalescing around efforts to bring social practices more in-line with biocentric worldviews include those led by ecological feminists seeking to bring about a more balanced relationship between humankind and the Earth (Johnson, 2011). Central to this feminist approach is an expectation of the ability to care (LaFollette, 2002), the promotion of the idea of interconnectedness (Besthorn & McMillen, 2002), and a rejection of dualistic hierarchical worldviews that divide humans and nature. In other words, ecological feminism is rejecting the ability expectation of "dominance, competition, materialism, and technoscientific exploitation inherent in modernist, competition-based social systems" (Besthorn & McMillen, 2002, p. 226) and nourishing the ability expectation of "caring and compassion and the creation and nurturing of life" (Besthorn & McMillen, 2002, p. 226). Applicable to all of these movements are insights already expressed within disability studies theories and methodologies – notions derived from a discipline whose corpus of thought is built upon the experiences of living within disabling contexts and from knowledge gained through working hard to change policies and practices that disable and exploit the "other."

I close this section with some open questions. What could the scope, possibilities, and limitation be of approaches that would be built around ideas such as: a medical health of nature, a social well-being approach to nature, a medical health of human-nature relationships, or a social well-being approach to human-nature relationships? How would animals fit into such projects? How could disabled people fit into these discourses? Could a medical or social model of disablement better deal with EcoHealth issues? What are the consequences for disabled people if the medical understanding of disabled people continues to be so pervasive in the EcoHealth discourse? How does this medical understanding limit how the field engages with human-nature relationships and human-animal-nature relationships? Certainly, having a clear understanding of the meaning and relationship between health, disease, disablement, and well-being and the ability expectation and ableism dynamics that drive the making of the meaning of health, disease, disablement, and well-being are important for EcoHealth research and practice. Ability expectations also shape the sustainable development agenda which, in turn, impacts emerging EcoHealth agendas. How the questions are answered are also relevant to the formation of the mandates and practices of government agencies, academic and nonacademic funding agencies, policy makers, national and international nongovernmental agencies, civil society organizations, and others interested in developing integrated approaches to promoting health and well-being for all, including disabled people.

CONCLUSION

Moving through the world with a body that is perceived as sub-species-typical – as impaired – teaches firsthand that perceptions shape practice, both in terms of how one personally navigates the challenges of disabling environments (both natural and social) and how society enables or disables health and well-being. From this vantage point, environmental health problems as well as societal views that enable or disable the ability of society to embrace the health of nature as part of the health of humans are important issues to reconcile. Relevant also is knowledge gained through the lived reality of being made sick by medical and social environments as well as from navigating through sick environments as someone labeled as lacking essential abilities because at some level these are interconnected phenomena. In short, there are myriad insights the disability studies and ability studies disciplines have to offer to work on health at the society-ecology interface such as how to develop a more nuanced, equitable, and sustainable research agenda.

This book chapter used the lenses of disability studies and ability studies to raise the issues of whose values count, introduce new values, and imagine future scenarios. I hope that (a) the framework of ableism and eco-ableism will be seen as useful by the reader in their EcoHealth endeavors and that more cross-fertilization takes place between the fields of ability studies and EcoHealth; (b) the visibility of disabled people increases in environment and ecology of health discourses and that more cross-fertilization takes place between disability studies and EcoHealth, and (c) the investigation of the influence of science and technology, in general, and science and technology–enabled ability expectations and ableism, in particular, be considered within future gazing EcoHealth research.

NOTE

1. When I use the term health I use it with the meaning of medical and social health.

REFERENCES

8th Disabled People's International (DPI) World Assembly. (2011). *Durban Declaration. Disabled Peoples' International Asia-Pacific Region*. Retrieved from http://www.dpiap.org/resources/article.php?id = 0000866&year = &genreid = 20

Abashian, P. (1998). GeoEngineering and the Internet: Collaborative computing finds its connective environment. *Geographic information systems development webpage*. Retrieved from http://www.gisdevelopment.net/proceedings/gita/1998/iote/iote001pf.htm

Ayim, M. (1997). Crimes against the deaf: The politics of ableism. *Canadian Journal of Education/Revue Canadienne de l'Education*, *22*, 330–335. JSTOR.

Besthorn, F. H., & McMillen, D. P. (2002). The oppression of women and nature: Ecofeminism as a framework for an expanded ecological social work. *Families in Society*, *83*(3), 221–232.

Blume, S. S. (2010). *The artificial ear: Cochlear implants and the culture of deafness*. New Brunswick, NJ: Rutgers Univ Pr.

Boyd, P. W. (2009). Geopolitics of geoengineering. *Nature Geoscience*, *2*(12). (812).

Butler, C. D., & Weinstein, P. (2011). Global ecology, global health, ecohealth. *Ecohealth*, *8*(3), 253–254.

Campbell, F. A. K. (2008). Exploring internalized ableism using critical race theory. *Disability & Society*, *23*(2), 151–162.

Carlson, L. (2001). Cognitive ableism and disability studies: Feminist reflections on the history of mental retardation. *Hypatia*, *16*(4), 124–146.

Chan, S. (2009). Should we enhance animals? *Journal of Medical Ethics*, *35*(11), 678–683. doi:10.1136/jme.2009.029512

Charron, D. F. (Ed.). (2012). *Ecohealth research in practice*. Ottawa, Canada: IDRC.

Choudhury, A. A. K., Conlan, J. V., Racloz, V. N., Reid, S. A., Blacksell, S. D., Fenwick, S. G., ... Whittaker, M. (2013). The economic impact of pig-associated parasitic zoonosis in northern Lao PDR. *Ecohealth*, *10*(1), 54–62.

Cleland, D., & Wyborn, C. (2010). A reflective lens: Applying critical systems thinking and visual methods to ecohealth research. *Ecohealth*, *7*(4), 414–424.

Corner, A., & Pidgeon, N. (2010). Geoengineering the climate: The social and ethical implications. *Environment: Science and Policy for Sustainable Development*, *52*(1), 24–37.

Dakubo, C. (2004). Ecosystem approach to community health planning in Ghana. *Ecohealth*, *1*(1), 50–59.

Davies, K. (2006). Economic costs of childhood diseases and disabilities attributable to environmental contaminants in Washington State, USA. *Ecohealth*, *3*(2), 86–94. doi:10.1007/s10393-006-0020-1

de Clerck, G. A. M. (2010). Deaf epistemologies as a critique and alternative to the practice of science: An anthropological perspective. *American Annals of the Deaf*, *154*(5), 435–446.

Ebi, K. L., Woodruff, R., von Hildebrand, A., & Corvalan, C. (2007). Climate change-related health impacts in the Hindu Kush–Himalayas. *Ecohealth*, *4*(3), 264–270.

Gardiner, S. M. (2011). Some early ethics of geoengineering the climate: A commentary on the values of the Royal Society Report. *Environmental Values*, *20*(2), 163–188.

Glickman, N. S., & Carey, J. C. (1993). Measuring deaf cultural identities – A preliminary investigation 69. *Rehabilitation Psychology*, *38*(4), 275–283.

Imrie, R. (1996). Ableist geographies, disablist spaces: Towards a reconstruction of Golledge's 'Geography and the disabled'. *Transactions of the Institute of British Geographers*, *21*(2), 397–403.

Jalaludin, B. B., & Garden, F. L. (2011). Does urban sprawl impact on self-rated health and psychological distress? A multilevel study from Sydney, Australia. *Ecohealth*, *8*(3), 268–276.

Johnson, S. (2011). Ecofeminism and Equality. *Feminist eZine*. Retrieved from http://www.feministezine.com/feminist/ecofeminism/Ecofeminism-and-Equality.html

Khan, R., Phillips, D., Fernando, D., Fowles, J., & Lea, R. (2007). Environmental health indicators in New Zealand: Drinking water – A case study. *Ecohealth*, *4*(1), 63–71.

Kickbusch, I. (1989). Approaches to an ecological base for public health. *Health Promotion International*, *4*(4), 265–268.

Kone, B., Feagan, M., Houenou, Y. A., Brou, N., Houenou, P. V., Fayomi, B., & ... Kouassi, E. (2012). Facilitating the relationship between researchers and policy-makers: Experiences from three ecohealth projects in West and Central Africa. *Ecohealth*, *8*(4), 413–417.

LaFollette, H. (2002). *Ethics in practice: An anthology* (3rd ed.). Oxford, UK: Wiley-Blackwell.

Lane, H. (1995). Constructions of deafness. *Disability & Society*, *10*(2), 171–189.

Langlois, E. V., Campbell, K., Prieur-Richard, A. H., Karesh, W. B., & Daszak, P. (2012). Towards a better integration of global health and biodiversity in the new sustainable development goals beyond rio+ 20. *Ecohealth*, *9*(4), 381–385.

Leipoldt, E. (2006). Disability experience: A contribution from the margins. Towards a sustainable future. *Journal of Futures Studies*, *10*(3), 15–32.

Liao, S. M., Sandberg, A., & Roache, R. (2012). Human engineering and climate change. *Ethics, Policy & Environment*, *15*(2), 206–221.

Loja, E., Costa, M. E., Hughes, B., & Menezes, I. (2013). Disability, embodiment and ableism: Stories of resistance. *Disability & Society*, *28*(2), 190–203.

McLaren, L., & Hawe, P. (2005). Ecological perspectives in health research. *Journal of Epidemiology and Community Health*, *59*(1), 6–14.

Miller, P., Parker, S., & Gillinson, S. (2004). *Disablism: How to tackle the last prejudice*. London, UK: Demos.

Monath, T. P., Kahn, L. H., & Kaplan, B. (2010). One health perspective. *ILAR Journal*, *51*(3), 193–198.

Naess, A. (1973). The shallow and the deep, long-range ecology movement: A summary. *Inquiry*, *16*(1/4), 95–100.

Narrod, C., Zinsstag, J., & Tiongco, M. (2012). A one health framework for estimating the economic costs of zoonotic diseases on society. *Ecohealth*, *9*(2), 150–162. doi:10.1007/s10393-012-0747-9

Noga, J., & Wolbring, G. (2012). The economic and social benefits and the barriers of providing people with disabilities accessible clean water and sanitation. *Sustainability*, *4*(11), 3023–3041.

Preston, P. (1995). Mother father deaf – the heritage of difference. *Social Science & Medicine*, *40*(11), 1461–1467.

Republic of Ecuador. (2008). Constitution of the Republic of Ecuador. Retrieved from http://pdba.georgetown.edu/Constitutions/Ecuador/english08.html

Schneider, S. H. (2008). Geoengineering: Could we or should we make it work? *Philosophical Transactions of the Royal Society A: Mathematical, Physical and Engineering Sciences*, *366*(1882), 3843–3862.

Stanford Encyclopedia of Philosophy. (2008). Environmental Ethics. *Stanford Encyclopedia of Philosophy*. Retrieved from http://plato.stanford.edu/entries/ethics-environmental/#EarDevEnvEth

Thomson, M. C., Connor, S. J., Ward, N., & Molyneux, D. (2004). Impact of climate variability on infectious disease in West Africa. *Ecohealth*, *1*(2), 138–150.

Tserendorj, A., Anceno, A. J., Houpt, E. R., Icenhour, C. R., Sethabutr, O., Mason, C. S., & Shipin, O. V. (2011). Molecular techniques in ecohealth research toolkit: Facilitating estimation of aggregate gastroenteritis burden in an irrigated periurban landscape. *Ecohealth*, *8*(3), 349–364.

Verhagen, F. C. (2008). Worldviews and metaphors in the human-nature relationship: An ecolinguistic exploration through the ages. *Language and Ecology*, *2*(3), 1–18.

Wernham, A. (2007). Inupiat health and proposed Alaskan oil development: Results of the first integrated Health Impact Assessment/Environmental Impact Statement for proposed oil development on Alaska's North Slope. *Ecohealth*, *4*(4), 500–513.

Wilcox, B. A., Aguirre, A. A., Daszak, P., Horwitz, P., Martens, P., Parkes, M., & … Waltner-Toews, D. (2004). EcoHealth: A transdisciplinary imperative for a sustainable future. *Ecohealth*, *1*(1), 3–5.

Wolbring, G. (2005). HTA Initiative #23 The triangle of enhancement medicine, disabled people, and the concept of health: A new challenge for HTA, health research, and health policy. Alberta Heritage Foundation for Medical Research (AHFMR) webpage: Alberta Heritage Foundation for Medical Research (AHFMR).

Wolbring, G. (2008a). Ableism, enhancement medicine and the techno poor disabled. In P. Healey & S. Rayner (Eds.), *Unnatural selection: The challenges of engineering tomorrow's people*. London, UK: Earthscan.

Wolbring, G. (2008b). The politics of ableism. *Development*, *51*(2), 252–258.

Wolbring, G. (2008c). Why NBIC? Why human performance enhancement? *Innovation: The European Journal of Social Science Research*, *21*(1), 25–40.

Wolbring, G. (2009). A culture of neglect: Climate discourse and disabled people. *Journal Media and Culture*, *12*(4). Available at http://dspace.ucalgary.ca/bitstream/1880/47465/1/wolbringclimate.pdf

Wolbring, G. (2011a). Ableism and energy security and insecurity.. *Studies in Ethics, Law, and Technology*, *5*(1). (Article 3).

Wolbring, G. (2011b). Disability, displacement and public health: A vision for Haiti. *Canadian Journal of Public Health*, *102*(2), 157–159.

Wolbring, G. (2011c). People with disabilities and social determinants of health discourses. *Canadian Journal of Public Health*, *102*(4), 317.

Wolbring, G. (2011d). Water discourse, Ableism and disabled people: What makes one part of a discourse? *Eubios Journal of Asian and International Bioethics*, *21*(6), 203–207.

Wolbring, G. (2012a). Eco-ableism. *Anthropology News*, September 14.

Wolbring, G. (2012b). Expanding ableism: Taking down the ghettoization of impact of disability studies scholars. *Societies*, *2*(3), 75–83.

Wolbring, G., & Burke, B. (2013). Reflecting on education for sustainable development through two lenses: Ability studies and disability studies. *Sustainability*, *5*(6), 2327–2342.

Wolbring, G., Leopatra, V., & Yumakulov, S. (2012). Climate change, water, sanitation and energy insecurity: Invisibility of people with disabilities. *Canadian Journal of Disability Studies*, *1*(3), 66–90.

Yanggen, D., Cole, D. C., Crissman, C., & Sherwood, S. (2004). Pesticide use in commercial potato production: Reflections on research and intervention efforts towards greater ecosystems health in northern Ecuador. *Ecohealth*, *1*(2), SU72–SU83.

Zinsstag, J., Mackenzie, J. S., Jeggo, M., Heymann, D. L., Patz, J. A., & Daszak, P. (2012). Mainstreaming One Health. *Ecohealth*, *9*(2), 107–110.

Zinsstag, J., Schelling, E., Bonfoh, B., Fooks, A. R., Kasymbekov, J., Waltner-Toews, D., & Tanner, M. (2009). Towards a 'One Health' research and application tool box. *Veterinaria Italiana*, *45*(1), 121–133.

STRUCTURAL VULNERABILITY AND NARRATIVE: SENSITISING CONCEPTS FOR UNDERSTANDING THE HEALTH IMPACTS OF CLIMATE CHANGE

Rebecca Moran, Julie Hollenbeck and Cassandra Phoenix

ABSTRACT

Purpose – *This chapter proposes a way to deepen our understanding of the health impacts of climate change. It explores how and why individuals and communities may experience the climate change-human health interface in different ways.*

Design/methodology/approach – *We suggest that the concepts of structural vulnerability and narrative inquiry can provide a thick (ethnographic) description of how and why individuals and communities experience and give meaning to the health impacts of climate change. We begin by defining the two concepts before bringing them together to explore the relationship between climate change and health.*

Ecological Health: Society, Ecology and Health
Advances in Medical Sociology, Volume 15, 109–124

ISSN: 1057-6290/doi:10.1108/S1057-6290(2013)0000015009

Findings – *The combination of these two concepts offers the potential to advance our knowledge in two key ways. Firstly, they facilitate a critical and interpretive approach to both the notion of agency and the public health paradigm of the 'rational-actor'. Secondly, they reveal how vulnerability to climate change is embodied at the level of the mundane and everyday.*

Social implications – *These concepts, when applied to the climate change–human health interface, can help demonstrate how vulnerability is often a social construction, and, with sufficient political will, may be ameliorated. We see the combination of the concepts discussed here as an opportunity for research to address inequality and justice.*

Originality/value – *This paper takes two innovative and established concepts in medical anthropology (structural vulnerability) and social science (narrative inquiry) and invites their application to our understanding of climate change and human health. Research analysed via these concepts will provide a clearer understanding of the impacts of climate change and experiences of vulnerability.*

INTRODUCTION

Experts predict that impending environmental impacts of climate change will affect a majority of the global population, putting billions of lives at increased risk of negative health effects. The occurrence of infectious diseases, physical harm from natural disasters and psychological impacts stemming from these effects are expected to rise in the coming years. Yet these direct and indirect health-related climate change impacts are (increasingly) experienced disproportionately since developing countries and vulnerable groups in developed countries have the weakest capacity to adapt (Costello, Abbas, Allen, Ball, & Bell, 2009; Crate & Nuttall, 2009; Raleigh, 2010).

Given its centrality to the topic of climate change, not least within discourses around adaptation and mitigation, the notion of vulnerability features regularly within the literature (e.g. Bohle, Downing, & Watts, 1994; Füssel, 2007; Füssel & Klein, 2006). However, much of this work presents either descriptions of likely vulnerabilities or provides tools for assessing levels of vulnerability. Currently there is a paucity of research seeking to understand the 'human face' of climate change (Finan, 2009). Research on climate change has explored, amongst other things, the human causes of global warming and the best ways 'to navigate along the twisting trail of impacts' (Finan, 2009, p. 175). Natural scientists have addressed important technical issues, such as agricultural land lost to climate change impacts, temperature rises and decreases in biodiversity. This, according to Finan, is climate change without a human face.

Researching the lived dimensions of climate change is part of a larger call to move beyond the 'old-fashioned idea that scientific facts build the appropriate foundation for knowing how to act in the world' (Sarewitz, 2004, p. 385), to explore what else contributes to building knowledge and how to enhance our understanding of why and how people act in the world (Geoghegan & Leyshon, 2012). This knowledge is particularly relevant to establishing the approach global policy makers will take when working with populations who are vulnerable to the human health impacts of climate change. While not drawing on empirical data in this conceptual chapter, we provide future researchers in this area with suggestions of how the human face of climate change may be revealed, by applying the concepts of structural vulnerability and narrative inquiry. We look at how narratives are used as tools to shape behaviour, influence policy and occasionally mislead. These concepts are 'sensitising' in that they help bring the human health impacts of climate change to life and provide context and perspective into 'how individuals and communities respond to climate change based on their needs, values, cultures, capacities, institutional forms and environmental factors' (Geoghegan & Leyshon, 2012, p. 1). We argue that structural vulnerability and narrative can work together to provide a thick (ethnographic) description of how and why individuals and groups experience and give meaning to the health impacts of climate change. Together they can help explore the interface between pathology and social, cultural and/or political exclusion in efforts to lead to a more multi-faceted understanding of the relationship between climate change and health.

We begin by defining structural vulnerability, outlining its key characteristics and demonstrating its importance for advancing knowledge of climate-related health impacts. We then proceed to advocate narrative inquiry as one methodology for studying the human face of climate change and place narrative inquiry within a structural vulnerability framework. Combined these methods provide a powerful tool for advancing knowledge of how climate change is experienced in two key ways. Firstly, it facilitates a critical and interpretive approach which provides a nuanced and detailed demonstration of social structures and the possibilities and limitations of agency. Secondly, these methods can be used to show how vulnerability to climate change impacts are embodied by social actors at the level of the everyday and mundane. As a result this framework has the analytical potential to connect every day, embodied realities with climate change, human health policy, public engagement and global governance.[1]

There are wider connections that can be drawn from these concepts. In this research we use a variety of examples to support our discussion, drawing mainly from policy and academic research. However, we acknowledge that the drivers of human behaviour go far beyond the realm of policy and governance. Future discussions may consider broader social influences on human behaviour using our conceptual framework, including the role of social media, community, culture and family.

STRUCTURAL VULNERABILITY

In its earlier iteration, structural vulnerability was termed 'structural violence'. The concept was developed by French sociologist Johan Galtung (1969) to help explain the impact of the Cold War on Third World states, subsumed in post-colonial politics and tensions between Western powers. The concept supposes that violence can be indirect and insidious, shaping behaviours and social norms, weaving them into the very fabric of society. As such, it has been argued that structural violence can perpetuate inequality, for example, by restricting some individuals' access to healthcare and educational opportunities through marginalisation. Elaborating further, Galtung writes:

> [W]hen one husband beats his wife there is a clear case of personal violence, but when one million husbands keep one million lives in ignorance there is structural violence. Correspondingly, in a society where life expectancy is twice as high in the upper as in the lower classes, violence is exercised even if there are no concrete actors one can point to directly attacking others, as when one person kills another. (1969, p. 171)

Physician and anthropologist Paul Farmer (2003), considers social inequality the heart of structural violence and has used structural violence to frame his research on health disparities. In doing so, his was one of the first analyses to connect structural violence explicitly with health. Exploring how the poor and marginalised experience health disparities, his work sheds light on how a host of factors, or 'insults', dispose individuals and groups towards ill health. The experience of structural violence not only disposes individuals and groups to particular pathogens, but causes health to be experienced differently, along socially structured patterns of disease and distress across population groups. For example, an individual's experience of infectious disease or mental health in a rural developing country context may be markedly different to that of one of the authors of this chapter. As Brown and Inhorn write: 'Exposure to a pathogen [is] a necessary but not sufficient cause of disease ... the progression from exposure to disease depend[s] in part on the health of the exposed person, in which ... an individual's vulnerability to a complex of insults, is never constant' (1997, p. 39).

The concept of structural violence enables the researcher to focus analytically on the behaviour, lifestyle, attitudes and environment of an individual or community, and ask why and how does this person/community act in the world? Structural violence, amongst other things, requires analysis of the socio-political and economic forces that have a bearing on human behaviour. It is also an implicit critique on the notion of agency. Specifically, a focus on structural violence calls for an analysis of the forces that constrain decision-making, frame choices and limit life options (Quesada, Hart, & Bourgois, 2011).

Structural violence forms the conceptual foundation of the notion structural vulnerability, a phrase coined by medical anthropologists analysing the health burden on Latino migrants in the United States (Quesada et al., 2011). Although analyses of structural violence may include cultural factors such as gender inequality, 'in practice the concept [of structural violence] has been used primarily to invoke materialist forces calling attention to class oppression and economic injustice' (2011, p. 341). Thus in order to extend the factors for analysis and attenuate the potentially provocative nature of the word 'violence', Quesada and fellow anthropologists employed the term 'structural vulnerability'. In doing so, their analysis extends beyond the socio-political and economic considerations to encompass cultural and idiosyncratic sources of physical and psychodynamic distress (2011, p. 341). Structural vulnerability entails a relative and contextual relationship of suffering (social and physical; direct and indirect) between a person or population, their history and the 'greater' social, political, economic, cultural and idiosyncratic forces within which they are embedded.

With this in mind, analysing structural vulnerability in the context of climate change and its human health impacts takes on an important role in understanding how and why people act in the world, as well as the apparent lack of a 'human face', in this emerging arena of study. An analysis of the 'greater' forces has the potential to help address this gap, thereby extending our understanding of why people might act in ostensibly counter-productive or 'illogical' ways. We shall now introduce one way of conducting this analysis before exploring how this concept can help develop our understanding of the health impacts of climate change.

NARRATIVE INQUIRY

Narrative inquiry is a school of thought that perceives 'people as active, socially constructed beings who live and lead storied lives' (Sparkes & Smith, 2008, p. 296). Scholars stress that our understanding of human life will always remain deficient if we do not also examine the narrative mode (Bruner, 2002; Frank, 2010; Freeman, 2010). A growing appreciation for the value this school of thought brings to the scientific body of knowledge has led some social scientists to explore narrative as a means of understanding how people make sense of the world. For example, in their research on narratives of ageing, Phoenix and Smith write: 'Narrative is a pervasive and crucially important form of human activity. It uniquely acts in human consciousness, shaping human conduct, projecting possible futures, affecting who we are and can come to be' (2011, p. 630).

'Who we are and can come to be' is intricately connected to the notion of habitus. The concept of habitus was developed by French sociologist Pierre Bourdieu and refers to 'a system of dispositions' (Bourdieu, 2005, p. 43). He explains how 'social agents are endowed with habitus, inscribed in their bodies by past experiences' (2006, p. 138). The agent knows the world, 'takes it for granted, precisely because he is caught up in it, bound up with it; he inhabits it like a garment (un habit) or a familiar habitat' (2006, p. 143). Habitus is what affirms being the kind of person we believe ourselves to be. It is also a 'tacit resource that structures how life is experienced' (Frank, 2010, p. 104), or as Bourdieu refers to a 'structuring structure'. Building upon the work of Bourdieu, Arthur Frank develops the concept of a 'narrative habitus'. That is 'knowing a corpus of stories; feeling comfortable telling and hearing certain stories (and not others); and sharing with others a sense of where events in a story are likely to lead' (2010, p. 105). These stories are embedded in our bodies, or what Bayard refers to as our 'inner library' (2007) and, for Bayard 'we are the sum of these accumulated books' (or stories) (p. 73). Thus narrative habitus provides us with a repertoire of stories that, in turn, shapes our taste in stories and helps predispose our preferred resolution to the stories as they unfold in our lives. These stories do not form in a social vacuum. Rather they are drawn from a range of dimensions that frame our everyday life.

According to Somers (1994), narratives can and should be analysed across different dimensions. This analysis can reveal the relationship between institutional and individual narratives as well as the structured and structuring nature of habitus. Somers offers a framework of narrativity as a means of distinguishing three layers of narrative in social life: ontological narratives, public narratives and meta-narratives. Firstly, ontological narratives are described as:

> the stories that social actors use to make sense of – indeed act in – their lives. Ontological narratives are used to define who we are; this in turn is a precondition for knowing what to do … ontological narratives make identity and the self something that one becomes. Thus narrative embeds identities in time and spatial relationships. (Somers, 1994, p. 618)

Ontological narratives are drawn from a wider but limited source, or corpus, of narrative resources that are 'social and interpersonal', specifically friends, family members, neighbours, co-workers, etc. Moreover, they are shaped by – and in some instances shape – the second dimension, public narratives. These refer to 'those narratives attached to cultural and institutional formations larger than the single individual, to intersubjective networks of institutions, however local or grand' (Somers, 1994, p. 619). In turn, public narratives are drawn from a third dimension outlined in Somer's framework of narrativity: the meta-narrative. These are 'master narratives' in which 'we are embedded as contemporary actors in history and as social scientists. Our sociological theories and concepts are encoded within aspects of these master narratives' (1994, p. 619). For instance, meta-narratives can comprise the ideology of our time, such as 'capitalism versus communism', 'healthy aging versus decline' and 'environmentalism versus consumerism'.

Embodiment

Both structural vulnerability and narrative inquiry involve an analysis of what is identity forming: that is an exploration of how and why people act in the world. The narrative habitus – that is the embedding of stories in the body – shapes 'not only expectations for how plots develop in stories, but also expectations for how people ought to emplot their lives' (Frank, 2010, p. 105). Bayard's concept of the 'inner library', goes further, suggesting that people spend their lives seeking a story that can match the inner story which, in turn, shapes perceptions and explains experiences of the world we inhabit.

This view connects strongly to a structural vulnerability analysis which helps explain vulnerability as a 'structured and structuring' experience. The vulnerability of an individual is 'produced by his or her location in a hierarchical social order and its diverse networks of power relationships and effects' (Quesada et al., 2011, p. 341). Bourdieu writes:

> … habitus is the basis of an implicit collusion among all the agents who are products of similar conditions and conditionings, … of a practical experience of the transcendence of the group, of its ways of being and doing, each agent finding in the conduct of all his peers the ratification and legitimation ('the done thing') of his own conduct, which, in return ratifies and, if need be, rectifies the conduct of others. (2006, p. 145)

A structural vulnerability analysis speaks to the notion of embodiment since the experience of vulnerability shapes the perceptions, behaviours, practices and self-conceptions that help form the habitus. This embodiment of marginalised and subordinated status produces a form of symbolic power (Bourdieu, 2006) whereby 'the everyday violence of imposed scarcity and insecurity is understood as natural and deserved' (Quesada et al., 2011, p. 342).

The dimensions of narrative identified by Somers (1994) circulate in different places are present in our social worlds in various forms (including fictional written works, government policies, social media, cultural and community norms, documentaries, films and journalistic accounts). Therefore, in piecing together the drivers of individual and group behaviour, it is important to recognise that narratives themselves are social constructions. One example is governmental policies and recommendations that are negotiated and distilled by social agents (such as policy makers, scientists, elected officials, media, community) around climate change (see Sarewitz 2004 for additional examples). Within these narratives, stories circulate as the narrative habitus, which enables, connects and teaches us ways of knowing and being in the world vis-à-vis changing social-ecological conditions.

CLIMATE CHANGE ANALYSIS

So far, we have introduced the concepts of structural vulnerability and narrative inquiry, and shown how they are used to critically analyse dimensions of individual and group vulnerability, as well as provide insights into how and why people act in and make sense of the social and the natural world. This final section discusses how these concepts can work together to help unravel the complex experiences encountered at the climate change–human health interface, and explore how and why this interface is experienced in specific but also diverse ways. Additionally these concepts show that variability of human health effects of climate change and the experiences of climate change in these contexts can be usefully thought of as social constructions. As such, research which takes this interplay into consideration will be highly influential in shaping strategies and encouraging political actors to take action for individuals and populations whose health is disproportionately affected by climate change. In short, if we can better understand the construction of vulnerability, we will be better placed to address any sense of its inevitability.

Two main points will be made in this section. The first is to show how these concepts combined pose a conceptual challenge to the current hegemonic thinking underlying approaches to public health and climate change. That is, the dominant biomedical paradigm of the 'rational-actor' that underpins many public health initiatives. This is questioned here through a critique of the notion of agency. The second point outlines the contribution narrative inquiry and structural vulnerability can make to research which explores the 'human face' of climate change. They can reveal how vulnerability to climate change impacts are embodied by social actors at the level of the everyday and mundane; and connect these everyday embodied realities to the impacts of climate change and human health policies and global governance practices.

Agency

The notion that habitus limits agency underpins narrative inquiry and structural vulnerability. Habitus implies a critique as it suggests that a 'set of dispositions' constrains decision-making capabilities and frames the choices presented, limiting available options. Therefore, one is a free agent within one's habitus. Referring back to Bayard's concept of the 'inner library', we draw from our finite inner library, which is based on our own accumulation of knowledge and stories through our personal experiences. Arguably this is an important consideration for individuals and groups subject to marginalisation, disenfranchisement, dispossession and uncertainties in their physical environment, particularly given uneven power structures that exist between individuals, groups, societies and cultures.

This is in line with Bourdieu's concept of 'symbolic violence' where dominion works based on the misrecognition of power structures by the dominated, who unknowingly become complicit in their own domination. For instance, Auyero and Swistun (2009) conducted an ethnography of

'Flammable', an Argentinian shantytown situated by a large petrochemical compound, which was responsible for severely polluting the town's soil, air and ground water. Despite nearly 5,000 sick or frail inhabitants, residents doubted or denied that petrochemical pollution was responsible for the illnesses they were experiencing. In this study, the authors found that 'physical and psychological suffering is compounded by doubts, disagreements, suspicions, fears, and endless waiting' (2009, p. 4). The experiences of the polluted reality were 'socially and politically produced; the meanings of contamination are the outcome of power relations between residents and outside actors. These produced meanings … shape those very same unequal relationships' (2009, p. 5). Many different actors (such as teachers, journalists, lawyers, Shell personnel, doctors and government officials) were part of the residents' everyday life in Flammable and all helped contribute to what the residents knew about the settlement. They also influenced what the residents ignored, misidentified and what they wanted to know.

Auyero and Swistun found that the concept of symbolic violence helped explain why injustices can become normalised and are not resisted. Instead they are subsumed into everyday life. It is important to note Bayard's observation that habitus can be predisposing, although predisposition does not necessarily determine belief or behaviour. Additionally, Bourdieu notes that 'Habitus change constantly in response to new experiences. Dispositions are subject to a kind of permanent revision … on the basis of the premises established in the previous state' (2006, p. 161). Therefore, the residents of Flammable were not necessarily destined to doubt or deny the source of their suffering. Indeed, one of the authors of the ethnography was a Flammable resident. However, the manner in which knowledge was produced and shared, and the uncertainties of the narratives in circulation, formed an empirical example of Bourdieu's idea of symbolic violence. The confusion and the doubts that became part of Flammable's narrative were shaped by 'the interacting "invisible elbows" of external power forces and of everyday routine survival struggles' (Auyero & Swistun, 2009, p. 6).

Structural vulnerability and narrative inquiry can help examine how and why the human health impacts of climate change are experienced. Armed with this data, policy makers and managers can better address the vulnerabilities in a manner that takes the specifics of these vulnerabilities and experiences into consideration. Recognising the limits to agency is crucial as it helps understand why people may behave in seemingly idiosyncratic or 'illogical' ways. For instance, in 2007 central Mozambique experienced heavy rains. Having invested in its flood response capacity after the floods in 2001, the Mozambique government was in a position to facilitate the evacuation of villages (Artur & Hilhorst, 2012). During the evacuation some villagers refused to leave. The army was assigned to forcibly remove them. The behaviour of the villagers appeared unreasonable given the extent and human cost of the floods in 2001 (where hundreds were killed). However, as Artur and Hilhorst found 'people have their own rationalities for dealing with the flood' (2012, p. 534). They discovered that evacuation was a strategy of last resort since protecting their personal assets like livestock, fishing nets and tools, their only buffer from abject poverty, was a priority. Villagers would only accept rescue when their lives were in definite and immediate danger. Furthermore, given that evacuation separates families, many preferred to stay together. Behaviour that seems idiosyncratic or illogical at first, may rest on the mundane and every day concerns of individuals who experience a structural vulnerability that makes evacuating on a helicopter, or into a canoe, a less reasonable choice than facing the possibility of a life-threatening flood. This is not an factor necessarily restricted to those living in poverty, both those living close to subsistence and those enjoying wealth and excess may be led to act in ways that seem contrary to survival. Clearly, there are more nuanced, personal and socially complex factors at play that a different approach to the study of behaviour change could reveal.

Structural vulnerability entails a contextual relationship of a person's or group's social and physical suffering with their history and greater economic, social, political and cultural forces within which they are inscribed. The result is a narrative habitus – an embodiment of these structures and the basis of implicit collusion by the dominated with the dominant (Bourdieu, 2006). Thus a narrative analysis of structural vulnerability entails the critique of agency and questions the traditional biomedical and socio-economic model of the rational-actor. In practice, climate change impacts compound existing inequalities and are built on everyday issues such as an individual's dependency on livestock and tools. By putting seemingly idiosyncratic or illogical behaviours into context, policy-makers, managers and emergency responders can begin to address predispositions and inequalities that exist within social groups and communities, as well as the ignorance that circulates at the institutional level.

Everyday Realities: Local and the Global

This brings us to the second point, which explores the contribution these concepts can make to research on the impacts of climate change on human health. Structural vulnerability and narrative analysis report a detailed thick description of the local: the lived experience of everyday life. This is the interface where policy and material (climate) change meet, where resources are contested and where everyday life is negotiated and enacted. The insights offered by narrative inquiry and structural vulnerability draw attention to how social inequalities are embodied and how they cause actual physical suffering. Although Farmer (2003) has demonstrated this in relation to infectious disease, it has yet to be dealt with in relation to climate change impacts. Given that research suggests the impacts of climate change will be worse for those already living in poverty and exclusion, it seems the time has come to address this research gap. A structural vulnerability analysis lends itself to the thick descriptive work of an ethnography. Fine-grained ethnographic accounts of lives in the context of the local forces that press on them are the foundation for a clearer understanding of the impacts of climate change on human health on vulnerable communities. This detailed description, this 'human face', can be used to highlight inequalities, injustices or even a complete oversight of particular needs in certain populations around the world.

The forces which we allude to regularly throughout this discussion as the 'political, social, economic, cultural and idiosyncratic' are all to some degree shaped by the circulating narratives which script governance and policy. That is to say that the resources and conditions available to communities around the world are tied into their histories, and their networks, which are also embedded in various strands of governance. As a term, 'global governance' refers to the changing nature of power whereby the geographical territory of the nation-state has had its sovereignty eroded through a transnational network of business, military intervention, regional economic agreements, transnational non-governmental organisations and other governments. Within all these different networks of power, narratives circulate and shape policies, which mediate or exacerbate conditions of vulnerability in marginalised communities.

One way to illustrate these observations is to turn to Farbotko and Lazrus (2012) who offer some interesting insights into the narrative of population displacement and mobility. They observe that the term 'climate refugee' is used to describe populations who are perceived as being forced to move due to environmental factors that precede or create a disaster. As a term, it evokes a sense of drifting dispossessed people. But this obfuscates both the institutional and human response, and the economic and social circumstances of a marginalised population, that can turn an environmental event, such as a flood, into a disaster (Ribot, 2010). Furthermore, 'climate refugee' narratives can be used to protect the interests of national security in the west by mobilising racist fears of a dangerous poor (Hartmann, 2010). The Australian Green Party proposed a bill to legislate a climate refugee visa in 2007, partly justified by a media report that New Zealand 'is accepting 75 climate refugees' from a Pacific island (Tuvalu) per year (Farbotko & Lazrus, 2012, p. 386). The scheme, which governed the intake of migrants into New Zealand, operated as part of the Pacific Access Category and was an economic migration policy, not a humanitarian one. Yet it was framed by the media as a policy for climate refugees, presumably because it was more newsworthy. In turn, this narrative was then used by a political party to help influence policy. Farbotko and Lazrus note that Pacific islanders have often exercised a narrative and practice of mobility between the islands as a way of life – this movement is not a new phenomenon practiced by 'climate refugees'. This example helps illustrate the importance of observing the purpose of a narrative and to identify who or what is being represented. The value for this discussion lies particularly in noting that narratives circulate, shape and are proactive in moulding perceptions and behaviours.

CONCLUSION

Vulnerability is socially constructed and once the threads have been identified, and the narratives in circulation noted, it becomes possible to visualise the foundation that helps us understand how people act in the world (Geoghegan & Leyshon, 2012). In medical anthropology, structural vulnerability has been used to highlight socially structured patterns of distress and disease, such as mental health, infant mortality and infectious disease (Quesada et al., 2011). For climate change we see this as an opportunity for research to address inequality and justice. Climate change cannot be treated as an exercise in disaster relief. Research shows how disaster relief is politicised and can often further depreciate agency in vulnerable communities (Artur & Hilhorst, 2012; Raleigh, 2010). Our approach analyses the narratives that help shape policy, everyday practices, histories and all that links behaviours with being in the world. A structural vulnerability analysis places the individual into the global context and does it in detail. The ethnographic 'thick description' (Quesada et al., 2011) of structural vulnerability analysis also helps to explore the purpose of narratives as they are scripted and circulated. Public and meta-narratives serve as ways of reinforcing behaviours and beliefs, or as Bourdieu refers to they emphasise 'the done thing' (2006). Sometimes these narratives are deliberately evocative; Farbotko and Lazrus provide an example that helps clarify this point. In their research on the Pacific island of Tuvalu, they describe the experiences of the islanders being appropriated by journalists and environmental NGOs to further the aims of renewable energy promotion:

> the image of the climate refugee is sustained as a sort of victim-commodity, providing news value, political point scoring, and a human embodiment of climate change 'evidence' for western environmental activists concerned with saving the planet. (Farbotko & Lazrus, 2012, p. 386)

The authors also describe how, in meetings and lobbying exercises, environmental NGOs circulated narratives about Tuvaluans to use them as evidence for reducing carbon emissions, rather than as the focus of adaptation policies. However, the 'owners' of the narratives being circulated wished to promote their stories in regard to equity and climate justice, but were not represented as such.

In this chapter we have explored how structural vulnerability and narrative inquiry can help offer insights into the experience of the impact climate change on human health. We began by outlining the concept of structural vulnerability and then introduced the dimensions of narrativity. Our discussion showed how combining these analytical tools can provide insights into the limits of individual and group agency, as well as provide a thick description of the lived experience of climate change. The effects of climate change exacerbate and rearrange the landscape of poverty, justice and inequality, as part of a network of elements that contribute to the shaping of everyday life. Research has demonstrated that a combination of socially constructed conditions, such as political instability, conflict, extreme poverty and corruption can go hand-in-hand with environmental degradation and vulnerability to extreme weather events, to create conditions for displacement, poor health and inadequate disaster recovery (Black, 1998; Farbotko & Lazrus, 2012; Raleigh, 2010). Research which continues to identify the narratives in circulation, and relate them to how the health impacts of climate change are experienced through an analysis of structural vulnerability, will be enormously powerful. It will unmask the purpose of narrative at its different levels and relate them to the experience and construction of vulnerability, and promise to be an invaluable tool when developing adaptation and mitigation strategies around the world.

NOTE

1. Throughout this discussion we will refer to the 'impacts of climate change' and in so doing we are referring to the direct and indirect threats to health (noted by Costello et al., 2009).

REFERENCES

Artur, L., & Hilhorst, D. (2012). Everyday realities of climate change adaption in Mozambique. *Global Environmental Change*, *22*, 529–536. doi:10.1016/j.gloenvcha.2011.11.013

Auyero, J., & Swistun, D. (2009). *Flammable: Environmental suffering in an Argentinian Shantytown*. Oxford: Oxford University Press.

Bayard, P. (2007). *How to talk about books you haven't read*. New York, NY: Bloomsbury.

Black, R. (1998). *Refugees, environment and development*. Harlow: Addison Wesley Longman.

Bohle, H. G., Downing, T. E., & Watts, M. J. (1994). Climate change and social vulnerability: Toward a sociology and geography of food insecurity. *Global Environmental Change*, *4*, 37–48.

Bourdieu, P. (2005). Habitus. In J. Hillier & E. Rooksby (Eds.), *Habitus: A sense of place*. Aldershot: Ashgate Publishing Limited.

Bourdieu, P. (2006). *Pascalian meditations*. Cambridge, UK: Polity Press.

Brown, P., & Inhorn, M. (1997). *The anthropology of infectious disease: International health perspectives*. Amsterdam: Gordon and Breach.

Bruner, J. (2002). *Making stories*. Cambridge, MA: Harvard University Press.

Costello, A., Abbas, M., Allen, A., Ball, S., Bell, S., Bellamy, R., … Patterson, C. (2009). Managing the health effects of climate change. *The Lancet Commissions*, *373*, 1693–1733.

Crate, S., & Nuttall, M. (Eds.). (2009). *Anthropology and climate change: From encounters to actions*. Walnut Creek, CA: Left Coast Press Inc.

Farbotko, C., & Lazrus, H. (2012). The first climate refugees? Contesting global narratives of climate change in Tuvalu. *Global Environmental Change*, *22*, 382–390. doi:10.1016/j.gloenvcha.2011.11.014

Farmer, P. (2003). *Pathologies of power: Health, human rights, and the new war on the poor*. Berkley, CA: University of California Press.

Finan, T. (2009). Storm warnings: The role of anthropology in adapting to sea-level rise in Southwestern Bangladesh. In S. Crate & M. Nuttall (Eds.), *Anthropology and climate change: From encounters to actions*. Walnut Creek, CA: Left Coast Press Inc.

Frank, A. (2010). *Letting stories breathe: A socio-narratology*. Chicago, IL: The University of Chicago Press.

Freeman, M. (2010). *Hindsight, the promise and peril of looking backward*. New York, NY: Oxford University Press.

Füssel, H.-M. (2007). Vulnerability: A generally applicable conceptual framework for climate change research. *Global Environmental Change*, *17*, 155–167.

Füssel, H.-M., & Klein, R. J. T. (2006). Climate change vulnerability assessments: An evolution of conceptual thinking. *Climatic Change*, *75*, 301–329.

Galtung, J. (1969). Violence, peace, and peace research. *Journal of Peace Research*, *6*(3), 167–191.

Geoghegan, H., & Leyshon, C. (2012). On climate change and cultural geography: Farming on the Lizard Peninsula, Cornwall, UK. *Climatic Change*, *113*(1), 55–66.

Hartmann, B. (2010). Rethinking climate refugees and climate conflict: Rhetoric, reality and the politics of policy discourse. *Journal of International Development*, *22*, 233–246.

Phoenix, C., & Smith, B. (2011). Telling a (good?) counterstory of aging: Natural bodybuilding meets the narrative of decline. *The Journals of Gerontology Series B: The Psychological Sciences and Social Sciences*, *66B*(5), 628–639. doi:10.1093/geronb/gbr077

Quesada, J., Hart, L., & Bourgois, P. (2011). Structural vulnerability and health: Latino migrant laborers in the United States. *Medical Anthropology*, *30*(4), 339–362. doi:10.1080/01459740.2011.576725

Raleigh, C. (2010). Political marginalization, climate change, and conflict in African Sahel states. *International Studies Review*, *12*, 69–86.

Ribot, J. (2010). Vulnerability does not fall from the sky: Toward multiscale, pro-poor climate policy. In R. Mearns & A. Norton (Eds.), *Social dimensions of climate change*. Washington, DC: The World Bank.

Sarewitz, D. (2004). How science makes environmental controversies worse. *Environmental Science and Policy*, *7*, 385–403.

Somers, M. (1994). The narrative constitution of identity: A relational and network approach. *Theory and Society*, *23*, 605–649. doi:10.1007/BF00992905

Sparkes, A., & Smith, B. (2008). Narrative constructionist inquiry. In J. Holstein & J. Gubrium (Eds.), *Handbook of constructionist research* (pp. 295–314). New York, NY: The Guildford Press.

PART III
RESEARCHING HEALTH IN THE NEXUS BETWEEN THE SOCIAL AND THE ECOLOGICAL

HEALTH AND ENVIRONMENTAL POLITICS IN THE UNITED STATES: A HISTORICAL PERSPECTIVE

Jennifer Thomson

ABSTRACT

Purpose – *This chapter examines the historical development of different conceptions of health among environmental activists in the postwar United States.*

Methodology/approach – *The historical analysis combines archival research with oral history interviews.*

Findings – *This study argues that applications of "health" to describe the environment are more diverse than generally acknowledged, and that environmental activists were at the forefront of connecting the two terms within broader public discourse.*

Originality/value of chapter – *This study provides a historical context for understanding the contemporary diversity of perspectives on the links between ecology and health. It illustrates the cross-fertilization between scientists, philosophers, and environmental activists in the 1970s that led to this contemporary diversity.*

Keywords: Environmental politics; biocentrism; Rachel Carson; Gaia theory; United States; Friends of the Earth

Ecological Health: Society, Ecology and Health
Advances in Medical Sociology, Volume 15, 127–146

ISSN: 1057-6290/doi:10.1108/S1057-6290(2013)0000015010

A language of health characterizes contemporary environmental politics in the United States. In Environmental Protection Agency (EPA) announcements and mission statements of environmental nonprofits, newspaper editorials and activist journals, claims about the health of ecosystems, the health of the planet, the health of animal species, and the health of humans abound. Yet despite this rhetorical ubiquity, modern environmental politics are ideologically and organizationally fractured along issues of whose health is at stake and just how that health should be protected.

In five distinct case studies, this chapter highlights the intellectual and political history of these competing conceptions of health. Each study documents how a varied group of postwar environmental activists appropriated concepts of health from ecologists, philosophers, conservation biologists, epidemiologists, and nuclear physicians in order to conceptualize and confront mounting environmental changes. This chapter explores this process of appropriation. In so doing, it demonstrates the historical complexity of the relationship between the ecology and health movements in the late twentieth century United States, a complexity which informs and frustrates contemporary attempts to further environmental protection and transform social practices more broadly.

Historians and scholars of environmentalism have often described the postwar period as a transition from conservation to environmentalism: in other words, a shift away from protecting wilderness and toward defending the health of humans and their standards of living (Hays & Hays, 1987). By contrast, this chapter presents five case studies which illustrate how the significance of health for environmental activists was far more complex:

(1) In *Silent Spring* (1962), Rachel Carson used the pressing problem of indiscriminate pesticide application to detail the *symbiotic* relationship between humans and their environment.
(2) Following the first Earth Day in April of 1970, many environmentalists began to espouse a vision of a *healthy planet* as one in which human population and economics achieved a steady, no-growth state. The newly formed environmental organization Friends of the Earth (1969) was instrumental in the development of this vision. By the early 1980s, however, the organization had transformed this systemic conception into a focus on saving the planet through informed consumer choices.
(3) In the 1970s, a new cohort of biocentric activists envisioned a planet whose health would only be ensured through *ecological egalitarianism*. Drawing upon deep ecological philosophy, civil disobedience, and conservation biology, biocentrists advocated the immediate reversal of

human demographic and economic growth and the "rewilding" of animals and humans.

(4) In 1978, with the discovery of 21,000 tons of buried chemical waste under their neighborhood, residents of Love Canal, New York, came to perceive their *environment as toxic* for human and animal health. This understanding came from their integration of epidemiological, anecdotal, and toxicological knowledge.

(5) Physicist James Lovelock proposed the Gaia hypothesis in the 1960s, positing that all life on Earth, from bacteria to the atmosphere, constituted a single integrated organism. By 1991, he had developed a system of *planetary medicine* to diagnose Gaia's ailments, chief among which was her steadily climbing fever.

This chapter offers a partial historical narrative of the evolving relationship between health, environment, and ecology in the United States. It focuses on environmental activists and independent scientists, rather than policy-makers or international organizations, and offers an overlapping narrative, one which demonstrates how politics of environmental management and protection existed alongside and interacted with politics of ecosystem approaches. In this nonprofessional and layered approach, the chapter offers a history of the *possibilities* for the emergence of the field of EcoHealth. It illustrates how several of the core principles of EcoHealth were enabled by the political actions of environmental activists and affiliated scientists in the late twentieth century. The conclusion will suggest several ways in which the following historical case studies shed light on the field of EcoHealth.

RACHEL CARSON

Rachel Carson is frequently portrayed as a point of origin for contemporary understandings of a symbiotic relationship between human and environmental health (Lear, 1997; Lytle, 2007; Souder, 2012). Her 1962 book *Silent Spring*, which detailed how the spraying of pesticides impacted all living creatures, is canonical among many environmental activists, scientists, and scholars.

Silent Spring tackled the question of pesticides progressively, addressing in turn the damages which their indiscriminate application had done to soil, water, air, birds, and humans. Carson opened the book with a comparison to the global, totalizing, intergenerational threat of nuclear war,

> Along with the possibility of the extinction of mankind by nuclear war, the central problem of our age has therefore become the contamination of man's total environment with such substances of incredible potential for harm – substances that accumulate in the tissues of plants and animals and even penetrate the germ cells to shatter or alter the very material of heredity upon which the shape of the future depends. (Carson, 1962, p. 8)

Carson reviewed the present and future consequences of heavy and indiscriminate pesticide application, illustrated the global reach of environmental toxins, and argued that the best opposition to the current state of affairs would come from an informed and engaged citizenry.

Silent Spring was responsive to ongoing and widespread popular unease regarding dispersed environmental toxicity. Throughout the writing of the book and following its publication, Carson was in communication with local conservation, gardening, and birding clubs across the United States and Europe, groups which had been fighting against aerial insecticide spraying since the mid-1950s. Research for the book was prompted by a letter from Boston resident Olga Owens Huckins, a friend of Carson's, whose bird sanctuary in Duxbury, Massachusetts, had suffered severely after being sprayed for mosquitoes in the summer of 1957 (Lear, 1997). Carson's mailbox was also filled with letters from doctors confronting unfamiliar patient symptoms. As Dr. Harry Lillie of Scotland wrote, "we have got to the stage when if a patient complains of pains anywhere, the first thing is to find out what insecticide or weed killing sprays he has been using in the house or garden or farm land, or what airosols [sic] a woman has been messing around with."[1]

Although *Silent Spring* is often remembered as a book about the deleterious effects of pesticides on wildlife, its most lasting message pertained to human health; specifically, the carcinogenic and mutagenic properties of pesticide exposure. As historian Linda Nash has argued, Carson popularized an understanding of environmental health based upon personal experience that stood in marked contrast to the laboratory-based definitions of environmental toxicity then espoused by the U.S. Public Health Service and the American Medical Association. In *Silent Spring*, Carson articulated a conception of the interconnection of the land, air, animals, water, and people that was accessible to a broad audience and supported by scientific research in multiple disciplines (Nash, 2006).

There is a paradox at the heart of *Silent Spring*'s legacy. While the book encouraged many traditional conservation organizations like the Sierra Club to become more ecological by including humans within their definition of the natural world, this expansion of focus tipped many organizations toward an anthropocentric environmentalism. As we will see in the following case study, one influential cohort of environmental activists in the 1970s was primarily interested in human health and social justice issues (Foreman, 1991).

FRIENDS OF THE EARTH

In 1971, environmental organization Friends of the Earth (FOE) wrote a letter to Senator Warren Magnuson, Chairman of the Senate Subcommittee on Labor, Health, Education and Welfare. The letter urged Magnuson to approve full funding of a program for the early detection and treatment of children's poisoning from lead-based paint. FOE explained its intervention on this question of public health,

> We have decried the damage being done to the environment by lead pollution of our air, soil, water, and food supplies, but there is another, equally as serious, form of lead pollution which is directly affecting thousands of children. The environment in this case is the ghetto and the pollution source is lead-based paint.[2]

Echoing Rachel Carson, FOE's letter framed the relationship between humans and their environment symbiotically. One part of this relationship (in this case, children) might evidence harm and require remediation more immediately; however, harm to one part would result in harm to the entire system. FOE redefined the environment to include cities and their poor and marginalized communities, and it laid claim to a set of issues about urban sanitation and socioeconomic inequality traditionally the province of public health and city planning officials.

The letter drew on a century of reform efforts targeting urban pollution and occupational hazards. From the Progressive Era reformism of Jane Addams and Alice Hamilton to the mid-century workplace activism of the Oil, Chemical and Atomic Workers and the United Farm Workers, middle class professionals and labor unions endeavored to protect the health of socioeconomically vulnerable human populations from hazards in the environment (Dewey, 1998; Montrie, 2000; Sellers, 1997). The letter was also informed by post-WWII grassroots attempts to counter the environmental effects of suburbanization, including smog from expanding automobile ownership and groundwater pollution from phosphate detergent use (Dewey, 2000; Rome, 2001).

Equally, the letter embodied a different era of environmental politics which emerged in the wake of the first Earth Day. On April 22, 1970, upwards of 20 million Americans had participated in a range of events, including university teach-ins, rallies in front of government buildings and corporate headquarters, and acts of ecotage.[3] Although the enormity of the event's participation diluted its message, Earth Day rhetoric was decidedly holistic, connecting the protection of the whole planet with that of human society (Rome, 2003; Turner, 2012). Speakers across the nation argued for a revolution in consciousness toward the environment, and indicted capitalism for the degradation of humans and their planet. The environmental discourse that emerged employed holistic, life-based, and planetary terms, linked participatory democracy with ecology, and relied upon health as a metaphor for assessing the success of this linkage. As poet Gary Snyder recalled two decades later,

> That first Earth Day was not exactly a beginning, but it was a hinge, a turn around a corner. It marked the gradual waning of the need for anti-war activism and the swinging of our energies toward the struggle for the *health* of the earth. (Snyder, 1995, p. 57)

FOE was part of this new activism. Throughout its first decade it brought a host of issues into the environmental fold: nuclear power, supersonic transport, Agent Orange, pesticide spray drift, and genetic engineering. In each campaign it argued that these practices threatened the present and future health of humans, wildlife, and wilderness. Illustrative of this expansive perspective was "Nuclear Man," a cartoon published in FOE's monthly journal *Not Man Apart* in June of 1977. The full-page graphic, modeled after Leonardo Da Vinci's iconic *Vitruvian Man*, illustrated where the strontium-90, plutonium-238, and cadmium-137 released by the routine and accidental emissions of nuclear power plants settled in the human body. The caption described the carcinogenic reality of these elements to present and future generations, "if the half life of the element is longer than the human life, the radioactive elements could cause cancer in someone else when the first host body returns to dust."[4] The graphic illuminated just how individual bodily circulation was connected to planetary circulation: radioactive fallout produced in one place could exert disastrous and long-lasting effects in places far removed. By the early 1980s, however, FOE had shifted its focus from planetary circulation to the consumer. This study concludes with the anti-pesticides work of FOE organizer Bob Scowcroft to understand how this transformation happened.

In 1978, Scowcroft began to organize health food stores and distributors on the West Coast around the issues of spray drift of 2,4,5-T and the phenoxy herbicides. His efforts paid off quickly. In August 1979, *The Corn Husk: Your Ear on the Whole Foods Industry*, afforded FOE a full page to make its case. Scowcroft's article contrasted pesticide use with organic farming, explaining how FOE was lobbying the United States Department of Agriculture to give organic farmers the same recognition and support as chemical farmers. Pointing toward the consumer, he argued that "Pesticides not only threaten farmers' and agricultural workers' health, but the health of everyone who eats fruits and vegetables, drinks the water, or breathes the air."[5] In the *Erewhon Monthly* for January 1980, Scowcroft made his case for why stores should get involved. "You, the whole food store, are essential to our work. You are our personal link with the people of your community."[6]

Scowcroft targeted the consumer concerned about his or her personal health. By organizing fundraisers and educational evenings through businesses that catered to an identity of achieving health through proper consumption, he portrayed opposition to pesticides as a lifestyle choice. This approach differed from FOE's fight against nuclear power, which apprehended human health in the aggregate, as public health. Although pesticides were understood to strike indiscriminately, thereby endangering the health of the human population at large, political outreach through health-based consumption made the true significance of pesticides personal and elective.

Scowcroft's "pesticide status report" for Fall 1981 surveyed three pesticide cases from different regions of the country: Endrin sprayed on wheat in Montana; Sevin sprayed in the Northeast to control the gypsy moth; and Malathion used in California to control the fruit fly. In the report, he emphasized three issues: the lack of scientific information on the long-term effects of exposure to these pesticides; the probability that unintended spray drift had exposed populations far from the site of actual spraying; and, the secrecy surrounding regulatory procedures. He repeatedly emphasized the public's inability to trust that either the government or industry held the health of the public in high regard.

Scowcroft's pesticides work for FOE connected individual health to informed citizenship. Like Rachel Carson, he asserted an individual's "right to know" what lurked in the environment and framed democratic citizenship as a process of demanding that government and industry be held responsive to the individuals who were affected by their actions. However, there was a crucial difference between FOE's focus on consumer health and Rachel Carson's message in *Silent Spring*. FOE's emphasis on individual health suggested that an individual could protect the health of the planet through knowledgeable consumption choices: in other words, that buying foods not sprayed with Endrin would save the fields and their inhabitants. From the holistic perspective it espoused in its early years, FOE had moved a great distance. Ultimately it took Carson's message about the importance of an informed citizenry to its logical capitalist conclusion: that the most effective way to convince people to act was to appeal to their identity as consumers.

BIOCENTRISM

FOE was just one manifestation of environmental activism in the 1970s. Throughout the decade, a motley crew of disgruntled environmental lobbyists, academic philosophers, back-to-the-landers, and conservation biologists strove to construct a politics *of* the wild. Despite dramatically different geographic and occupational niches, this sprawling group identified as *biocentrists*, crafted a politics intended to represent, diagnose, and ultimately restore the health of the wild.

Biocentrists employed a large set of terms to diagnose the planet's ills and prescribe a right course of action. On the diagnostic side, ecologist Raymond Dasmann spoke of humanity's assault on biodiversity as the "Third World War"; conservation biologist Michael Soulé described it as the "end of evolution," the "Sixth Great Extinction," and the "Anthropocene Era"; and Earth First! founder Dave Foreman, borrowing from interwar ecologist Aldo Leopold, compiled a list of human-inflicted "wounds" to the earth. On the prescriptive side, Dasmann advised adoption of a "future primitive" state; Gary Snyder and activist Peter Berg urged humans to "reinhabit" the planet; Soulé, Foreman, and conservation biologist Reed Noss developed "rewilding" proposals; and Earth First! practiced nonviolent direct action. While the terms employed differed, the message did not: humanity lived on the brink of disaster, and only immediate attention to the health of the earth and its nonhuman inhabitants could help ameliorate the impending downfall.

Biocentrism found its first practical expression in bioregionalism. As students and social activists left university campuses and the disintegrating countercultural community in San Francisco and elsewhere in the late 1960s to resettle in remote and sparsely populated regions, rural communities were infused with the social activism and countercultural lifestyles of these new residents, many of whom were rich in ecological ideals but lacked any concrete knowledge of the ways of life best suited to the region they had chosen to inhabit (Aberley, 1999). Poet Gary Snyder, himself a part of this diaspora, described the process of learning to live in a place in accordance with its ecological and cultural possibilities as *bioregionalism.*

Bioregionalists believed that human-drawn boundaries were ecologically detrimental, and should be replaced by regions "governed by nature, not legislature" (Sale, 1985, p. 43). Northern California bioregionalist Jim Dodge took this definition even further,

> Bioregionalism means life territory, place of life, or perhaps by reckless extension, government by life. If you can't imagine that government by life would be at least 40 billion times better than government by the Reagan administration, or Mobil Oil ... then your heart is probably no bigger than a prune pit. (Dodge, 1981, p. 6)

A central practice of bioregionalism was ecological restoration. Reinhabitants studied and attempted to reconstruct their region's historical ecological community. The idea of the Earth as a living entity, whose body could be approached as a doctor would approach a patient, animated much of this restoration work. In 1977, bioregional organization Frisco Bay Mussel Group articulated this conception,

> Watershed is a living organism: rivers and streams and underground flows are veins and arteries; marshes are the pollution-removing kidneys; water is the cosmic sense organ of the earth, the dimpled skin between above and below; water rhythms show us moon, season, shape and sense of land. (Frisco Bay Mussel Group, 1977, original document unpaginated)

This analogy between the planetary and human body encouraged an emotional connection with the nonhuman world; by comparing planetary processes to the familiar pathways of the human circulatory system, the organization provided its readers with a tangible way of imagining and comprehending the natural world. Moreover, the anatomical imagery offered humans a template for assessing and treating the planet as they would a sick human.

Stephanie Mills, a Michigan bioregionalist, observed that health was the dominant metaphor for the ecological restoration which bioregionalism entailed.[7] This metaphor relied upon the work of early twentieth century ecologist Aldo Leopold; in particular, his concepts of land health as the "capacity of the land for self renewal," and of the ecologist as a "doctor" (Leopold, 1949). Echoing Leopold, Mills wrote,

> Even when attempts at domesticating, or subjugating, wild systems or organisms appear to have succeeded, the results are a little perverse ... In contrast to these anomalies are the qualities of wildlife and wild places: authenticity, indigeneity, specificity and spontaneity; resilience and health. (Mills, 1994, p. 47)

Mills understood health as a gauge of ecological stability and historical continuity. She judged an area of land to be healthy insofar as it had maintained its historical ecological community, self-regulated through fire and large carnivores, and existed relatively untrammeled by human affairs. The proper role of humans, Mills felt, was to facilitate the land's reclamation of these inherent patterns.

Health has three valences for bioregionalists. First, it describes the optimal state for a particular bioregion, usually through reference to a past state of being. Second, it assesses the quality of relationship between humans and the environment they inhabit. A healthy inhabitory relationship is one in which humans understand and accept their role within the lifecycles of a particular place. Third, it functions as a heuristic device. Comparing the planetary to the human body fosters an understanding of an ecosystem's disease. This last valence stems most directly from the biocentric premise that humans are one with their environment. If, as Snyder wrote, "we are all composite beings, not only physically but intellectually, whose sole individual identifying feature is a particular form or structure changing constantly in time" (Snyder, 1995, p. 189), then there is no substantive difference to be found between one's own health and that of the neighboring river.

In another quarter of the biocentric community, ongoing debates about defining ecosystem health and the proper role of science vis-à-vis politics led to the creation of the Wildlands Project in 1991. Cofounded by a mixture of former Earth First! members and conservation biologists, the Project debuted at the Society for Conservation Biology's annual meeting in 1993. The Project's primary goal was to reorganize North America into a system of large core wilderness areas, from which humans would be excluded, and to connect these areas with wildlife corridors that would enable the passage of large predators and the migration of species. The Project's work was based upon the premise that "healing the land" meant allowing for the recovery of "whole ecosystems and landscapes" across North America (Davis & Foreman, 1993).

The philosophical framework for the Wildlands Project had been outlined by Earth First! co-founder Dave Foreman in *Confessions of an EcoWarrior* (1991), in which he proclaimed, "In our decimation of biological diversity, in our production of toxins, in our attack on the basic life-support system of Earth, in our explosive population growth, we humans have become a disease – the Humanpox" (Foreman, 1991, p. 57). If the human species had become a disease of planetary proportions, Foreman believed that the planet would produce antibodies – conservationists – to defend itself, "a new race of ... humans who love the wild, whose primary loyalty is to earth and not to *Homo sapiens* ... [who] will fight like antibodies and phagocytes for the wild, for the previous native diversity of our planet" (Foreman, 1991, p. 58)

The Wildlands Project fought for this diversity on both local and national levels. It convinced the United States Department of Fish and Game to reintroduce the Mexican wolf to the New Mexican deserts in 1994; contributed to the 1992 Northern Rockies Ecosystem Protection Act, the first federal legislation based upon core ecosystem reserve areas and connecting corridors; and developed an ambitious and idealistic proposal for a North American Wildlands Network (Foreman, 2004).

By the close of its first decade, the Wildlands Project had explicitly reframed its work as "land medicine." Working closely with pathologist Robert Howard, and influenced by Aldo Leopold, it identified seven human-inflicted "wounds" to the land, and spoke of the conservationists' proper role as a land doctor.

> Just as a medical doctor seeks not only to treat the symptoms and the disease but also to understand the root cause(s) of the illness, so do ecological "doctors" seek both to heal the wounds of the land and to understand their underlying causes. (Foreman, 2004, p. 63)

The doctor's job was both historical and preventative: healing the wounds that had already been inflicted, while attempting to prevent future harm. The stakes were high, for just as in medicine, the "end point" of the human pathology was death. For the earth, this end point would manifest as a loss of resilience and an inability to maintain normal ecological processes such as climate regulation.

Yet all was not harmonious within the conservation biology family.[8] Fearing that the discipline and its affiliated political projects had tilted so exclusively toward the biosphere as to exclude humans, by the late 1990s some conservation biologists began to engage with medical and veterinary professionals. The result was conservation medicine, a discipline which, according to practitioner Gary Meffe (1999), aimed to place biodiversity within the framework of global health. Meffe was explicitly pragmatic, arguing that the inclusion of human health within biodiversity discussions was a sure way to win over individuals who would otherwise remain opposed to the protection of biodiversity. Others, including Reed Noss, David Pimentel, and Laura Westra, formed the Global Integrity Project, which connected human well-being, ecosphere, and economics (Noss, Pimentel, & Westra, 2000). These various offshoots of conservation biology, among related initiatives such as One Health, would directly contribute to the emergence of the field of EcoHealth.

THE LOVE CANAL HOMEOWNER'S ASSOCIATION

In 1978, following the discovery of 21,000 tons of buried chemical waste under its homes and school, the small suburb of Love Canal, New York, was inundated with health and environmental investigators. The New York State Health Department and Department of Environmental Conservation, as well as the EPA, the Department of Health and Human Services, and the Centers for Disease Control, performed scores of epidemiological and environmental studies on the residents, homes, gardens, and creeks of Love Canal. These studies ranged from adverse pregnancy outcomes, home basement air testing for seven "Love Canal indicator chemicals," autopsies on three dogs, one blackbird, and two gulls, and medical examinations of 112 construction workers from the Canal remedial construction project.[9]

From a public health standpoint, the results were alarming. Elevated levels of toluene, chloroform, and benzene vapors were found in many homes; 20 school children were diagnosed with severe liver problems; and in May 1980, the EPA announced that a statistically significant number of residents suffered from chromosome breakage. Residents expected these results to prompt swift government intervention. They were deeply mistaken. Once provided with their test results, residents were frequently told to make their own informed decisions. As Eileen Matsulavage testified in 1979 about receiving the results of her basement air testing,

> I was told by Dr. Nancy Kim that the readings were quite high and I should not use the basement ... she suggested that I seal the bottom of the doorway with rags. When I questioned whether the house was safe to live in, she stated that it was a personal decision.[10]

Similarly, despite a Department of Health study that found increased rates of miscarriage and low birth weight among Canal residents, the Department announced that it had "provided women of childbearing age with sufficient information as to the risks before making a conscious and voluntary decision to become pregnant."[11]

Love Canal, at which a shifting population had been exposed to more than 200 chemicals at indeterminate strengths and for indeterminate lengths, provoked an epistemological crisis for the fields of epidemiology and toxicology. While scientists acknowledged elevated rates of illness and death, their analytic tools could not establish precise causal mechanisms for morbidity and mortality. Epidemiologists and toxicologists, accustomed to studying occupational exposure and tracking discrete chemical vectors, had only recently begun to study chronic environmental exposure (Nash, 2006;

Sellers, 1999). Scientists lacked substantive data on the health effects of most of the Canal chemicals, much less their interaction, and tended to minimize or negate residents' diseases. A 1978 statement by Mrs. Walters illustrated the health crisis many families experienced:

> My husband and I moved into our home in 1964 with no major health problems. During the following two years, I had two miscarriages and our daughter, Michelle, was born in July of 1966. A year later, when we celebrated her first birthday and she was not attempting to walk, we were concerned because her right foot was turned inward. I took her to an orthopedic doctor and he prescribed corrective shoes. One morning, not long after her first birthday, I went in to get her out of her crib and there was something the matter with her leg. It was drawn up toward her back ... This was the beginning of a nightmare.[12]

Speaking as a former resident who had fled the Canal shortly after the discovery of the chemical wastes, she pointed to the drastic improvement in Michelle's health following the family's departure. "She has grown and her health is excellent. We don't need health testing to prove that once you move away from a toxic dump, your health can improve." For Mrs. Walters, life at Love Canal had meant prolonged and unresolved illness. She was far from alone in making this equation.[13]

In September 1978, led by resident Lois Gibbs, roughly 500 residents formed the Love Canal Homeowner's Association (LCHA). One of the LCHA's first actions was to attack the discrepancy between official health data and conclusive action on relocation by conducting its own health survey. To do so, it recruited the help of Beverly Paigen, a cancer research scientist at Buffalo's Roswell Park Memorial Institute, whose main research interest was genetic susceptibility to environmental toxins.

Paigen and the LCHA devised a health questionnaire which focused on length of residency and illnesses, only counting those health effects that had been diagnosed by a physician and were known to residents by name. After collecting the health data, residents plotted it on a map of the neighborhood waterways. They observed a strong correlation between disease incidence and proximity to underground streambeds. The health survey concluded that the homes closest to the waterways demonstrated a threefold increase in miscarriages, a 20 percent increase in birth defects, and elevated rates of asthma, central nervous system toxicity, and urinary disease (Paigen, Goldman, Highland, Magnant, & Steegman, 1985).

The LCHA asserted that scientific studies should begin from the premise that residents were threatened by the uncertain hazards present in their environment, rather than from the need to identify discrete causal links between Canal chemicals and resident ailments. In the summer of 1980, having failed to alter either the methodology of scientific testing or government policy priorities, the LCHA boycotted all further government-sponsored testing.

The boycott represented residents' certitude that the chemicals were causing their illnesses. As they saw it, the proper role for epidemiological investigation was to identify causal links between the contents of the Canal and residents' specific diseases. On these grounds, Gibbs asked Joseph Highland, head of the Toxic Chemicals Program at the Environmental Defense Fund, to perform an independent epidemiological analysis of the neighborhood's children (Gibbs, 1982). In cooperation with Gibbs and Paigen, Highland determined that the best way to investigate a connection between the chemicals and the children's health was to study children's growth and nerve conduction (Paigen et al., 1985; Paigen, Goldman, Highland, Magnant, & Steegman, 1987). In all, they studied 523 Love Canal children, and concluded that seizures, learning problems, hyperactivity, eye irritation, skin rashes, abdominal pain, and incontinence were more prevalent among these children than in a control population of 440 (Paigen et al., 1987).

Love Canal residents took their personal experience of disease as the basis for understanding the relationship between the human body and its environment. They believed that accurate knowledge regarding health and disease arose from the immediate lived experiences of specific communities, rather than from abstract measurements of probability or causality. Furthermore, they believed that the human body should constitute the normative site against which to judge scientific practices.

GAIA

In 1961, physicist James Lovelock began work as a consultant with the NASA Jet Propulsion Laboratory in Pasadena, California. The main objective of his team was to invent technologies for detecting life on Mars and other planets. Convinced that whatever life did exist on Mars would be identical to life on Earth, the team designed experiments to test the Martian soil for the exact bacteria, microbes, and fungi found on Earth.

Lovelock grew skeptical of this approach. Believing that life on Mars would not necessarily resemble life on Earth, he asserted that the key thing to assess Mars for was "entropy reduction." With the help of colleague Dian Hitchcock, he examined how the chemical composition of the Earth's atmosphere provided evidence for life, concluding that the composition of gases in the Earth's atmosphere was "improbable" by at least one hundred orders of magnitude. Building on this evidence, they hypothesized that the atmosphere was being "manipulated on a day-to-day basis from the surface, and that the manipulator was life itself" (Lovelock, 1979, p. 6).

Lovelock nicknamed this hypothesis the "Gaia hypothesis" after the Greek Earth goddess. He formally presented Gaia to the 1968 meeting of the American Geological Union. The hypothesis was poorly received, with the exception of biologist Lynn Margulis, who thenceforth became Lovelock's closest collaborator. The two would define Gaia as,

> A complex entity involving the Earth's biosphere, atmosphere, oceans, and soil; the totality constituting a feedback or cybernetic system which seeks an optimal physical and chemical environment for life on this planet. The maintenance of relatively constant conditions by active control may be conveniently described by the term "homeostasis." (Lovelock, 1979, p. 10)

By 1991, Lovelock had developed a system for diagnosing and treating Gaia's ills, which he published in *Gaia: The Practical Science of Planetary Medicine*. The book opened provocatively,

> I speak as the representative, the shop steward of the bacteria and the less attractive forms of life who have few others to speak for them. My constituency is all life other than humans. (Lovelock, 1991, p. 9)

Lovelock intended the book to be a home medical encyclopedia – something which everyone could refer to when "some new or frightening phenomenon afflicts the region of the Earth where you live." In it, he argued that the time was ripe for abandoning concerns about human rights and human sufferings. Rather, the true patient was the Earth, of which humans were an integral part. "We are so tied to the Earth that its chills or fevers are our chills and fevers also, p. 18."

Lovelock's comparison of the human and planetary bodies bore strong resemblance to the work of the biocentrists. However, he took a decidedly more material and less romantic approach by enumerating the technical procedures for diagnosing the patient. While a family practitioner measures human temperature with a thermometer, a planetary physician uses a satellite radiometer; similarly, while a stethoscope measures human breathing, atmospheric carbon dioxide and oxygen monitors attend to the planet's respiration.

Lovelock concluded the book with an extended analysis of the "people plague": Gaia's current affliction. He invited readers to consider that Gaia was a patient arriving at a clinic with disturbing medical records from the pathologist and dermatologist.

> Atmospheric CO_2 and methane are above the patient's normal range, and there is a suspicion of fever. Some skin damage is apparent – the land surface shows a number of bare patches. Most revealing are certain abnormal chemicals in the air – CFCs, substances that are never made by the natural chemistry of living organisms. (Lovelock, 1991, p. 153)

However, Lovelock argued, the primary pathology was not the presence of people but the effect which their presence exerted on Earth's natural functions. He surveyed the ways in which humanity had so far disrupted Gaia's homeostasis: industrial agriculture, deforestation, acid rain, the ozone hole, and global warming. He charted a case history for Gaia's fever, arguing that whereas projected warming was less than previous episodes of "interglacial fever," the real concern was that humans had stripped Gaia of the forest ecosystems which normally performed the cooling action.

What constituted a suitable planetary prescription? Lovelock cautioned that humans knew virtually nothing about the Earth. "Keeping a healthy planet with an equable climate is as difficult for us as was keeping a healthy body free of disease for our forebears (p. 174)." Rather, the best cure, or beginning of a cure, was to live healthily and lightly, without excessive use of "powerful medicines" such as geo-engineering or antibiotics, respecting the body's innate ability to heal itself.

Lovelock concluded with a plea to seriously consider whether humans wished to become stewards of the Earth; a fate which he argued was inevitable if Gaia's natural cycles were disrupted to such a degree as to prevent her self-regulation. "There can be no worse fate for people than to be conscripted for such a hopeless task – to be made forever accountable for the smooth running of the climate, the composition of the oceans, the air, and the soil (p. 186)." Rather, he mused, humans should understand themselves as representatives for the rest of life on the planet, life which in the current world was extremely angry at the "diabolical liberties" which humans had taken with the planet. Ultimately, Lovelock's planetary medicine attempted an uneasy synthesis of the biocentric belief in ecological egalitarianism with a positivist hope that humans would eventually exercise the intelligence to self-correct.

CONCLUSION

This chapter has illustrated the diverse and historically intertwined conceptual and political linkages made between health and the environment within five iconic social movements in the United States. The 1970s were a formative decade for the articulation of a range of issues at play today, including pervasive environmental toxicity, climate change, the necessity of limiting human demographic and economic growth, and the argument that first world consumers can protect the planet and its vulnerable inhabitants through proper consumption patterns. The politics, science, and public discourses surrounding each of these dilemmas are the products of the specific postwar history this chapter has presented.

The field of EcoHealth, which pursues integrative and ecosystem approaches to health and sustainability (Bunch, Morrison, Parkes, & Venema, 2011), emerged in the 1990s from the interactions of professionals from diverse fields, including conservation medicine, medical geology, health and sustainability, and ecosystem health (Wilcox et al., 2004). Although this chapter has not endeavored to trace the specific history of the field, as this has been done by other authors (Bunch et al., 2011; Parkes, 2012), it has illuminated philosophical and political developments which contributed in different ways to the emergence of EcoHealth. In particular, the chapter has noted the active participation of nonprofessionals in the shaping of scientific and public discourse regarding the connections between human, animal, and ecosystem health. In its first decade, Friends of the Earth gave political expression to Rachel Carson's articulation of the symbiotic relationship between humans, animals, and ecological context, thereby helping to shift wilderness-oriented conservation politics toward a more ecological, integrated perspective. The residents of Love Canal and the environmental justice movement they helped to shape have encouraged a resurgent focus on the environment as the primary determinant of human and animal health, as well as an emphasis on the connection between environmental degradation and socioeconomic status. Bioregionalists, with their carefully articulated concepts of bioregion and watershed, along with their practices of environmental restoration and reinhabitation, recognized human communities as embedded within a broader ecological context, and believed that human community could only be maintained in harmony with the ecological possibilities of place. Conservation biologists recognized the integrity of ecosystems, and endeavored to protect biodiversity through preserving large interconnected wilderness areas. Gaia theory, which comprehends the Earth as a living, breathing whole, gives the most profound expression to the insight of interconnectedness by bringing even the atmosphere into the equation. The field of EcoHealth has global, interdisciplinary, and institutional roots which stretch far beyond the scope of this chapter (Parkes, 2012; Wilcox et al., 2004). Yet it is equally rooted in the health and environmental activism of these diverse communities and the ways they brought the links between humans and the natural world to life in their work throughout the United States.

NOTES

1. Lillie, H. (October 18, 1962). Letter to Rachel Carson. Yale Beinecke Library, Rachel Carson papers.

2. Friends of the Earth. (July 9, 1971). Letter to Warren Magnuson. David Ross Brower Papers, The Bancroft Library, University of California, Berkeley.

3. A neologism combining "ecological" and "sabotage," ecotage refers to deliberate acts of property destruction performed against perpetrators of environmental harm: for example, pouring sugar into the gasoline tank of a bulldozer.

4. Parkinson, A. (June 1977). Nuclear Man. *Not Man Apart*. David Ross Brower Papers, The Bancroft Library, University of California, Berkeley.

5. Friends of the Earth. (August 1979). Friends of the Earth. *The Corn Husk: Your Ear on the Whole Foods Industry*. David Ross Brower Papers, The Bancroft Library, University of California, Berkeley.

6. Friends of the Earth. (January 1980). Meet Friends of the Earth and Join the Fight Against Pesticide Abuse!. *Erewhon Monthly*. David Ross Brower Papers, The Bancroft Library, University of California, Berkeley.

7. Author's telephone interview with Stephanie Mills, February 2012.

8. For a thorough history of conservation biology, its influences and cross-currents, see Meine (2010).

9. New York State Department of Health. (June 1980). Love Canal Studies by the New York State Department of Health. Ecumenical Task Force of the Niagara Frontier Records, 1946–1995 (1978–1990 bulk), University Archives, The State University of New York at Buffalo.

10. Matsulavage, E. (April 5, 1979). Love Canal Testimony. Ecumenical Task Force of the Niagara Frontier Records, 1946–1995 (1978–1990 bulk), University Archives, The State University of New York at Buffalo.

11. Love Canal Homeowners Association. (January 1980). Love Canal Chronological Report, April 1978 to January 1980. Center for Health, Environment, and Justice, records, 1978–2001, Digital Collections and Archives, Tufts University.

12. Walters, Mrs. (1978). Testimony. Ecumenical Task Force of the Niagara Frontier Records, 1946–1995 (1978–1990 bulk), University Archives, The State University of New York at Buffalo.

13. Cerillo, D. & Hale, J. (November 15, 1983). Testimony to the Joint Public Hearing on the Need for Community Health Information on Toxic Substances and Hazardous Waste, convened by the State of New York Office of the Assistant Majority Leader, Standing Committee on Health and Standing Committee on Environmental Conservation. Ecumenical Task Force of the Niagara Frontier Records, 1946–1995 (1978–1990 bulk), University Archives, The State University of New York at Buffalo.

REFERENCES

Aberley, D. (1999). Interpreting bioregionalism: A story from many voices. In M. V. McGinnis (Ed.), *Bioregionalism*. New York, NY: Routledge.

Bunch, M. J., Morrison, K. E., Parkes, M. W., & Venema, H. D. (2011). Promoting health and well-being by managing for social-ecological resilience: The potential of integrating ecohealth and water resources management approaches. *Ecology and Society*, *16*(1), 6.

Carson, R. (1962). *Silent Spring*. New York, NY: Houghton Mifflin.

Davis, J., & Foreman, D. (1993). The Wildlands project. *Earth Island Journal*, *8*(2).

Dewey, S. H. (1998). Working for the environment: Organized labor and the origins of environmentalism in the United States, 1948–1970. *Environmental History*, *3*(1), 45–63.

Dewey, S. H. (2000). *Don't breathe the air: Air pollution and U.S. environmental politics, 1945-1970*. College Station, TX: Texas A&M University Press.

Dodge, J. (1981). Living by Life: Some bioregional theory and practice. *Co-Evolution Quarterly*, *32*, 6–12.

Foreman, D. (1991). *Confessions of an eco-warrior*. New York, NY: Harmony Books.

Foreman, D. (2004). *Rewilding North America: A vision for conservation in the 21st century*. Washington, DC: Island Press.

Frisco Bay Mussel Group. (1977). *Living here*. San Francisco, CA: Frisco Bay Mussel Group.

Gibbs, L. M. (1982). *Love canal: My story*. New York, NY: Grove Press.

Hays, S. P., & Hays, B. D. (1987). *Beauty, health, and permanence: Environmental politics in the United States, 1955–1985*. New York, NY: Cambridge University Press.

Lear, L. J. (1997). *Rachel Carson: Witness for nature*. New York, NY: Henry Holt.

Leopold, A. (1949). *A Sand County Almanac. With other essays on conservation from Round River*. New York, NY: Oxford University Press.

Lovelock, J. (1979). *Gaia: A new look at life on Earth*. New York, NY: Oxford University Press.

Lovelock, J. (1991). *Gaia, the practical science of planetary medicine*. London: Gaia Books.

Lytle, M. H. (2007). *The gentle subversive: Rachel Carson, Silent Spring, and the rise of the environmental movement*. New York, NY: Oxford University Press.

Meine, C. (2010). Conservation biology: Past and present. In P. R. Ehrlich & N. S. Sodhi (Eds.), *Conservation biology for all* (pp. 7–27). New York, NY: Oxford University Press.

Meffe, G. K. (1999). Editorial. *Conservation Biology*, *13*(5), 953–954.

Mills, S. (1994). The wild and the tame. In D. C. Burks (Ed.), *Place of the wild: A wildlands anthology*. Washington, DC: Island Press.

Montrie, C. (January 2000). Expedient environmentalism: Opposition to coal surface mining in Appalachia and the United Mine Workers of America, 1945–1977. *Environmental History*, *5*(1), 75–98.

Nash, L. L. (2006). *Inescapable ecologies: A history of environment, disease, and knowledge*. Berkeley, CA: University of California Press.

Noss, R., Pimentel, D., & Westra, L. (2000). *Ecological integrity: Integrating environment, conservation, and health*. Washington, DC: Island Press.

Paigen, B., Goldman, L., Highland, J., Magnant, M. M., & Steegman, A. T. (1985). Prevalence of health problems in children living near Love Canal. *Hazardous Waste and Hazardous Materials*, *2*(1), 23–43.

Paigen, B., Goldman, L., Highland, J., Magnant, M. M., & Steegman, A. T. (1987). Growth of children living near the hazardous waste site, Love Canal. *Human Biology*, *59*(3), 489–508.

Parkes, M. (2012). Diversity, emergence, resilience: Guides for a new generation of ecohealth research and practice. *EcoHealth*. doi:10.1007/s10393-011-0732-8

Rome, A. W. (2001). *The bulldozer in the countryside: Suburban sprawl and the rise of American environmentalism*. New York, NY: Cambridge University Press.

Rome, A. W. (2003). 'Give Earth a Chance': The Environmental Movement and the Sixties. *The Journal of American History*, *90*(2), 525–554.

Sale, K. (1985). *Dwellers in the land: The bioregional vision*. San Francisco, CA: Sierra Club Books.

Sellers, C. (1997). *Hazards of the job: From industrial disease to environmental health science*. Chapel Hill, NC: University of North Carolina Press.

Sellers, C. (1999). Body, place and the state: The makings of an "environmentalist" imaginary in the post-World War II U.S. *Radical History Review*, *74*, 31–64.

Snyder, G. (1995). *A place in Space: Ethics, aesthetics, and watersheds*. Washington, DC: Counterpoint Press.

Souder, W. (2012). *On a farther shore: The life and legacy of Rachel Carson*. New York, NY: Crown Publishers.

Turner, J. (2012). *The promise of wilderness: American environmental politics since 1964*. Seattle: University of Washington Press.

Wilcox, B. A., Aguirre, A. A., Daszak, P., Horwitz, P., Howard, J., Lannigan, R., … Waltner Towes, D. (2004). Introduction. *EcoHealth*, *1*(1), 1–2.

Archival Sources

Rachel Carson Papers. Yale Collection of American Literature, Beinecke Rare Book and Manuscript Library, Yale University.

Center for Health, Environment, and Justice, records, 1978–2001, Digital Collections and Archives, Tufts University.

David Ross Brower Papers, BANC MSS 79/9 c, The Bancroft Library, University of California, Berkeley.

Friends of the Earth Records, BANC MSS 82/98 c and BANC MSS 86/104 c, The Bancroft Library, University of California, Berkeley.

Love Canal Collections, State University of New York, Buffalo.

Interviews

Joseph Highland, Skype, October 2011.

Stephanie Mills, Telephone, February 2012.

EXPLORING THE LINKS BETWEEN HIV/AIDS AND FORESTS IN MALAWI: MORBIDITY, MORTALITY, AND CHANGING DEPENDENCE ON FOREST RESOURCES

Joleen Timko

ABSTRACT

Purpose – *By bringing together aspects of sustainable forest management, population health, and local livelihoods, the purpose of this study was to characterize how household dependence on forest resources changes through three phases: the period before HIV became a problem in the household, the period during HIV-related morbidity, and after AIDS-related mortality.*

Methodology/approach – *Sixty semi-structured interviews were conducted with members of unaffected and HIV/AIDS-affected households in four case study districts in Malawi.*

Ecological Health: Society, Ecology and Health
Advances in Medical Sociology, Volume 15, 147–171

ISSN: 1057-6290/doi:10.1108/S1057-6290(2013)0000015011

Findings – *This study demonstrates that the relationship between HIV/AIDS and dependence on specific forest resources appears to correspond closely with the stage of the disease. Firewood and water were consistently ranked as being one of the three most important resources, regardless of HIV-affectedness. During the morbidity phase, respondents reported their need for medicinal plants increased substantially, along with other resources. The importance of timber increased significantly after HIV-related mortality.*

Social implications – *Interview respondents themselves suggested key interventions that would assist households in the HIV/AIDS-mortality phase, in particular, to obtain the forest resources they require. These interventions could address the impacts of HIV/AIDS on the sustainability of important resources, compensate for a decreased availability of household labor, and foster greater access to these resources for vulnerable households in the four study sites.*

Originality/value of chapter – *In spite of the fact that forest resources can play a crucial role in enabling a household to control and adapt to the disease, research on the environmental dimensions of HIV/AIDS remains limited. This chapter helps to address this knowledge gap, suggests practical, innovative interventions that could alleviate some of the disease burden on rural Malawian households, and offers insight into potential areas of further inquiry in this research domain.*

Keywords: HIV/AIDS; Malawi; forest resource; dependence

INTRODUCTION AND PURPOSE

Forests are fundamental to the health and well-being of rural people in developing countries. Forests are important as they are the providers of medicinal plants; animal food sources (including insects, mollusks, fish, crustaceans, amphibians, and bushmeat); plant food sources (including mushrooms, seeds, edible fruits, vegetables, and root crops); gums and resins; charcoal; and honey. It has been estimated that over two-thirds of Africa's 600 million people rely on forest resources for both subsistence and to supplement their cash incomes (Arnold, 2001; CIFOR, 2005; Kaimowitz, 2003; Sunderlin et al., 2005). In addition, approximately 70% of the energy consumed in southern Africa is in the form of fuelwood or charcoal

(Syampungani, Chirwa, Akinnifesi, Sileshi, & Ajayi, 2009). Sub-Saharan Africa has the highest prevalence of the human immunodeficiency virus (HIV) in the world, with 35 of the 45 most highly affected countries located in this region (Drimie & Gandure, 2005). The acquired immunodeficiency syndrome (AIDS) is a leading cause of death in Africa and worldwide for prime-aged adults – ages 15–49 – the very people who work to support families, those responsible for building society, and those usually the most productive economically (Drimie, 2002; Hunter, Twine, & Johnson, 2008; Torell et al., 2006).

The HIV/AIDS pandemic is having not only devastating and tragic social, economic, and political impacts throughout Sub-Saharan Africa, but also ecological ones (Barany, Hammett, Stadler, & Kengni, 2004). Across this region, health care is predominantly a forest-based service (Barany, Hammett, Sene, & Amichev, 2001). Traditional healers are the dominant providers of medical care in forested areas, often providing between 70% and 95% of primary health care (Colfer, Sheil, & Kishi, 2006). Wild foods or forest resources can improve health and be used to detoxify the effects of AIDS-related drug treatments (ABCG, 2002), and can play a key role in enabling an afflicted person and their household to control and adapt to HIV/AIDS (Villarreal, Holding-Anyonge, Swallow, & Kewsiga, 2006). Wild foods, fruits, berries, and leaves can boost the immune system and help protect against opportunistic disease (Villarreal et al., 2006) while "wild meats, insects and fruit kernels provide high quality protein and wild leafy vegetables are key sources of micronutrients" (Kaschula, 2008, p. 163). Some non-timber forest resources have been found to be high in key nutrients required by people living with HIV/AIDS, particularly protein, fat, iron, zinc, and vitamins A and C (Barany et al., 2004).

In many developing countries, the collection of natural resources for subsistence purposes can be very time consuming. Bishop-Sambrook (2003) notes that the collection of firewood and water are two of the most burdensome tasks that take place on a daily basis for women who must often walk for several hours before starting their collection, which itself can take several hours. With natural resources already acting as critical safety nets for rural households in times of crisis (i.e., as a source of plant- and animal-based foods (bushmeat, seeds, fruits, mushrooms, etc.) and medicinal plants), rural and poor HIV/AIDS-affected households with limited income opportunities often increase their reliance on forest resources as a consistent livelihood strategy to minimize the socioeconomic burden of HIV/AIDS (Lopez, 2008; Shackleton & Shackleton, 2004; Villarreal et al., 2006). More timber is harvested to build coffins; more firewood is used for cooking, funerals, and ceremonies; there is increased experimentation and use of medicinal plants to treat the side effects of HIV/AIDS (e.g., shingles, diarrhea); there is a greater reliance on bushmeat and charcoal for alternative income purposes; and more forest lands are converted to cemeteries (Barany, Holding-Anyonge, Kayambazinthu, & Sitoe, 2005; Bolton & Talman, 2010; Holding-Anyonge, Rugalema, Kayambazinthu, Sitoe, & Barany, 2006; Mauambeta, 2003; Topouzis, 2007). In some regions of Sub-Saharan Africa, HIV/AIDS has had "insidious and negative impacts on the environment" as the pandemic has spread (De Souza et al., 2008, p. 240). Zambia, Tanzania, Malawi, and Mozambique are countries in which HIV/AIDS is not only particularly prevalent, but are now among the countries with the highest rates of deforestation (Barany et al., 2005; Frank & Unruh, 2008).

There are gains to be made by protecting these forest ecosystems that provide essential ecosystem services (Weinstein, 2005) as there are growing health implications of, and risks to, human well-being from environmental degradation including deforestation (Keesing et al., 2010; McMichael, 2002; Myers & Patz, 2009; Plowright, Sokolow, Gorman, Daszak, & Foley, 2008; Tabor, 2002). The loss of forests and biodiversity can, in the case of many infectious diseases, exacerbate disease emergence and resurgence (Patz et al., 2004; Weiss & McMichael, 2004). This can occur both directly, such as by reducing predation and competition on reservoir hosts (Keesing et al., 2010), and indirectly, through increasing forest fragmentation, pollution, and poverty (Patz et al., 2004). It is therefore imperative to better understand the interaction between anthropogenic changes to environmental resources (Parkes, Panelli, & Weinstein, 2003), in this case forest resources, and the corresponding impacts on human health (McMichael, 1999; Ostfeld, Meffe, & Pearl, 2002).

Despite the extent of the pandemic in many developing countries, the links between HIV/AIDS and forest resources have not been well studied. Research on the interactions between the dependence on forest biodiversity and contemporary epidemics in general, and on the environmental dimensions of the HIV/AIDS pandemic in particular, remains surprisingly limited (Holding-Anyonge et al., 2006; Hunter et al., 2008). There has been a dearth of inquiry into the links between the livelihoods of HIV/AIDS-affected households and forest resources (Barany et al., 2005; Torell et al., 2006), and on the long-term impacts of people living with HIV/AIDS on the management of forest resources (Dwasi, 2002). Evidence suggests that HIV/AIDS-affected households increase their reliance on forest resources – especially related to the increasing use of fuelwood, medicinal plants, and charcoal (Barany et al., 2005; Holding-Anyonge et al., 2006; Lopez, 2008; Topouzis, 2007; Torell et al., 2006). Indeed, while there has been minimal empirical investigation on this topic, a recent systematic analysis of the limited literature in this domain revealed an increased dependence by HIV/AIDS-affected households on forest resources (Timko, Kozak, & Innes, 2010).

To grapple with these kinds of complex socio-ecological problems, we require innovative ideas and scholarship that combine social scientific, ecological, and population health research (Parkes et al., 2005). Health, conservation, and agricultural projects must work together to develop an understanding among the many disciplines spanning the natural and social sciences at work in this field (Tabor, 2002; Wilcox et al., 2004). Indeed, it was through comparing a map of forest cover in Africa with a map of the continent's HIV prevalence (such as those in Barany et al., 2001) that inspired me to pursue the line of inquiry described in this chapter. By bringing together aspects of sustainable forest management, population health, and local livelihoods, the purpose of this study was to characterize how household dependence on forest resources in rural Malawi changes through three phases: the period before HIV became a problem in the household, the period during HIV-related morbidity, and after AIDS-related mortality. This is an important research gap given the extent of prime-age adult mortality attributable to the HIV/AIDS pandemic, particularly in Sub-Saharan Africa (Hunter et al., 2008).

Two key terms are used repeatedly in this chapter. Fisher (2004) defined forest dependence in terms of the share of earnings derived from forest-based activities. This definition does not, however, adequately reflect how forest resources are used to meet the fundamental, daily, nonmonetary livelihood requirements of local respondents in this study. In the context of the analysis presented here, I define *dependence* specifically as the frequency with which a forest resource is collected or purchased. Therefore, responses indicating an increased frequency of use are taken to mean an increased dependence on that forest product in that time period. The term *forest resource* in this chapter refers to "any wild biological resource (animal or plant) harvested from forested lands ... by rural households for domestic consumption or small-scale trade, with no, or limited capital investment" (Shackleton, Shackleton, Buiten, & Bird, 2007, p. 560). These include roots, fruits, medicinal plants, resins and essential oils, and fibers such as bamboos, rattans, and other palms used for weaving and structural applications (Belcher, 2005).

METHODS

Selection of Case Study Sites

I selected Malawi as the country in which to situate this study as it can shed crucial insight into the linkages between HIV/AIDS and forest resources. Malawi is a small (land area of 118, 480 km^2) landlocked country bordered to the south, east and west by Mozambique, to the west by Zambia, and to the north by Tanzania (Kachule, Tchale, & Mataya, 2003) (Fig. 1). The majority (61%) of Malawi's land is under agriculture while forests occupy about 38% (*ibid.*). Miombo woodlands, characterized by species of *Brachystegia*, *Julbernadia*, and *Isoberlinia*, account for approximately 70% of the country's total forest area and 97% of all forest species (Government of Malawi, 1996). These woodlands are generally used for grazing and as sources of wood and non-wood forest products, while also providing cultural, social, and other forest services (Kayambazinthu, Matose, Kajembe, & Nemarundwe, 2003).

In Malawi, there are essentially two types of miombo woodlands: those woodlands located on public lands and protected by the government, and village woodlands on customary land (Masangano, Kayambazinthu, & Mwabumba, 2003). Protected woodlands are managed strictly by the Departments of Forestry and Parks and Wildlife, and local communities are not allowed to collect forest products in these areas without a permit. In comparison, woodlands on customary land consist of thousands of small areas of woodland usually designated as village forest areas (VFAs) under the control of village headmen. Community members living in close proximity to these areas are normally allowed access to collect specified resources such as firewood, thatch grass, fruits, and mushrooms, but cutting of trees is not normally allowed. Malawi currently advocates a policy of communal management of miombo woodlands (Masangano et al., 2003). National legislation and guiding policies are set mainly by the Forestry Department and provide the broader institutional framework within which lower level government institutions operate. These lower level structures (the community or village) implement national legislation. And it is at the local level that traditional chiefs "are the encapsulation of all institutions, formal and informal" and are at the center of all user-group rules, sacred controls, and civic controls (Kayambazinthu et al., 2003, p. 50). Local community development institutions include village natural resource management committees (VNRMCs) and village forest committees (VFCs). While these are less powerful usually than traditional leaders over the management of

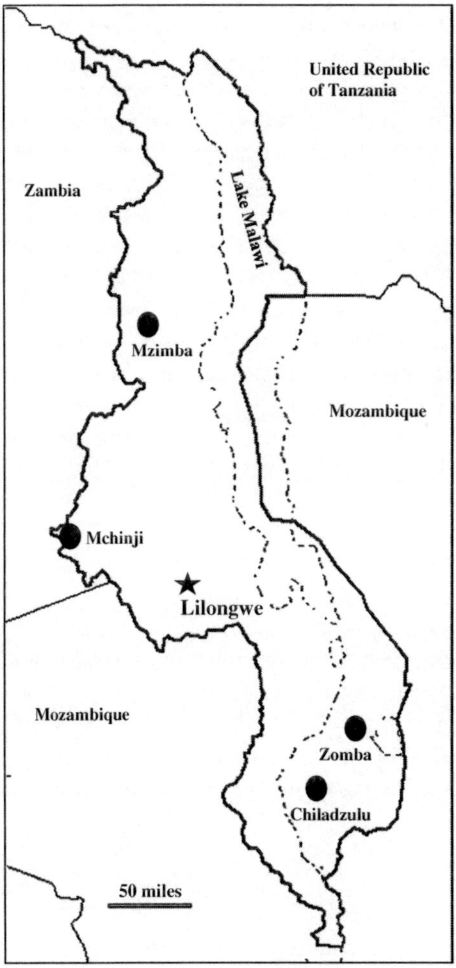

Fig. 1. Map of Malawi Indicating the Approximate Location of the Four Case Study Districts in Relation to the Capital City of Lilongwe.

indigenous resources, their existence has sometimes resulted in divided loyalty between the committees and traditional leaders (Kayambazinthu et al., 2003). However, often traditional and modern institutions have been observed to work side by side.

For the purposes of this study, Malawi was selected specifically because it has: a largely rural population with a high dependence on forest resources, high HIV/AIDS prevalence, and governmental openness about the impacts of HIV/AIDS on forest resources and forest management institutions (Government of Malawi, 2007). I selected the case study sites based on variation according to three main characteristics: proximity to forest resources in either protected forest reserves or VFAs under customary land tenure, HIV prevalence, and regional representation. National HIV prevalence in Malawi is considered high at around 12%, and has been estimated at 6.5% in the north, 8.6% in the center, and 16.5% in the south (National AIDS Commission, 2008). Case study sites were located within four districts (rural study site names appear in brackets): Zomba (Domasi) and Chiladzulu (Milepa) in the south, Mzimba (Mbalachanda) in the north, and Mchinji (Kamboni) in the central region of the country (Fig. 1). Two of the study sites (Chiladzulu and Mzimba) rely on forest resources from VFAs, while the other two (Zomba and Mchinji) rely on protected forest reserves.

Interviews

The results presented in this chapter stem from a larger study in this domain which employed both focus groups and semi-structured interviews. The data were specifically obtained from semi-structured interviews conducted with 5 respondents from each of the 12 focus groups (3 focus groups per study district). Sixty semi-structured interviews were conducted between June and September 2010 in ChiChewa, the national language of Malawi, by two Malawian field assistants at a time and location deemed safe and convenient to the respondents. Prior to conducting focus groups and interviews, all questions were approved by the *Behavioural Research Ethics Board* at the author's university. Consent was obtained from each respondent before they were allowed to participate in the study.

Each respondent was characterized according to age, gender, ethnic group, number and composition of people living in the household, and whether their household qualified as "affected" or unaffected. Households were broadly deemed *affected* if one of the following criteria was met: a prime-age adult member of the household was presently infected or chronically ill; widow- and widower-headed households where the spouse died from HIV; a household which had taken in an orphaned child or children (Barany et al., 2005). Households were deemed *unaffected* provided prime-aged adults (ages 19–49) were present part of year and not chronically ill and there was no prime-aged illness or death within the last five years.

Fifteen semi-structured interviews were conducted in each of the four case study districts, for a total of 60 interviews. Of these, 20 involved respondents from unaffected households and 40 involved respondents from HIV-affected households. Interviews were exploratory in nature; they were not intended to be inferential, but rather to provide insight into the issue of forest resources dependence by HIV/AIDS-affected households.

Each interview was comprised of a characterization of the respondent and their forest resource use. This first section of each interview asked all respondents to think back to a time before HIV was a problem in their region to identify the forest resources (e.g., firewood, medicinal plants, honey, bushmeat, etc.) deemed to be most important to the household including: a rank order preference for each resource; whether these resources were collected (from the forest/river/home gardens) or purchased; the frequency with which they were obtained; and the perceived availability of each resource. All of the respondents (unaffected and HIV/AIDS-affected) answered these questions.

Only HIV/AIDS-affected respondents were asked to answer further questions regarding the impacts of HIV/AIDS-related morbidity and mortality on their forest resource use. These questions resembled the first section, but sought to differentiate how the presence of having a prime-aged member of the household fall ill (morbidity) or die (mortality) from HIV/AIDS affected their use of forest resources (e.g., which forest resources became more or less important, if their use increased, decreased, or did not change, if the perceived availability of each resource increased, decreased, or did not change). Respondents were also asked about any new and innovative strategies they employed to obtain or use these important forest resources.

Data Analysis

All interviews were recorded on digital voice recorders and transcribed as soon as possible afterwards into individual Microsoft Excel spreadsheets. Once completed, these spreadsheets were then compiled into one master spreadsheet. The data were summarized to assess forest resources according to the proportion of times they were listed as the first, second, or third most important resource, changes in their frequency of use, etc. Data were grouped according to whether the respondents were unaffected or affected. For the latter designation, the impacts of HIV were considered in three phases: pre-HIV (the time before HIV became a problem in that respondent's household), HIV-morbidity phase (while someone in the household or the respondent themselves were ill), and HIV-mortality phase (after a household member had died from HIV).

RESULTS

I found that while 19 different forest resources were mentioned by respondents as being important, 10 resources stood out as more important (i.e., they were ranked as one of the three most important by at least 10% of respondents in at least one of the HIV phases): firewood, water, medicinal plants, (thatch) grass, mushrooms, fruits, poles, timber, bushmeat, and honey. Other resources mentioned in the interviews but excluded from further discussion in this chapter included: bamboo, rope fiber, construction materials (e.g., wood materials from the forest used as rafters, purlins, cross-bars), edible caterpillars, and elephant grass.

Respondents were first asked to list the most important types of forest resources used in their households for each HIV-phase. Firewood, medicinal plants, grass, mushrooms, fruits, and poles were mentioned by more than 50% of unaffected and pre-HIV phase respondents (Fig. 2). Respondents were then asked to rank the first, second, and third most important forest

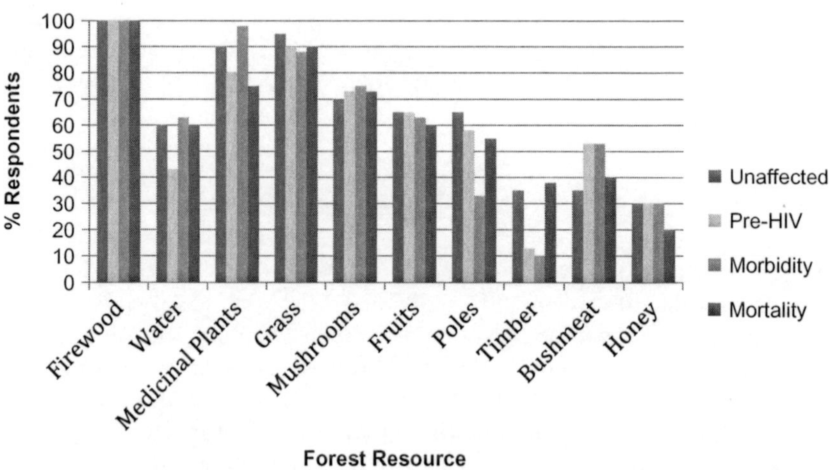

Fig. 2. Important Forest Resources Mentioned According to Household Phase of HIV/AIDS.

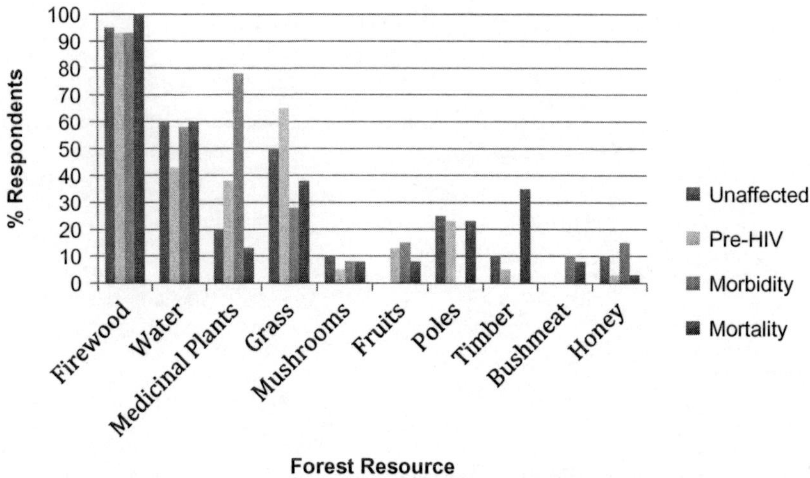

Fig. 3. Important Forest Resources Ranked by Respondents as Being One of the Three Most Important According to Household Phase of HIV/AIDS.

resources they had mentioned. Firewood and water were consistently ranked as being one of the three most important resources, regardless of HIV-affectedness, while the need for medicinal plants increased substantially for morbidity-affected households (Fig. 3). Many of the respondents also reported an increased need for fruit, bushmeat, and honey during the morbidity phase, with a decreased need post-mortality (Fig. 3). The importance of timber increased significantly after HIV-related mortality. The remainder of this section will predominantly discuss the results for these forest resources in greater detail, with less discussion paid to the other important forest resources.

Firewood

Firewood was consistently ranked by most respondents as one of the three most important forest resources (Fig. 3), and its importance is reflected in the results in Table 1. Response rates to the questions about firewood collection, frequency, and availability (see Table 1) were outstanding with 100% of respondents providing a response on all but one question. On the remaining question, 98% (59/60) provided a response. Firewood is a forest resource that is mainly collected, however a small proportion of unaffected

Table 1. Data Depicting the Importance of Firewood to Unaffected and Affected Respondents.

	Collected/Bought		Frequency of Use				Availability			
	Collected	Bought	dy	w	m	f	e	d	vd	nd
Unaffected (%)	91	9	15	80	5	0	37	58	5	0
Affected – pre-HIV phase (%)	100	0	22.5	75	2.5	0	100	0	0	0
	Collected	Bought	inc	dec	nc	nd	inc	dec	nc	nd
Affected – morbidity phase (%)	95	5	92.5	5	2.5	0	2.5	92.5	5	0
Affected – mortality phase (%)	98	0	0	100	0	0	0	100	0	0

Data represent the proportion of respondents (%) who indicated a response, and include whether the resource was bought or collected, frequency of use, and availability (legend: dy = daily, w = once or a few times per week, m = monthly, f = a few times per year, y = once per year or less, s = seasonally, o-opportunistically as needed, e = easy, d = difficult, vd = very difficult, nd = no data, inc = increased, dec = decreased, nc = no change).
The above legend also applies to Tables 2, 3, and 4.

and morbidity-affected respondents purchase some firewood. Most respondents (80% of unaffected and 75% of affected (pre-HIV phase) respondents)[1] collect firewood on a weekly basis; 93% of morbidity-affected respondents reported an increase in its use, while 100% of mortality-affected respondents decreased their use of firewood. Unaffected respondents were roughly split on whether firewood was easy or difficult to source, while 100% of affected (pre-HIV phase) respondents reported firewood's ease of availability. The vast majority of morbidity and mortality-affected respondents reported a decreased availability of firewood.

Firewood would be expected to be important to HIV-morbidity households as many respondents noted that more calorically dense foods (such as cassava) needed to meet the caloric requirements of ill household members require cooking. Firewood is also required as a heat source for ill members, and is sold as a source of income, perhaps to purchase medicines. Possibly due to a decreased availability of labor in the household for collecting such resources, 100% of mortality-affected respondents decreased their use of firewood (Table 1), in spite of the fact that firewood was even more important to mortality-affected respondents (Fig. 3). One affected 35-year-old widow commented that she has to use other pieces of wood found around her homestead "because I cannot work much more due to less energy in my body." Other common, short-term strategies used by respondents to cope with limited quantities of firewood include: substituting other products for cooking, heating, and lighting such as: pigeon pea stems, cassava stems, reeds, maize and sorghum stalks, tobacco stems, plastic paper, and rubbish; walking further to get firewood; purchasing firewood from a market or from others; or eating less food or eating foods that don't require cooking.

Water

Though water was only mentioned as an important forest resource by 60% of unaffected and 43% of affected (pre-HIV phase) respondents (Fig. 2), it was ranked as one of the three most important resources, particularly by respondents from morbidity- and mortality-affected households (Fig. 3). Table 2 presents the results for water. Of the four most important forest resources discussed in this chapter, water garnered the least responses with only 50–63% of respondents providing answers to the questions about water collection, frequency, and availability (see Table 2). Of these, all respondents reported that water is collected daily. Ngwenya and Kgathi (2006) found that affected households expressed a need for more water when caring for

Table 2. Data Depicting the Importance of Water to Unaffected and Affected Respondents.

	Collected/Bought		Frequency of Use				Availability			
	Collected	Bought	dy	w	m	f	e	d	vd	nd
Unaffected (%)	100	0	100	0	0	0	50	50	0	0
Affected – pre-HIV phase (%)	100	0	100	0	0	0	100	0	0	0
	Collected	Bought	inc	dec	nc	nd	inc	dec	nc	nd
Affected – morbidity phase (%)	100	0	88	4	8	0	0	68	32	0
Affected – mortality phase (%)	100	0	10	80	10	0	0	90	10	0

Data are listed as proportions of respondents (%) who indicated a response, and include whether the resource was bought or collected, frequency of use, and availability.

the sick, and indeed, 88% of morbidity-affected respondents in my study reported an increased use of water for tasks such as cooking, increased laundry requirements, and bathing of the affected person. Eighty percent of mortality-affected respondents decreased their use of water as would be expected because these same needs would not be present post-mortality. Unaffected respondents were evenly split on whether water was available in such quantities as to be easy or difficult to find, while 100% of affected (pre-HIV phase) respondents reported that water was easy to find. The majority of morbidity- (68%) and mortality-affected (90%) respondents perceived a decreased availability of water. Many respondents noted having to cope with limited availabilities of water by either walking longer distances to find water; using wetland/stagnant water for other purposes to save borehole water for drinking; reducing water use by washing less clothes, eating less boiled foods, or leaving domestic chores undone; and taking grandchildren out of school to collect water.

Medicinal Plants

Medicinal plants were the third most mentioned forest resource by both unaffected and affected (pre-HIV phase) respondents (Fig. 2). While medicinal plants were not highly ranked by unaffected and pre-HIV phase respondents, they were the second most important forest resource for morbidity-affected respondents (Fig. 3). This is unsurprising given that medicinal plants are a primary response to HIV-related illness in Malawi for the treatment of symptoms such as diarrhoea and shingles (Barany et al., 2005). While medicinal plants tend to be collected opportunistically as needed when HIV is not a factor, the data show an increased need for these resources during morbidity with a decreased need after mortality (Table 3). The vast majority of respondents provided answers to the questions about medicinal plant collection, frequency, and availability (see Table 3), with most questions garnering a response from 98 to 100% of the respondents. The lowest response rate on any of the questions was still very good, with 78% of respondents providing an answer. The majority (78%) of unaffected respondents reported that medicinal plants were difficult to find, while 90% of affected (pre-HIV phase) respondents reported they were easy to find (Table 3). The vast majority of morbidity- and mortality-affected respondents reported that medicinal plants were more difficult to find. This supports the findings of Barany et al. (2005) and Kayambazinthu, Barany, Mumba, and Holding-Anyonge (2005) who note that medicinal plants used

Table 3. Data Depicting the Importance of Medicinal Plants to Unaffected and Affected Respondents.

	Collected/Bought		Frequency of Use				Availability			
	Collected	Bought	w	f	y	o	e	d	vd	nd
Unaffected (%)	75	25	0	0	0	100	22	78	0	0
Affected – pre-HIV phase (%)	96	4	3	7	3	87	90	7	3	0
	Collected	Bought	inc	dec	nc	nd	inc	dec	nc	nd
Affected – morbidity phase (%)	75	25	85	15	0	0	0	95	5	0
Affected – mortality phase (%)	72	28	10	90	0	0	0	97	3	0

Data are listed as proportions of respondents (%) who indicated a response, and include whether the resource was bought or collected, frequency of use, and availability.

in the treatment of HIV/AIDS-related illnesses were becoming less available in their study sites due to overharvesting.

Thatch Grass

Thatch grass was the second most mentioned important forest resource by both unaffected and pre-HIV respondents (Fig. 2), and was consistently ranked as being one of the three most important resources (Fig. 3). For HIV-affected respondents, it was more important in pre-HIV years (65%) than during morbidity (28%) and after mortality (38%) (Fig. 3), and is commonly used as a roofing material. Table 4 presents the results for grass. The vast majority of respondents provided answers to the questions about thatch grass collection, frequency, and availability (see Table 4), with 88–92% of all questions garnering a response. Grass is mostly collected rather than purchased, and is generally collected seasonally. A slight majority of morbidity- and mortality-affected respondents noted there was no change in the use of thatch grass. Unaffected respondents were almost evenly split on whether grass was available in such quantities as to be easy or difficult to find, while 100% of affected (pre-HIV phase) respondents reported that it was easy to find. Seventy-four percent of morbidity-affected respondents and 89% of mortality-affected respondents perceived a decreased availability of grass.

Table 4. Data Depicting the Importance of Grass to Unaffected and Affected Respondents.

	Collected/Bought		Frequency of Use				Availability			
	Collected	Bought	dy	s	y	o	e	d	vd	nd
Unaffected (%)	100	0	0	95	5	0	58	42	0	0
Affected – pre-HIV phase (%)	100	0	3	83	14	0	100	0	0	0
	Collected	Bought	inc	dec	nc	nd	inc	dec	nc	nd
Affected – morbidity phase (%)	97	28	31	11	58	0	3	74	23	0
Affected – mortality phase (%)	3	3	17	29	54	0	0	89	11	0

Data are listed as proportions of respondents (%) who indicated a response, and include whether the resource was bought or collected, frequency of use, and availability.

Other Important Forest Resources

Fruits were the fifth most mentioned forest resource (Fig. 2) and appear to increase in importance for HIV-morbidity affected households (Fig. 3). These resources are both consumed locally and are used to generate an income for the purchase of other commodities such as sugar. Bushmeat was mentioned by roughly 30–50% of the respondents, depending on HIV-phase (Fig. 2), but appears to be ranked as important to HIV-mortality affected households in particular (Fig. 3). Honey and timber were arguably the least important of the most important forest resources mentioned (Fig. 2). However, the importance of timber increased significantly after HIV-related mortality (Fig. 3), supporting the assertion by Mauambeta (2003) that the amount of wood required for coffins in Malawi has increased, constraining sustainable harvesting of timber.

DISCUSSION AND CONCLUSION

The results presented in this chapter appear to support the assertion by Holding-Anyonge et al. (2006) that the relationship between HIV/AIDS and dependence on specific forest resources appears to correspond closely with the stage of the disease. Many of the respondents in this study demonstrated an increased need for medicinal plants, fruit, bushmeat, and honey during the morbidity phase, with a decreased need for these same resources post-mortality. Timber was the exception in that its importance increased significantly after HIV-related mortality. Firewood remained the most important forest resource, regardless of a household's affectedness. Whether these resources are used to treat opportunistic infections and symptoms associated with HIV/AIDS, or to decrease household vulnerability to the disease through either direct consumption or sale, or both, remains a knowledge gap.

Due to the breadth of the topic covered in this study, two main limitations are acknowledged. First, there is a consistent discrepancy in responses provided by the unaffected and pre-HIV phase respondents on questions regarding resource availability. All pre-HIV phase respondents reported that the resource in question was easy to find in the phase before HIV became a problem, while the unaffected respondents were often split between easy and difficult responses for the same resource. These discrepancies could be attributed to the reality of the disease burden on affected households as compared to unaffected households. The unaffected respondents have likely not suffered to the same extent that affected respondents have. Yet, respondents from HIV-affected households were required to continually differentiate their answers between "then" (pre-HIV) and "now" (post-HIV). This could possibly have enabled them to reflect back to a time when life was "easier," providing an indication of how HIV-affected respondents perceive the added burden and hardship caused by HIV-morbidity and mortality. A larger sample size could have helped to clarify why this discrepancy arose. The second limitation to this study is its small sample size. A much larger sample would have provided a clearer understanding about the relationship between HIV/AIDS and dependence on specific forest resources. In particular, it would have enabled me to disaggregate the data by study site so as to better understand how household dependence on forest resources is affected not only by HIV/AIDS, but also by institutional access (whether access is on federally protected lands or customary lands).

Based on the results presented, it is pertinent to ask if HIV/AIDS is a unique shock to households that justifies specific interventions. Twine and Hunter (2008) found that AIDS mortality is not necessarily a unique household shock with regard to its impact on household food security. Households may develop coping strategies during the protracted period of illness, suggesting that policies and interventions should target vulnerable households more broadly. However, the results presented here indicate an increased need for food and medicines during the morbidity phase, particularly medicinal plants, with a decreased need for these same resources during the post-mortality phase. Likewise, the sudden requirement for timber in the post-mortality phase could have implications from the perspectives of both household labor availability and forest sustainability.

This is an important finding given that a positive feedback loop could occur whereby impoverished people living with HIV/AIDS increasingly depend on forest resources for their food security and livelihoods, resulting in increased forest degradation and deforestation that compromises their abilities to continue to meet these critical needs (Lopez, 2008). The negative health impacts of an impaired environment for HIV/AIDS-affected households could be catastrophic: decreased stamina, increased vulnerability to further infections and other diseases, and less livelihood and food security are all possible impacts (Timko et al., 2010). For instance, if important medicinal plant species become scarce, an ill individual could suffer more from a loss of stamina and a compromised immune system. If firewood or foodstuffs are harder to come by, one potential coping mechanism could be to skip meals, and this was indeed a strategy used by several respondents in this study. A decreased availability of firewood, as was perceived by HIV-affected households in this study, could result in the loss of income-generating opportunities, and an increase in labor spent collecting harder to find forest products (*sensu* Barany et al., 2005). Forest fragmentation and deforestation could also increase the risk of other disease transmissions (Keesing et al., 2010; Wolfe, Daszak, Kilpatrick, & Burke, 2005), furthering the disease burden on HIV-affected people.

The complex interactions between the health and well-being of an HIV/AIDS-affected population and environmental resources means that solutions cannot be sought from a purely medical standpoint. Instead, an interdisciplinary approach based on combined ecological, sociological, and health perspectives is required. We must look beyond our existing frameworks which limit our ability to understand diseases (*sensu* Farmer, 1996), to deal with complex environmental problems, and to develop adequate policy responses in these domains. For instance, specific policy options to safeguard the forest resource-related needs of HIV/AIDS-affected households in particular settings could include those that: foster the most promising coping strategies and innovations used by local people to obtain and manage important forest resources (Timko, 2013); ensure local values for forest resources (such as medicinal plants) are protected by forest management plans (Colfer et al., 2006; Holding Anyonge et al., 2006); engage local people in the preparation of forest management plans and in forest management decisions; promote agroforestry as a means to assist rural residents to establish and manage tree crops, particularly those possessing medicinal properties (Barany et al., 2001); promote collaboration to build a multi-sectoral approach to assist HIV-affected households (Shackleton et al., 2006); and examine rural health delivery to determine the environmental consequences of access, or a lack thereof, to health services.

Concerted attention to developing an understanding among the many disciplines spanning the natural and social sciences at work in this field is essential (Wilcox et al., 2004), including anthropology and human ecology, economics, epidemiology, ethnobotany, forest ecology, pharmacology, politics, and economics (Colfer et al., 2006). Synergies between the social and physical environment as determinants of health create opportunities for both health protection and health promotion strategies (Parkes et al., 2003). For the four case study sites in Malawi, policies and interventions that address the impacts of HIV/AIDS on the sustainability of these important resources, that compensate for a decreased availability of household labor (*sensu* Timko, 2013), and that foster greater access to these resources for vulnerable households, are justified. In particular, while not the focus of this study, institutional structures (such as federally protected lands and forest reserves and customary lands and VFAs) and any barriers they create to accessing important forest resources should be carefully examined for their unintended impacts on HIV/AIDS-affected households. The results of my study highlight the need for interventions that would assist households in the HIV/AIDS-mortality phase in particular to obtain the forest resources they require, especially medicinal plants, fruit, (bush)meat, and honey. Because there is a constant need for firewood and water by all households, regardless of HIV-affectedness, they are included in the recommended interventions. Respondents of the study offered many simple, practical, yet innovative solutions including:

Firewood: providing firewood directly to affected households; providing seedlings of fast-growing species of firewood trees (such as various species of *Eucalyptus* and *Pinus*, *Gmelina arborea*, *Senna siamea*, *Toona ciliata*, and Neem (*Azedrachta indica*) for planting in home gardens or on common lands; reforesting degraded lands to replenish firewood species; providing fuel-efficient wood stoves or stoves using alternative fuel types to all household types to alleviate pressure on the forest (*sensu* Holding-Anyonge et al., 2006);

Water: fostering local collective actions to install/providing households with storage tanks or rainwater harvesting systems; installing new or repairing degraded village boreholes and waterpumps;
Medicinal Plants: providing traditional healers with important medicinal plants; creating a medicinal plant community herbarium; training to ensure sustainable harvesting methods are used in the collection of medicinal plants in woodlands and on other common lands;
Fruit: investing in agroforestry projects (*sensu* Jayne, Villarreal, Pingali, & Hemrich, 2005) and the domestication of important medicinal plants, wild vegetables, and indigenous fruit trees; providing domesticated fruit tree and indigenous fruit tree seedlings (such as *Flacoutia indica*, *Amisophylla pomifera, Friesodielsia obovata* (Benth.) Verdc) for planting in home gardens or on common lands;
(Bush)meat: reforesting to restore wildlife populations; encouraging animal husbandry (particularly chickens, goats, dairy cattle, pigs, and guinea fowl) as a means to meet local protein requirements; building capacity around fish farming; and
Honey: providing beehive and honey harvesting materials to enable honey to be sustainably used locally as both a food and a medicine, and to be sold to generate an income.

Given the extent of prime-age adult mortality attributable to the pandemic across Sub-Saharan Africa and the links between HIV/AIDS and forest resources, there is ample scope for further research in this domain. Further exploratory studies that emerge from my study include: (1) replicating the methods used in this study with more households across Malawi to garner a greater understanding of the environmental dimensions of the disease; (2) assessing the health impacts of changes in the availability of the most important forest resources; (3) assessing how household resource use changes according to the phases of the disease, such as from onset of HIV, stabilization of HIV, early versus prolonged morbidity, progression into AIDS, immediate aftermath of mortality, and resource use 1, 2, and 5 years post-mortality; (4) analyzing the therapeutic effects of the most important medicinal plants identified in this study (but not reported in this chapter); and (5) testing the interventions recommended in this chapter to understand what, if any, role they play in enabling affected households to obtain important forest resources.

Throughout this chapter, I have attempted to characterize how rural household dependence on forest resources changes through three phases of household HIV-affectedness. By doing so, I hope to have contributed to the inquiry into the environmental dimensions of HIV/AIDS. I have suggested practical, innovative interventions that could alleviate some of the disease burden on rural Malawian households; and offered insight into potential areas of further inquiry in this research domain. Above I have listed some of the many ways for governments, NGOs, researchers, civil society organizations, and local people to contribute to policies and interventions that address the impacts of HIV/AIDS on the sustainability of important forest resources, that compensate for a decreased availability of household labor, and that foster greater access to these resources for vulnerable households. For when you are told that children and grandchildren are being pulled out of school to collect firewood or water, or that a woman's risk of rape increases as she is having to walk further to collect forest resources, the call to action is clear. I have been implored to try to contribute on even the smallest level, as the consequences of not trying to improve the situation, of not acting, are grave. And it is with this sense of responsibility, that in 2012 I initiated a follow-up study in two of the study sites: Zomba and Chiladzulu. In these sites and with my Malawian field assistant, I will carry out two small studies on interventions discussed in this chapter.

NOTE

1. All percentages refer to the proportion of respondents.

REFERENCES

Africa Biodiversity Collaborative Group (ABCG). (2002). HIV/AIDS and natural resource management linkages. *Workshop Proceedings*, September 26–27, Nairobi, Kenya. Retrieved from http://www.iucnbot.bw/pages_sub_dir/Proceedings%20-%20final.pdf

Arnold, J. E. M. (2001). *Forests, poverty and aid.* CIFOR Occasional Paper No. 33. CIFOR, Bogor, Indonesia.

Barany, M., Hammett, A. L., Sene, A., & Amichev, B. (2001). Nontimber forest benefits and HIV/AIDS in Sub-Saharan Africa. *Journal of Forestry*, *99*(12), 26–41.

Barany, M., Hammett, A. L., Stadler, K. E., & Kengni, E. (2004). Non-timber forest products in the food security and nutrition of smallholders afflicted by HIV/AIDS in Sub-Saharan Africa. *Forests, Trees and Livelihoods*, *14*(1), 3–18. doi:10.1080/14728028.2004.9752475

Barany, M., Holding-Anyonge, C., Kayambazinthu, D., & Sitoe, A. (2005). Firewood, food, and medicine: Interactions between forests, vulnerability and rural responses to HIV/AIDS. *Proceedings from the IFPRI conference: HIV/AIDS and food and nutrition security*, April 14–16, Durban, South Africa. Retrieved from http://www.fao.org/forestry/9719-088e4b8cdc0c6bd4b67eaa81e6427a705.pdf

Belcher, B. (2005). Forest product markets, forests and poverty reduction. *International Forestry Review*, *7*(2), 82–89. (doi:http://dx.doi.org/10.1505/ifor.2005.7.2.82)

Bishop-Sambrook, C. (2003). *Labour saving technologies and practices for farming and household activities in Eastern and Southern Africa: Labour constraints and the impact of HIV/AIDS on rural livelihoods in Bondo and Busia districts, western Kenya*. Rome, Italy: International Fund for Agricultural Development (IFAD)/Food and Agriculture Organization of the United Nations (FAO).

Bolton, S., & Talman, A. (2010). *Interactions between HIV/AIDS and the environment: A review of the evidence and recommendations for next steps*. Nairobi, Kenya: IUCN ESARO Office. Retrieved from http://cmsdata.iucn.org/downloads/hiv_aids_and_environment.pdf

CIFOR (Centre for International Forestry Research). (2005). *Contributing to African development through forests: Strategy for engagement in Sub-Saharan Africa*. Bogor, Indonesia: CIFOR. Retrieved from http://www.cifor.org/publications/pdf_files/Books/PCIFOR0501.pdf

Colfer, C. J. P., Sheil, D., & Kishi, M. (2006). *Forests and human health: Assessing the evidence*. CIFOR Occasional Paper No. 45. CIFOR, Bogor, Indonesia. Retrieved from http://www.cifor.org/publications/pdf_files/OccPapers/OP-45.pdf

De Souza, R. M., Heinrich, G., Senefeld, S., Coon, K., Sebanja, P., Ogden, J., & … Oglethorpe, J. (2008). Using innovation to address HIV, AIDS, and environment links: Intervention case studies from Zimbabwe, Uganda, and Malawi. *Population and Environment*, *29*, 219–246. doi:10.1007/s11111-008-0070-0

Drimie, S. (2002). The impact of HIV/AIDS on rural households and land issues in Southern and Eastern Africa. A background paper prepared for the Food and Agricultural Organization, Sub-Regional Office for Southern and Eastern Africa. Retrieved from ftp://ftp.fao.org/docrep/nonfao/ad696e/ad696e00.pdf

Drimie, S., & Gandure, S. (2005). *The impacts of HIV/AIDS on rural livelihoods in southern Africa: An inventory and literature review*. Harare, Zimbabwe: Food and Agricultural Organization of the United Nations. Retrieved from http://www.zimrelief.info/files/attachments/doclib/Impact%20of%20HIV%20&%20AIDS%20on%20Livelihood%20in%20Southern%20Africa.pdf

Dwasi, J. (2002). *HIV/AIDS and natural resource management in Africa: Findings from Kenya, Namibia, South Africa, and Uganda*. Unpublished report for the Africa Biodiversity Collaborative Group. Retrieved from http://www.frameweb.org/adl/en-US/2323/file/327/Final_Draft_of_Jane_Dwasi_%20Report.doc

Farmer, P. (1996). Social inequalities and emerging infectious diseases. *Emerging Infectious Diseases*, *2*(4), 259–269.

Fisher, M. (2004). Household welfare and forest dependence in southern Malawi. *Environment and Development Economics*, *9*(2), 135–154. doi:10.1017/S1355770X03001219

Frank, E., & Unruh, J. (2008). Demarcating forest, containing disease: Land and HIV/AIDS in southern Zambia. *Population and Environment*, *29*, 108–132. doi:10.1007/s11111-008-0067-8

Government of Malawi. (1996). *National forest policy of Malawi*. Lilongwe, Malawi: Ministry of Natural Resources.

Government of Malawi. (2007). *Forestry HIV and AIDS strategy 2007-2011*. Lilongwe, Malawi: Ministry of Energy and Mines.

Holding-Anyonge, C., Rugalema, G., Kayambazinthu, D., Sitoe, A., & Barany, M. (2006). Fuelwood, food and medicine: The role of forests in the response to HIV and AIDS in rural areas of southern Africa. *Unasylva*, *57*(224), 20–23.

Hunter, L. M., Twine, W., & Johnson, A. (2008). *Adult mortality and natural resource use in rural South Africa: Evidence from the Agincourt Health and Demographic Surveillance Site*. Institute of Behavioral Science Working Paper EB2005-0004. University of Colorado at Boulder, Boulder, CO. Retrieved from http://www.colorado.edu/ibs/pubs/eb/eb2005-0004.pdf

Jayne, T. S., Villarreal, M., Pingali, P., & Hemrich, G. (2005). HIV/AIDS and the agriculture sector: Implications for policy in eastern and southern Africa. *Journal of Agricultural and Development Economics*, *2*(2), 158–181. doi:10.1111/j/1467-9353.2006.00310.x

Kachule, R., Tchale, H., & Mataya, C. (2003). Management of miombo woodlands in Malawi: An application of system dynamics modelling. In G. Kowero, B. M. Campbell & U. R. Sumaila (Eds.), *Policies and governance structures in woodlands of Southern Africa* (pp. 423–438). Bogor: CIFOR.

Kaimowitz, D. (2003). Not by bread alone ... forests and rural livelihoods in sub-Saharan Africa. In T. Oksanen, B. Pajari, & T. Tuomasjukka (Eds.), *Forests in poverty reduction strategies: Capturing the potential* (pp. 45–63). EFI Proceedings No 47. European Forest Institute, Joensuu, Finland.

Kaschula, S. A. (2008). Wild foods and household food security responses to AIDS: Evidence from South Africa. *Population and Environment*, *29*, 62–185. doi:10.1007/s11111-008-0068-7

Kayambazinthu, D., Barany, M., Mumba, R., & Holding-Anyonge, C. (2005). *Miombo woodlands and HIV/AIDS interactions: Malawi country report*. Forest Policy and Institutions Working Paper – Food and Agriculture Organization 2005 No. 6. Retrieved from http://www.fao.org/docrep/008/j6038e/j6038e00.htm

Kayambazinthu, D., Matose, F., Kajembe, G., & Nemarundwe, N. (2003). Institutional arrangements governing natural resource management of the miombo woodland. In G. Kowero, B. M. Campbell & U. R. Sumaila (Eds.), *Policies and governance structures in Woodlands of Southern Africa* (pp. 45–64). Bogor: CIFOR.

Keesing, F., Belden, L. K., Daszak, P., Dobson, A., Harvell, C. D., Holt, R. D., & ... Ostfeld, R. S. (2010). Impacts of biodiversity on the emergence and transmission of infectious diseases. *Nature*, *468*, 647–652. doi:10.1038/nature09575

Lopez, P. (2008). The subversive links between HIV/AIDS and the forest sector. In C. J. P. Colfer (Ed.), *Human health and forests: A global overview of issues, practice, and policy* (pp. 221–237). Sterling, VA: Earthscan.

Masangano, C. M., Kayambazinthu, D., & Mwabumba, L. (2003). Conflicts over the miombo woodlands: The case of Blantyre, Lilongwe and Kasungu, Malawi. In G. Kowero, B. M. Campbell & U. R. Sumaila (Eds.), *Policies and governance structures in woodlands of Southern Africa* (pp. 65–79). Bogor: CIFOR.

Mauambeta, D. C. (2003). *HIV/AIDS mainstreaming in conservation: The case of wildlife and environmental society in Malawi*. Retrieved from http://www.frameweb.org/adl/en-US/2308/file/209/WESM_AIDS_Paper.pdf

McMichael, T. (1999). Prisoners of the proximate: Loosening the constraints on epidemiology in an age of change. *American Journal of Epidemiology*, *149*(10), 887–897.

McMichael, T. (2002). The biosphere, health, and "sustainability". *Science*, *297*(5584), 1093. doi:10.1126/science.297.5584.1093

Myers, S. S., & Patz, J. A. (2009). Emerging threats to human health from global environmental change. *Annual Review of Environment and Resources*, *34*, 223–252. doi:10.1146/annurev.environ.033108.102650

National AIDS Commission. (2008). *HIV and syphilis sero – Survey and national HIV prevalence and AIDS estimates report for 2007*. Lilongwe, Malawi: Ministry of Health, Department of Preventive Health Services.

Ngwenya, B., & Kgathi, D. (2006). HIV/AIDS and access to water: A case study of home-based care in Ngamiland, Botswana. *Physics and Chemistry of the Earth*, *31*(15–16), 669–680. doi:10/1016/j.pce.2006.08.041

Ostfeld, R. S., Meffe, G. K., & Pearl, M. C. (2002). Conservation medicine: The birth of another crisis discipline. In A. A. Aguirre, R. S. Ostfeld, G. M. Tabor, C. House & M. C. Pearl (Eds.), *Conservation medicine: Ecological health in practice* (pp. 17–26). Oxford, UK: Oxford University Press.

Parkes, M., Panelli, R., & Weinstein, P. (2003). Converging paradigms for environmental health theory and practice. *Environmental Health Perspectives*, *111*(5), 669–675. doi:10.1289/ehp.5332

Parkes, M. W., Bienen, L., Breilh, J., Hsu, L. N., McDonald, M., Patz, J. A., & … Yassi, A. (2005). All hands on deck: Transdisciplinary approaches to emerging infectious disease. *EcoHealth*, *2*(4), 258–272. doi:10.1007/s10393-005-8387-y

Patz, J. A., Daszak, P., Tabor, G. M., Aguirre, A. A., Pearl, M., Epstein, J., & … Bradley, D. J. (2004). Unhealthy landscapes: Policy recommendations on land use change and infectious disease emergence. *Environmental Health Perspectives*, *112*(10), 1092–1098.

Plowright, R. K., Sokolow, S. H., Gorman, M. E., Daszak, P., & Foley, J. E. (2008). Causal inference in disease ecology: Investigating ecological drivers of disease emergence. *Frontiers in Ecology and the Environment*, *6*, 420–429.

Shackleton, C., & Shackleton, S. (2004). The importance of non-timber forest products in rural livelihood security and as safety nets: A review of evidence from South Africa. *South African Journal of Science*, *100*(11&12), 658–664. (ISSN 0038-2353).

Shackleton, S., Kaschula, S., Twine, W., Hunter, L., Holding-Anyonge, C., & Petheram, L. (2006). *Forests as safety nets for mitigating the impacts of HIV/AIDS in southern Africa*. CIFOR: Bogor, Indonesia. Retrieved from http://www.cifor.org/publications/pdf_files/livebrief/livebrief0604e.pdf

Shackleton, C. M., Shackleton, S. E., Buiten, E., & Bird, N. (2007). The importance of dry woodlands and forests in rural livelihoods and poverty alleviation in South Africa. *Forest Policy and Economics*, *9*, 558–577. doi:10.1016/j.forpol.2006.03.004

Sunderlin, W. D., Angelsen, A., Belcher, B., Burgers, P., Nasi, R., Santoso, L., & Wunder, S. (2005). Livelihoods, forests, and conservation in developing countries: An overview. *World Development*, *33*(9), 1383–1402. doi:10.1016/j.worlddev.2004.10.004

Syampungani, S., Chirwa, P. W., Akinnifesi, F. K., Sileshi, G., & Ajayi, O. C. (2009). The miombo woodlands at the cross roads: Potential threats, sustainable livelihoods, policy gaps and challenges. *Natural Resources Forum*, *33*, 150–159. doi:10.1111/j.1477-8947.2009.01218.x

Tabor, G. M. (2002). Defining conservation medicine. In A. A. Aguirre, R. S. Ostfeld, G. M. Tabor, C. House & M. C. Pearl (Eds.), *Conservation medicine: Ecological health in practice* (pp. 8–16). Oxford: Oxford University Press.

Timko, J. A., Kozak, R. A., & Innes, J. L. (2010). HIV/AIDS and forests in Sub-Saharan Africa: Exploring the links between morbidity, mortality, and dependence on biodiversity. *Biodiversity*, *11*(1&2), 45–48. doi:10.1080/14888386.2010.9712646

Timko, J. A. (2013). Exploring forest-related coping strategies for alleviating the HIV/AIDS burden on rural Malawian households. *International Forestry Review*, *15*(2), 230–240. doi:http://dx.doi.org/10.1505/146554813806948495

Topouzis, D. N. (2007). HIV/AIDS: A risk to the social and economic sustainability of forestry in Sub-Saharan Africa? *International Forestry Review*, *9*(1), 558–562. doi:10.1505/ifor.9.1.558

Torell, E., Tobey, J., Thaxton, M., Crawford, B., Kalangahec, B., Madulu, N., & ... Sallema, R. (2006). Examining the linkages between AIDS and biodiversity conservation in coastal Tanzania. *Ocean & Coastal Management*, *49*, 792–811. doi:10.1016/j.ocecoaman.2006.08.007

Twine, W., & Hunter, L. (2008). *AIDS mortality and the role of natural resources in household food security in a rural district of South Africa*. HIV, livelihoods, food and nutrition security: Findings from Renewal Research (2007–2008), Brief 11. IFPRI, Washington, DC.

Villarreal, M., Holding-Anyonge, C., Swallow, B., & Kewsiga, F. (2006). The challenge of HIV/AIDS: Where does agroforestry fit in? In D. Garrity, A. Okono, M. Grayson & S. Parrott (Eds.), *World agroforestry into the future* (pp. 181–190). Nairobi, Kenya: World Agroforestry Centre.

Weinstein, P. (2005). Human health is harmed by ecosystem degradation, but does intervention improve it? A research challenge from the millennium ecosystem assessment. *EcoHealth*, *2*, 228–230. doi:10.1007/s10393-005-6329-3

Weiss, R. A., & McMichael, A. J. (2004). Social and environmental risk factors in the emergence of infectious diseases. *Nature Medicine Supplement*, *10*(Suppl. 12), 570–576.

Wilcox, B. A., Aguirre, A. A., Daszak, P., Horwitz, P., Martens, P., Parkes, M., & Waltner-Toews, D. (2004). EcoHealth: A transdisciplinary imperative for a sustainable future. *EcoHealth*, *1*, 3–5. doi:10.1007/s10393-004-0014-9

Wolfe, N. A., Daszak, P., Kilpatrick, A. M., & Burke, D. S. (2005). Bushmeat hunting, deforestation, and prediction of zoonotic disease emergence. *Emerging Infectious Diseases*, *11*(12), 1822–1827. doi:10.3201/eid1112.040789

DROPS AND HOT STONES: TOWARDS INTEGRATED URBAN PLANNING IN TERMS OF WATER SCARCITY AND HEALTH ISSUES IN LEH TOWN, LADAKH, INDIA

Daphne Gondhalekar, Adris Akhtar, Pascal Keilmann, Jenny Kebschull, Sven Nussbaum, Sonam Dawa, Phuntsok Namgyal, Lobzang Tsultim, Tsering Phuntsog, Stanzin Dorje, Tsering Mutup and Phunchok Namgail

ABSTRACT

Purpose – *This chapter studies the link between urban planning and health. Access to safe drinking water is already a very serious issue for large urban populations in fast-growing economies such as India. Water availability is further being impacted by climate change, leading to the drastically increased spread of water-related diseases.*

Design/methodology/approach– *Leh Town, which is located in an ecologically vulnerable semi-arid region of the Himalayas in Ladakh, has*

Ecological Health: Society, Ecology and Health
Advances in Medical Sociology, Volume 15, 173–193

ISSN: 1057-6290/doi:10.1108/S1057-6290(2013)0000015012

been considered for this study because it is undergoing large-scale transformation due to rapid growth in its tourism industry. In 2012–2013 our interdisciplinary group comprising researchers from Germany and India conducted field surveys, including geographic information system-based (GIS) mapping of point sources of water pollution, questionnaire surveys of 200 households and 70 hotels and guesthouses and semi-structured interviews. We also reviewed secondary medical data.

Findings – *We found that diarrhoeal incidence has increased in the local population in Leh in the past decade, which may be linked to water pollution: Further, we found that rapidly increasing water consumption coupled with a lack of adequate water and sanitation infrastructure is causing serious water pollution.*

Research limitations/implications – *Further, data is needed for causal connections between water pollution and health impacts to be conclusively drawn.*

Practical implications – *This study discusses the use of GIS to support a call for the need for more integrated urban planning and decision-making that holistically addresses water and health challenges in Leh and advocates the development of a decentralized or hybrid sanitation system to support water resources conservation as a central dimension of an integrated health management approach.*

Social implications – *GIS is also a very useful platform for supporting participatory urban planning in Leh.*

Originality/value – *With such an integrated urban planning approach, Leh would be a lighthouse example for other towns in the region.*

Keywords: Urban planning; water resources management; health; climate change; geographic information system (GIS); India

INTRODUCTION

Urban areas are undergoing huge transformations due to rapid economic growth especially in terms of water-related environmental challenges in developing economies such as India (Marcotullio, 2007). Urban infrastructure planning is often unable to keep up with the increasing water demands of growing urban populations. The resulting lack of access to safe drinking water and sanitation in these rapidly growing cities is helping to drive the spread of water-related diseases (Galea & Vlahov, 2005). This situation is being exacerbated by climate change-induced water scarcity (Vörösmarty, Green, & Lammers, 2000).

One of the leading water-related diseases, diarrhoea, to which children are especially vulnerable, is already a major public health concern in developing countries. Diarrhoea accounts for 16 percent of deaths of children under five years of age globally, one third of which occur in India (White Johansson & Wardlaw, 2009, pp. 5–7). Internationally, diarrhoea is the second leading cause of death in children under five (WHO, 2002, pp. 138–139) and kills more children every year than AIDS, malaria and measles combined (Plan, 2012). Diarrhoea can have a number of causes, but the main transmission route is through drinking water (Howard & Bartram, 2003; Sakdapolrak, Seyler, & Prasad, 2011, p. 88). The stark reality is that globally one in three people do not have access to adequate sanitation (Plan, 2012); therefore, adequate sanitation access is currently considered by some academics to be the most off-track of the Millennium Development Goals (e.g. Albuquerque, 2012). Lack of sanitation infrastructure contributes considerably to drinking water pollution (WHO, 2002, pp. 138–139). Nearly all deaths due to diarrhoea worldwide could be prevented through access to safe water, adequate sanitation and good hygiene (White Johansson & Wardlaw, 2009, pp. 10–13).

Sanitation has always been a consideration within urban planning and continues to be so in the contemporary moment where a focus is being placed on how to address the interplay between urbanization, urban planning and sustainable development. Increasing evidence shows how living in urban areas impacts human health (Barton & Pretty, 2010; Wilson et al., 1998); thus, urban areas promise opportunities for health research (Frumkin, 2003; Northridge, Sclar, & Biswas, 2006). In this chapter, we aim to link public health and urban planning through presenting a case study on Leh Town, India. Specifically, this chapter begins with a brief survey of the literature linking urban planning and health and then describes social and ecological drivers of increases in incidences of acute diarrhoea, a preventable disease which remains inadequately addressed globally, in this small city. The chapter then outlines the methodology used in this international, interdisciplinary, multi-partner research project. In the following two sections we then share some key results from the health data and water extraction and pollution analyses conducted and discuss their implications for urban planning. This chapter culminates in a conclusion that draws these threads together and advocates for the development of integrated approaches to social-ecological health research through collaborations between the disciplines of urban planning and public health.

LINKING URBAN PLANNING AND SOCIAL-ECOLOGICAL HEALTH ISSUES IN LEH

Although health issues do not directly drive urban design, they did provide the original impetus for the urban planning profession. One well-known example from history is the discovery in 19th-century London that cholera is a water-borne disease and that it was spreading from one particular contaminated water pump. This insight had huge implications for urban design which is increasingly considered a powerful tool for addressing new public health concerns (Jackson, L. E., 2003; Jackson, R. J., 2003). Today, several studies are advocating the development of new frameworks for linking public health and urban planning (Frank, Andresen, & Schmid, 2004; Heath et al., 2006; Hoehner, Brennan, Brownson, Handy, & Killingsworth, 2003; Naess, 2006), including this study which calls for a linking of disciplines through an integrated social-ecological approach. Offering evidence which substantiates the need for such a call are a number of studies on the relationship between the built-environment and health conducted in a number of academic fields yet these results are difficult to synthesize as each discipline uses its own theoretical frameworks, methodologies and terminologies (Dannenberg et al., 2003; Galea & Vlahov, 2005). Clearly, cross-disciplinary collaboration and further research is needed in order to strengthen the associations between urban design and health (Jackson, L. E., 2003). In particular, linking urban design and health to social practices in terms of water utilization, such as, cultural habits of water consumption, is key for operationalizing such research. Further interdisciplinary and international research cooperation is needed (Bork, Butsch, Kraas, & Kroll, 2009) as is enabling cross-country learning experiences.

When reviewing the literature for work that links urban planning, water, sustainable development and health, and when looking for exemplars of integrated, intersectoral social-ecological studies within this literature, a few trends became clear. Globally, there are more studies on the environmental impacts of urbanization than on related human health impacts (Jackson, L. E., 2003). Of the existing studies linking urbanisation and health, most are on developed countries, leaving a gap in the literature regarding developing countries such as India. When undertaken, research on urbanization in India in particular has focused on the complexity of current urban issues arising in the country and has resulted in a large number of studies on urbanization (Sandhu, 2003), urban land-use policy (Kulabkar,

2002; Sridhar, 2010), urban water management (Durga Rao, 2005; Gandy, 2008), urban governance (Dupont, 2007), urban planning history (Banerjee, 2009) and the impacts of climate change on urban areas (Revi, 2008) but relatively few studies have also looked at related human health impacts. When urban studies in India have focused on environmental challenges, they have also tended to be at a regional scale (Bose & Leitmann, 1996; Pacione, 2006), which makes the results difficult to translate into implementable solutions on the ground. In addition, many studies look at new urban developments (Bhattacharya & Sanyal, 2011; Vedula, 2007; Wang, Kundu, & Chen, 2010) and large cities while there is a dearth of information on water and health issues in small and medium-sized cities.

Recognizing and understanding the health benefits of urban design as a means to helping to devise solutions applicable to future urban development patterns (Frumkin, 2002) and addressing the need for research in small cities, we embarked on a study of Leh Town, which is considered one of the fastest-expanding small towns in India (Rieger-Jandl, 2005, p. 123). Located in a remote, ecologically sensitive semi-arid region in the Himalayas at an altitude of 3,500 metres above sea level, Leh is a green oasis of agricultural fields nestled between barren mountains and fringing a historic town centre (Fig. 1). Administratively, Leh Town is the capital of Leh District in the Ladakh Region of the Jammu and Kashmir State of India. The Ladakh Autonomous Hill Development Council (LAHDC), responsible for economic, health and land-use planning, has been governing Leh District since

Fig. 1. Geographical Location of Leh, Ladakh, India.

1995. According to the 2011 Census of Jammu and Kashmir State, Leh District has a population of 147,104 inhabitants. It is administratively divided into nine blocks, one of which is Leh Block with a population of 34,400 according to the Chief Medical Officer (CMO). Leh Town is a separate Municipality, with a population of 17,553 according to the 2011 Census. In addition, 40,000 army personnel live in Leh (Skeldon, 1985) and several tens of thousands of migrant workers come to Leh every summer.

Social factors are not only shaping the health setting of Leh Town, but are also environmental determinants of health. One significant environmental factor is the limited availability of water. The precipitation in Leh is only 115 mm per year (Owen, Caffee, Bovard, Finkel, & Sharma, 2006, p. 384); therefore, the sole source of surface water is a glacier above Leh. The apparent lushness of Leh is thus not a natural occurrence but the result of hundreds of years of careful irrigation and cultivation of a fertile desert. Since 1974, when Ladakh was opened to tourism, the number of visitors has increased exponentially. Particularly since 2002 there has been a large increase and in 2012, 179,000 tourists visited Leh (Fig. 2). Most tourists visit in the summer, between April and October, and during this period Leh is bustling with traders from other Indian States, Nepal and Tibet. In the harsh winter when the average temperature drops to $-16.5°C$ to $-5.5°C$ (Singh, 1998, p. 353), only the local population remains.

To cater to the huge increase in visitors, hundreds of guesthouses and hotels have been constructed in Leh. But although Leh has a masterplan, which was published in 1996, urban development in Leh is haphazard (Eichert, 2009, p. 77) and unsustainable. For example, traditional Ladakhi dry toilets do not require any water, but 99 percent of tourists prefer to use flush toilets, according to one survey conducted in July–August 2009 (Akhtar, 2010, p. 57), which is why increasingly tourist accommodations are constructing flush toilets and showers, pushing up the water demand. The

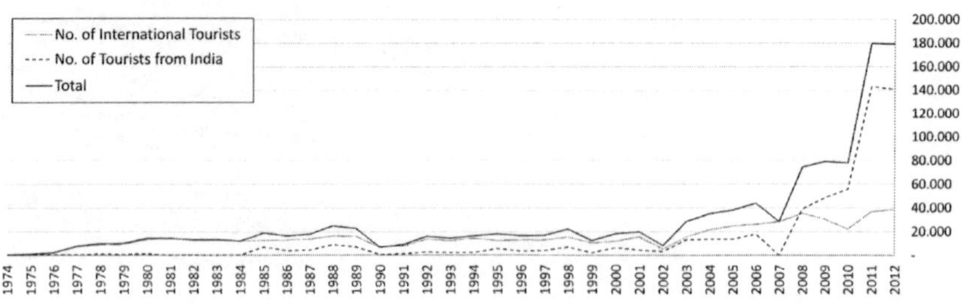

Fig. 2. Year-wise Number of Visitors to Leh (*Source:* Tourist Board Leh).

water supply and wastewater management systems have been unable to keep up with the resulting huge increase in demand. At the same time, surface water from glacial runoff is decreasing (Eichert, 2009, p. 53) possibly as a result of climate change, and is affecting groundwater recharge (LEDeG, 2010). Water-related environmental issues such as water pollution, lack of water and sanitation infrastructure, and rubbish dumping are already severe in Leh. Springs used to be the main source of household and drinking water in the past, but currently about 1.5 million litres of water daily is supplied by the Public Health Engineering Department (PHE) through groundwater extraction in the summer months (LEDeG, 2010). However, PHE provides water only for a few hours per day (Akhtar, 2010, p. 71), which is insufficient for flush toilets and showers, and so many guesthouses and hotels are constructing private bore wells. In 2005, the Leh Development Authority was established to monitor development projects in Leh District, but so far there is no political agency that monitors environmental issues; this work is only done by local non-governmental organizations (NGOs) (Akhtar, 2010, p. 34). Increases in water-borne diseases such as hepatitis and diarrhoea were already recorded in Leh over a decade ago (Bashin, 1999), and exposure to drinking water pollution and health impact may have profound implications for economic growth (Dasgupta, 2004). However, so far there has been no systematic study of the link between water and health in Leh. Taking an Ecohealth approach (Charron, 2012), therefore, the aim of our study has been to assess water-related health issues, water pollution factors and their interlinkages in Leh.

METHODOLOGY

Field surveys were conducted between July 2012 and February 2013 in collaboration with our research partner, the Ladakh Ecological Development Group (LEDeG), a local NGO. We obtained health data on various diseases from the CMO, including District Level yearly data from 1998 to 2012 and monthly data from 2001 to 2012 and monthly data at the Leh Block Level from 2010 to 2012. The CMO reported that some health records had been eaten by cows; thus, there is a data gap between 2008 and 2010. We used global positioning system (GPS) to map new hotels, guesthouses and restaurants, as well as 270 point sources of water pollution. The World Health Organization recommends a minimum distance of 100 metres of pollution sources from water resources to avoid water pollution (WHO, 2002, p. 139); thus, distribution of point sources of water pollution was measured within a 100 metre buffer of rivers and streams. A WorldView-2 very high-resolution satellite image (ground resolution 50 cm) from November 2011 served as a base map. Geographic information systems (GIS) data on rivers, roads, hotels and guesthouses mapped by Akhtar (Akhtar, 2010) were used. We also conducted a questionnaire survey of 200 households and 70 hotels and guesthouses, as well as semi-structured interviews with households, representatives of LAHDC, PHE, tourists and doctors at the central government hospital and the CMO. All research activities were carried out under the Urban Water Health Project Leh, a joint participatory research project of LEDeG and the Center for Development Research (ZEF), University of Bonn, Germany.

RESULTS

Health Issues

The CMO compiles health data from government hospitals and health centres; however, other or traditional medical institutions are not included. Health data has been compiled systematically in the Leh District since 1998, overlapping with the last decade when the largest increase in tourists occurred (Fig. 2) and thus the greatest environmental impacts are expected to have taken place. Of the diseases for which data was systematically compiled in 1998–2012, acute diarrhoeal diseases, typhoid, viral hepatitis, leprosy, tuberculosis and malaria are water-related. Only very few cases occurred throughout for typhoid, leprosy and tuberculosis, and in 2002–2005 only the army reported a few cases of malaria. Further, viral hepatitis on average affected only 0.1 percent of the population. However, acute diarrhoeal diseases has affected over 10 percent of the population yearly over the past decade, and affected 12 percent of the population in 2011–2012. Despite the fact that recurrent cases are included in the yearly data it seems that acute diarrhoea, the only disease with major impact that is water-related, is a major public health concern in Leh.

Looking at the monthly data, acute diarrhoeal disease seems to peak in summer. The dashed trend line seems to indicate a linear increase (Fig. 3). However, since health data include tourists, this could be due to the increasing numbers of tourists visiting Leh, who are highly susceptible to diarrhoea due to exposure to unfamiliar bacteria in water and food and because of altitude sickness. Therefore, we assessed incidences of diarrhoeal diseases in the local population in Leh during a time period when it was

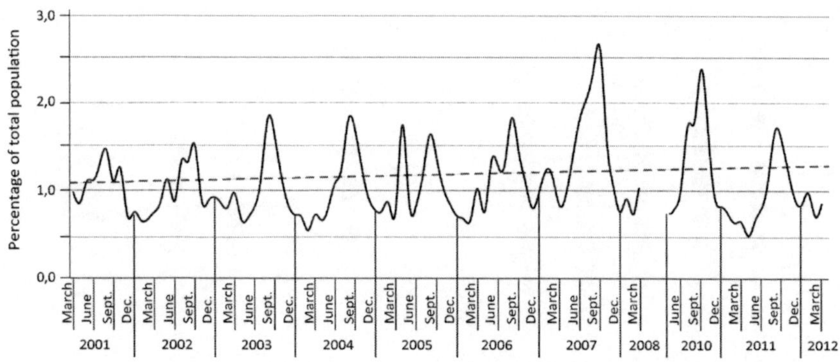

Fig. 3. Percentage of Population with Acute Diarrhoeal Diseases in Leh District, 2001–2012 (*Source:* CMO).

expected that a minimum of tourists would be present. According to the interview survey, this is between November and end of March. After October, the climate is too cold for tourism, and in early April the first big influx of domestic tourists escaping the high temperatures in the pre-monsoon Indian lowlands arrives in Leh. Thus, analysis of incidences of acute diarrhoea between November and March, which is expected to include predominantly the local population, seems to indicate that here also the monthly average has risen in the past decade, from 0.8 to 0.9 percent.

One of the most vulnerable segments of the population, children under five years of age, account for 8 percent of the total population of Leh. Most tourists visit Leh without small children due to the type of tourism and high-altitude conditions, and thus most of the under five-year olds cases of diseases recorded in Leh are expected to be of the local population. It seems that the percentage of under five-year olds afflicted by acute diarrhoeal disease in the past decade was consistently between 5 and 20 percent, and in 2011–12 was over 10 percent. Further, acute diarrhoea affects far more under five-year olds than tuberculosis, pneumonia or dysentery (Fig. 4). In addition, monthly data from April 2010 to March 2012 at the Leh Block level show that for under five-year olds, acute diarrhoeal diseases also seem to peak in summer (Fig. 5). If we assume that very few tourists are children below age five, this may also be an indicator for acute diarrhoeal disease as a health concern of the local population.

In 2011–2012, different types of acute diarrhoeal diseases started to be recorded in Leh District: with dehydration 23 percent, without dehydration 75 percent and with blood 2 percent. This may indicate an increase in institutional awareness of acute diarrhoea. To summarize, we found that

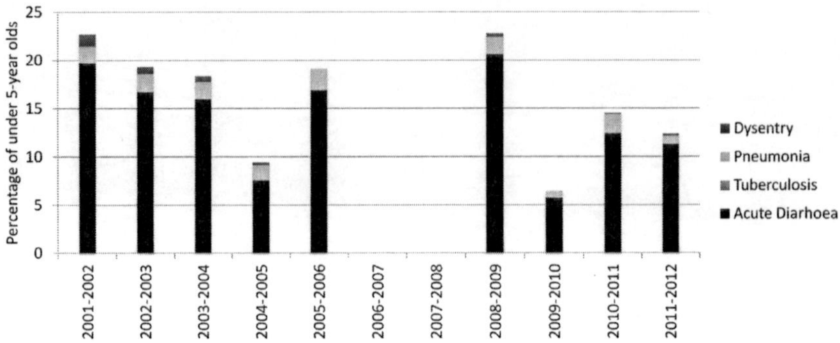

Fig. 4. Major Diseases Affecting Under Five-Year-Olds in Leh District, 2001–2012 (*Source:* CMO).

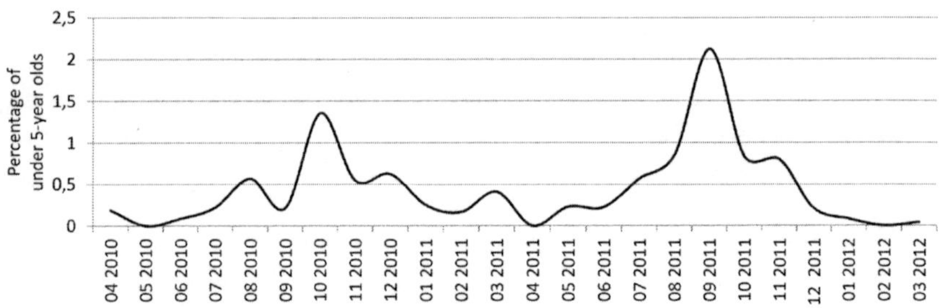

Fig. 5. Percentage of Under Five-Year-Olds with Acute Diarrhoeal Diseases in Leh Block (*Source:* CMO).

acute diarrhoeal disease seems to be a serious health issue in Leh meriting more in-depth investigation.

Water Extraction and Pollution

The number of hotels and guesthouses in Leh Town has increased very rapidly in the past decades: in 1980 there were only 24 hotels and guesthouses in Leh, but by 1990 there were 62, by 2000 there were 117, by 2010 there were 282 and just from 2010 to 2012, the number had increased to ca. 360 guesthouses and hotels in business, with another ca. 60 not yet in business or under construction (Fig. 6). Of 21 wards in Leh Town, 10 have predominantly agricultural land, whilst the others are dry and desert-like.

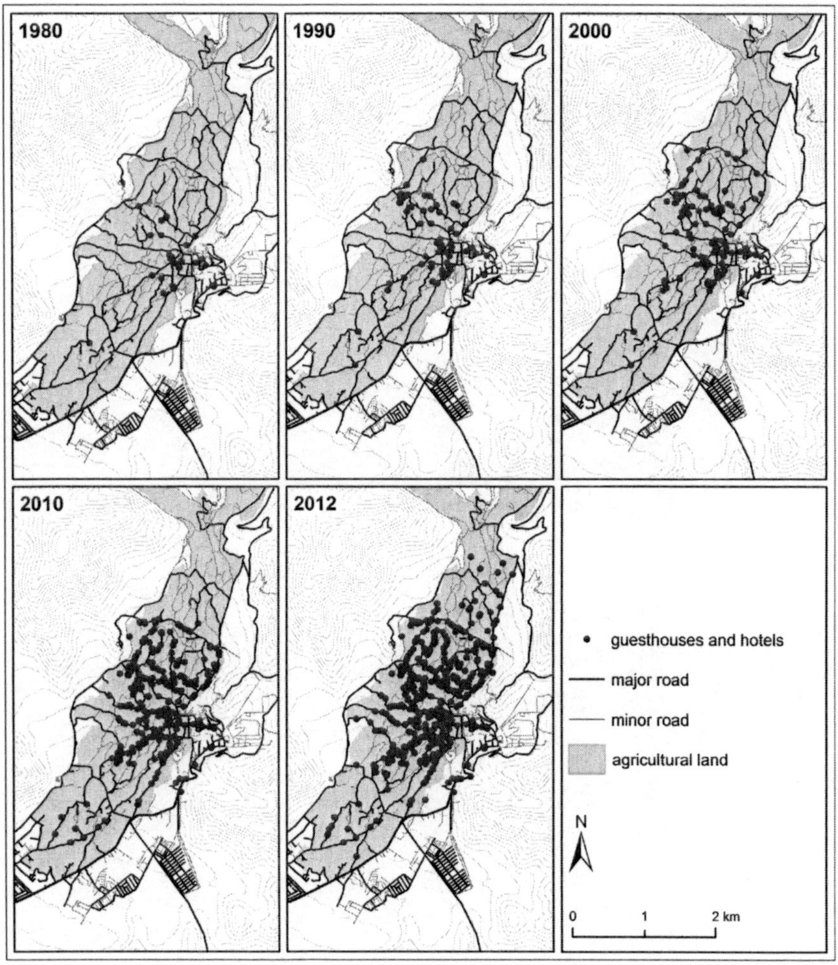

Fig. 6. Increase in Hotels and Guesthouses in Leh Town from 1980 to 2012.

We found that 90 percent of hotels and guesthouses in Leh are located in its predominantly agricultural land area (Fig. 6; Table 1).

The substantial increase in tourists in Leh signifies a huge increase in water demand. Hotels and guesthouses strive to provide flush toilets and showers to increase their rating and, thus, their overnight prices, according to our interview surveys. Consequently, on top of the official extraction of groundwater by PHE, the questionnaire survey revealed that over half of

Table 1. Distribution of Point Sources of Water Pollution in 2012 in Agricultural Wards.

Ward ID	Ward Name	Ward Type	No. of Hotels (*H*) or Guesthouses (*GH*)	No. of Restaurants (*R*)	No. of Pollution Points (*PP*)	No. of *PP* Within 100 Metres of a River/Stream	% of Total *PP* Within 100 Metres of a River/Stream	Relation of No. of *GH*/*H* + *R* to % of *PP* Within 100 Metres of a River/Stream
1	Sanker	Inner	57	0	26	19	73	1
2	Sheynam I	Inner	17	0	32	28	88	0
3	Sheynam II	Inner	22	4	55	24	44	1
4	Tukcha	Inner	103	18	97	66	68	2
5	Karzoo	Inner	110	17	36	24	67	2
6	Skara I	Outer	6	2	22	5	23	0
7	Gonpa I	Outer	1	0	0	0		
8	Gonpa II	Outer	4	0	0	0		
9	Gangles	Outer	0	0	0	0		
10	Horzey	Outer	0	0	0	0		
Total			320	41	270	168	62	

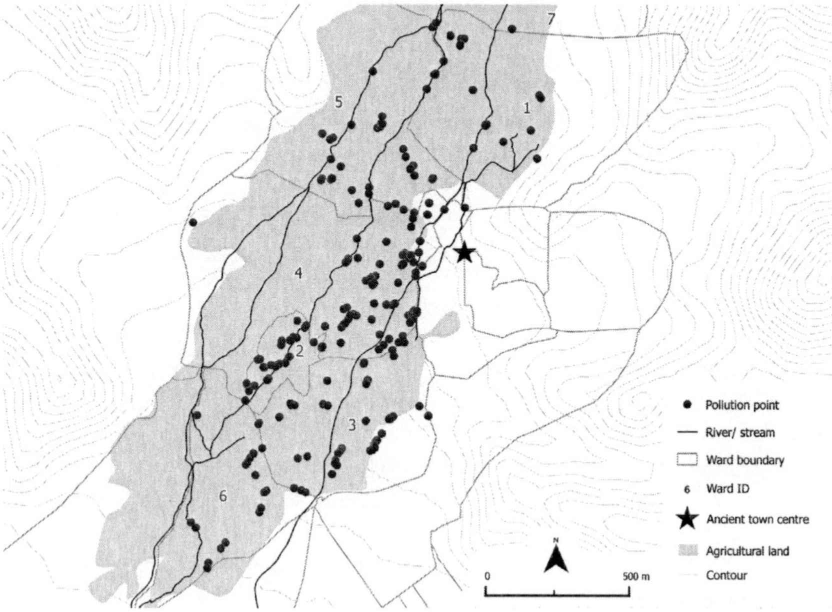

Fig. 7. Distribution of Point Sources of Water Pollution on Agricultural Land Area in 2012.

hotels and guesthouses are using private bore wells in order to extract several thousand litres of groundwater daily in the tourist season. For example, one hotel owner of a hotel with 18 en-suite rooms reported extracting up to 8,000 litres per day during the tourist season.

Focusing on water pollution factors in the 10 wards of Leh Town with predominantly agricultural land, we found the following. As there is currently no sanitation infrastructure or systematic wastewater management existing in Leh, hotels and guesthouses use septic tanks or soak pits to collect black and grey wastewater. Some guesthouses were found to have only soak pits, potentially representing a significant source of effluents and thus groundwater pollution threat. In addition, we mapped a total of 270 point sources of water pollution in the wards with predominantly agricultural land (Fig. 7), which can be divided into three categories (total numbers of each are in brackets):

1. Black water pollution sites including black water inlets (from toilets) (8), public toilets without septic tanks and foul-smelling empty lots

being used as toilets (8) and soak pits other than those of hotels and guesthouses (7).
2. Grey water pollution sites including grey water inlets (other bathroom and kitchen wastewater) (216), clothes (12) and car washing points (1).
3. Garbage dumps (18).

Of all point sources of water pollution, 80 percent are grey water inlets, which is of concern because increasing amounts of chemicals from grey water are being released into the water system in Leh, for example from detergents used for cleaning and washing purposes. Further, 41 restaurants were mapped in wards with predominantly agricultural land (Table 1). We found in our field survey that many restaurants in the agricultural land area of Leh Town are garden restaurants without sanitary infrastructure and only dug pit toilets, so that effluents from restaurants along rivers and streams signify both surface and groundwater pollution threats.

The wards with predominantly agricultural land in Leh Town can be divided into five inner and five outer wards, with inner wards being those directly adjoining the ancient town centre (Fig. 7); a total of 95 percent of hotels, guesthouses and restaurants are located in the inner wards. Consideration of the distribution of point sources of water pollution in terms of proximity to rivers and streams in these 10 wards showed that there are hardly any point sources of water pollution in the outer wards. Rather, the percentage of the total number of point sources of water pollution located within 100 metres of a river or stream was more than twice as high in the inner than in the one outer ward where we found point sources of water pollution. Further, the relation of the number of hotels, guesthouses and restaurants to the percentage of point sources of water pollution within 100 metres of a river or stream was clearly the highest in the two wards Tukcha and Karzoo, which have the largest number of hotels, guesthouses and restaurants. Thus, increase in water pollution may be directly linked to the tourism industry. Overall, 62 percent of point sources of water pollution in the wards with predominantly agricultural land in Leh Town are within 100 metres of rivers and streams (Table 1).

In terms of people's perception of water issues, the questionnaire survey revealed that 98 percent of households thought that drinking water quality is good and safe in Leh. However, 49 percent of households thought drinking water quality today is worse than 10 years ago. Thirty-five percent of households reported having problems with their drinking water in terms of smell, taste or colour: some households mentioned that tap water tastes bitter in the mornings, has a bad taste or smell or has a muddy or greenish colour. Several households reported that groundwater from bore wells is hard and dark in colour, and one household mentioned that groundwater darkens the boiling pot. In terms of drinking water pollution, 43 percent of households pointed out the relation to garbage dumping. Lack of adequate sanitation system, that is, septic tanks or soak pits, were thought by 31 percent of households to be the main source for groundwater pollution. Increased use of chemical fertilizer in agriculture was also perceived as a water quality threat. Forty percent of households thought drinking water pollution is related to diarrhoea. Thus, we found drinking water pollution to be a serious social-ecological concern of the local population.

DISCUSSION: IMPLICATIONS FOR URBAN PLANNING

The results provided above serve as a starting point for demonstrating that water pollution may be a serious environmental issue in Leh with expected health impacts. What is more, the interlinkages of factors influencing water pollution in Leh are quite complex. A better understanding of these interrelationships is needed in order to link urban planning and health concretely. Further, in addition to the actual water quality, the perceptions of water quality held by local inhabitants and tourists plays a vital role in how water is utilized and the challenges are understood. Importantly, over a third of the population thinks that drinking water quality is not as good as they would wish it would be (Section *Water Extraction and Pollution*). According to our interview survey, most tourists drink imported bottled water. As a result, plastic bottle waste is another of the major environmental issues in Leh.

Acute diarrhoeal disease seems to be a serious health concern in Leh. The data only portrays the cases that were reported to government health institutions and were categorized as acute (Section *Health Issues*) and thus the possibility exists that there are many other cases that are not categorized as acute and are not appearing in these figures. Overall, to date, concrete evidence is lacking that diarrhoeal disease can be linked to groundwater pollution in Leh. Acute diarrhoeal disease can have many causes besides water pollution including stress, diet, altitude sickness, basic hygiene practices such as washing ones hands and others. Further, what is termed diarrhoea is very much subject to cultural perception. In many development contexts, diarrhoea may not be regarded as a disease but as a fact of life.

Also, in the minds of the local population, diarrhoea may not be directly associated with water pollution. For instance, according to our interview survey, in winter, vegetables that are grown in greenhouses with too much fertilizer cause indigestion, and the same phenomenon occurs in the summer but in relation to imported fruit. However, overall we would expect to see a health issue like diarrhoea decrease and not increase with economic growth. In the light of such a pattern, a more in-depth spatial analysis of possible causes of water pollution may help explain existing patterns in the data.

Groundwater extraction is not regulated in Leh, and the total number of bore wells, rates of extraction and groundwater aquifer levels are currently not known. The main concern of PHE, according to the interview survey, is how to procure more water from glacial, groundwater and river water sources. An increasing percentage of hotels and guesthouses are installing showers and flush toilets (Section *Water Extraction and Pollution*); although, according to the interview survey, inhabitants think that some springs in Leh seem to have dried up because of high rates of groundwater extraction. There is so far no strategy tackling water issues in Leh from the demand side. Negligible rainfall, decrease in surface water and irrigated agricultural land due to hotel and guesthouse construction, added to the sum of PHE and private extraction by hotels and guesthouses, may mean that groundwater is being depleted faster than the rate of recharge in Leh. Further spatial investigation of distribution of water resources and water demand may help to estimate Leh's carrying capacity in terms of water resources. Overall, the type and distribution of water pollution in Leh (Section *Water Extraction and Pollution*) indicates further environmental planning is needed.

In order to deal with increasing amounts of wastewater, LAHDC plans to implement a centralized sanitation system by 2040. However, this may require even more water resources in order to flush long pipes, which will require more energy for extraction. Furthermore, the system will not be operable in the harsh winters due to pipes freezing and it may entail high maintenance costs due to the mountainous topography. Nonetheless, a centralized sanitation system is a large-scale infrastructure investment opportunity for the government. In-depth understanding of stakeholder power distribution is needed to approach water-related health impacts through alternative urban development scenarios. One hotel in Leh is currently already implementing a decentralized wastewater treatment plant in order to treat its own wastewater locally, due to environmental considerations. An overall decentralized sanitation system in Leh may help to conserve water resources as well as groundwater quality. For example wastewater could be treated and channelled back to the aquifer or used in agriculture locally with less seepage due to short pipes. However, to utilize a decentralized sanitation system effectively will require significant increase in awareness of inhabitants and tourists to initiate change in social behaviour, for example the use of Ecosan (urine separation) toilets, or using less water for showers.

Environmental challenges in Leh may increase dramatically if adequate steps to deal with the rising water demand and disposal of resulting wastewater are not taken soon, potentially representing a very serious ecological threat to this fragile Himalayan mountain region and a growing health threat to residents and visitors. Integrated urban planning is urgently needed to avert potential future bottlenecks in terms of safe drinking water provision. For this, GIS can be a powerful decision-making support tool in visualizing complex water and health interlinkages and help highlight possible paths of action for decision makers. For example, GIS can be used to systematically plan the distribution of drinking water access points and wastewater disposal. Based on the insights gathered from Phase One of our project, Urban Water Health Leh, we are now aiming to support the development of an integrated urban planning strategy with participatory discussions in Leh until April 2015. An inception workshop was held in August 2012 attended by representatives of the local population, LAHDC, the health sector, and other NGOs. Further, a senior advisory board of local stakeholders was set up with whom joint international peer-reviewed publications are being written. Regular stakeholder workshops and focus group discussions will be held to discuss project outcomes and potential implications. Our project aims to create a GIS-based decision-support system in Leh as a basis for participatory discussion on integrated urban planning. In order to help operationalize this, we are conducting GIS capacity building of LEDeG staff using the open ware Quantum GIS (QGIS; http://www.qgis.org). We hope that through this action-oriented process, the project can contribute to Eco-town planning in other towns and for small and medium-sized cities in the region.

One approach could be to analyse the interlinkages between water and health issues to help to demonstrate to decision makers the advantages of sanitation infrastructure investments enabling adequate and clean drinking water supplies. The premise is that the cost of infrastructure in order to prevent health impacts is lower than continuous investment into health impact mitigation, the basic public health argument. Thus, future phases of this study will take an interdisciplinary approach to linking urban planning and health through alternative water management scenarios in Leh. The intention is to empower people to "discus and action oriented ..." in order to discuss an action-oriented research approach to alleviating water related human health impacts in small and medium-sized towns in India.

CONCLUSION

Whereas there seems to be a link between the tourism industry and economic growth in Leh, the authors believe tourism may actually have not only positive but also direct negative impacts, for example by impeding access to safe drinking water. Further, environmental decline may eventually impact the tourism industry. Thus, if diarrhoea in Leh is indeed linked to water pollution, neglect of water pollution issues may be undermining the government's current goal of achieving optimal growth of the tourism industry. As long as short-term economic goals prevail, developing enough awareness of water issues in Leh to change water consumption patterns will be difficult.

The water and health interlinkages indicated in this study amount only to circumstantial evidence. However, as far as possible, causal connections of health impacts in Leh have to be identified in order to direct intervention strategically so that authorities can get the maximum return on their investments. To our knowledge, the centralized sanitation system is being planned in Leh without systematic study on whether this is the best option in terms of water resources conservation. There may be unexplored alternative development choices and opportunities in Leh as a water-deficient tourist destination. A groundwater quality assessment, not within the scope of our study, is currently being conducted in Leh, but the results have not yet been published. Further research is needed on how water resource availability is being impacted by climate change, the percentage of water-pollution-related health impacts on local inhabitants and tourists and their cost to the government health care system. This study calls for further research on whether a decentralized or hybrid sanitation system with appropriate sanitation technology could be a suitable option for Leh and draws attention to the practical importance of integrated, cross-sectoral approaches to water and health governance and urban planning.

ACKNOWLEDGEMENTS

We thank our research partner Ladakh Ecological Development Group (LEDeG), Leh, India, for organizing and carrying out data collection. We thank the editor and three anonymous reviewers for their very valuable comments. This research is supported by a Marie Curie International Reintegration Grant within the 7th European Community Framework

Programme (PIRG06-GA-2009-256555) and the German Research Foundation (DFG) (KE 1710/1-1), and is conducted in collaboration with the International Centre for Integrated Mountain Development (ICIMOD).

REFERENCES

Akhtar, A. (2010). *Tourism and water resources in Leh Town (NW-India): Analysis from a political ecology perspective*. Master thesis in Geography, Ruprecht-Karls-Universität Heidelberg, January 2010, pp. 116.

Albuquerque, C. de (2012). *Address on World Toilet Day*. United Nations Special Rapporteur on the human right to safe drinking water and sanitation. Retrieved from http://www.ohchr.org/EN/NewsEvents/Pages/DisplayNews.aspx?NewsID = 12806&LangID = E. Accessed on November 29, 2012.

Banerjee, T. (2009). U.S. planning expeditions to postcolonial India: From ideology to innovation in technical assistance. *Journal of the American Planning Association*, *75*(2), 193–208.

Barton, J., & Pretty, J. (2010). Urban ecology and human health and wellbeing. In K. J. Gaston (Ed.), *Urban ecology* (p. 319). Cambridge, UK: Camrbidge University Press.

Bashin, V. (1999). Leh – An Endangered City? *Anthropology*, *1*(1), 1–17.

Bhattacharya, R., & Sanyal, K. (2011). Bypassing the squalor: New towns, immaterial labour and exclusion in post-colonial urbanisation. *Economic & Political Weekly*, *xlvi*(31).

Bork, T., Butsch, C., Kraas, F., & Kroll, M. (2009). Megastaedte: Neue Risiken fuer die Gesundheit. *Deutsches Aerzteblatt*, *39*, 1609. (in German).

Bose, R. K., & Leitmann, J. (1996). Environmental profile of the Singrauli region, India. *Cities*, *13*(2), 71–77.

Charron, D. F. (2012). Ecohealth: Origins and approach. In D. F. Charron (Ed.), *Ecohealth research in practice: innovative applications of an ecosystem approach to health* (p. 282). Ottawa, Canada: International Development Research Centre.

Dannenberg, A. L., Jackson, R. J., Frumkin, H., Schieber, R. A., Pratt, M., Kochtitzky, C., & Tilson, H. H. (2003). The impact of community design and land-use choices on public health: A scientific research agenda. *American Journal of Public Health*, *93*(9), 1500–1508.

Dasgupta, P. (2004). Valuing health damages from water pollution in urban Delhi, India: A health production function approach. *Environment and Development Economics*, *9*(1), 83–106.

Dupont, V. (2007). Conflicting stakes and governance in the peripheries of large Indian metropolises – An introduction. *Cities*, *24*(2), 89–94.

Durga Rao, K. H. V. (2005). Multi-criteria spatial decision analysis for forecasting urban water requirements: A case study of Dehradun city, India. *Landscape and Urban Planning*, *71*, 163–174.

Eichert, D. (2009). *Die historische Altstadt von Leh: Wandel und Sanierung eines historischen Stadtquartiers im indischen Trans-Himalaya*. Master thesis in Geography, Ruprecht-Karls-Universität Heidelberg, Sept. 2009, pp. 117.

Frank, L. D., Andresen, M. A., & Schmid, T. L. (2004). Obesity relationships with community design, physical activity, and time spent in cars. *American Journal of Preventive Medicine*, *27*(2), 87–96.

Frumkin, H. (2002). Urban sprawl and public health. *Public Health Reports*, *117*, 201–217.

Frumkin, H. (2003). Healthy places: Exploring the evidence. *American Journal of Public Health*, *93*(9), 1451–1456.

Galea, S., & Vlahov, D. (2005). Urban health: Evidence, challenges and directions. *Annual Review of Public Health*, *26*, 341–365.

Gandy, M. (2008). Landscapes of disaster: Water, modernity, and urban fragmentation in Mumbai. *Environment and Planning A*, *40*, 108–130.

Heath, G. W., Brownson, R. C., Kruger, J., Miles, R., Powell, K. E., Ramsey, L. T., & the Task Force on Community Preventive Services. (2006). The Effectiveness of Urban Design and Land Use and Transport Policies and Practices to Increase Physical Activity: A Systematic Review. *Journal of Physical Activity and Health*, *3*(1), S55–S76.

Hoehner, C. M., Brennan, L. K., Brownson, R. C., Handy, S. L., & Killingsworth, R. (2003). Opportunities for integrating public health and urban planning approaches to promote active community environments. *American Journal of Health Promotion*, *18*(1), 14–20.

Howard, G., & Bartram, J. (2003). Domestic water quantity, service level and health. In *Publication of World Health Organization* (p. 33). Geneva, Switzerland: WHO Press.

Jackson, L. E. (2003). The relationship of urban design to human health and condition. *Landscape and Urban Planning*, *64*, 191–200.

Jackson, R. J. (2003). The impact of built environment on health: An emerging field. *American Journal of Public Health*, *93*(9), 1382–1384.

Kulabkar, P. (2002). The politics of implementing urban plans in India. *IDPR*, *24*(1), 77–103.

LEDeG. (2010). *Water in Leh: A drop in the desert*. Ladakh Ecological Development Group brochure, Leh, Ladakh, India.

Marcotullio, P. J. (2007). Urban water-related environmental transitions in Southeast Asia. *Sustainability Science*, *2*, 27–54.

Naess, P. (2006). *Urban structure matters* (p. 328). Abingdon, England: Routlegde.

Northridge, M. E., Sclar, E. D., & Biswas, P. (2006). Sorting out the connections between the built environment and health: A conceptual framework for navigating pathways and planning healthy cities. *Journal of Urban Health*, *80*(4), 556–568.

Owen, L. A., Caffee, M. W., Bovard, K. R., Finkel, R. C., & Sharma, M. C. (2006). Terrestrial cosmogenic nuclide surface exposure dating of the oldest glacial succesions in the Himalaya orogen: Ladakh Range, northern India. *Bulletin of the Geological Society of America*, *118*, 5/6.

Pacione, M. (2006). City profile Mumbai. *Cities*, *23*(3), 229–238.

Plan. (2012). *World Toilet Day: Children lead the way. Plan: Promoting child rights to end poverty organisation*. Retrieved from http://plan-international.org/about-plan/resources/news/world-toilet-day-children-lead-the-way. Accessed on November 27, 2012.

Rieger-Jandl, A. (2005). Living culture in the Himalayas. *Anthropological guidelines for building in developing countries*. Facultas, Wien, pp. 263.

Revi, A. (2008). Climate change risk: An adaptation and mitigation agenda for Indian cities. *Environment and Urbanization*, *20*, 207–229.

Sakdapolrak, P., Seyler, T., & Prasad, S. (2011). Measuring the local burden of diarrhoeal disease among slum dwellers in the megacity Chennai, South India. In A. Krämer, M. H. Khan & F. Kraas (Eds.), *Health in megacities and urban areas* (pp. 87–100). Berlin: Physica-Verlag.

Sandhu, R. S. (Ed.). (2003). *Urbanization in India* (p. 258.). New Delhi: Sage Publications.

Singh, H. (1998). Economy, society and culture – Dynamics of change in Ladakh. In I. Stellrecht (Ed.), *Karakoram – Hindukush – Himalaya: Dynamics of change, Part 2* (pp. 351–366). Cologne, Germany: RüdigerKöppe Publishers.

Skeldon, R. (1985). Population pressure, mobility, and socioeconomic change in mountainous environments: Regions of Refuge in comparative perspective. *Mountain Research and Development*, *5*, 233–250.

Sridhar, K. S. (2010). Impact of land use regulations: Evidence from India's cities. *Urban Studies*, *47*(7), 1541–1569.

Vedula, A. (2007). Blueprint and reality: Navi Mumbai, the city of the 21st century. *Habitat International*, *31*, 12–23.

Vörösmarty, C. J., Green, P., & Lammers, R. P (2000). Global water resources: Vulnerability from climate change and population growth. *Science*, *289*, 284–288.

Wang, L., Kundu, R., & Chen, X. (2010). Building for what and whom? New Town development as planned suburbanization in China and India. *Research in Urban Sociology*, *10*, 319–345.

White Johansson, E., & Wardlaw, T. (2009). *Diarrhoea: Why children are still dying and what can be done* (p. 68). Geneva: WHO Press.

Wilson, A., Seal, J. L., McManigal, L. A., Lovins, L. H., Cureton, M., & Browning, W. D. (1998). *Green development: Integrating ecology and real estate* (p. 552). New York, NY: Wiley.

World Health Organization (WHO). (2002). *Environmental health in emergencies and disasters: A practical guide*. In B. Wisner & J. Adams (Eds.), p. 272. Geneva, Switzerland: World Health Organization. Retrieved from http://whqlibdoc.who.int/publications/2002/9241545410_eng.pdf. Accessed on July 25, 2013.

World Health Organization (WHO). (2013). *Diarrhoeal disease*. Fact sheet N°330. April 2013. Retrieved from http://www.who.int/mediacentre/factsheets/fs330/en/index.html. Accessed on April 23, 2013.

THE ECOLOGY OF DYING: COMMODITY CHAINS, GOVERNANCE, AND THE MEDICALIZATION OF END-OF-LIFE CARE

Christine Vatovec, Laura Senier and Michael Mayerfeld Bell

ABSTRACT

Purpose – *Millions of people die of chronic diseases within inpatient settings annually in the United States, despite patient preferences for dying at home. This medicalization of dying has received social and economic critiques for decades. This chapter offers a further analysis to these critiques by examining the ecological impacts of inpatient end-of-life care on the natural environment and occupational and public health.*

Methodology – *We compare the ecological health outcomes of medical care in three inpatient units (conventional cancer unit, palliative care ward, and hospice facility) using ethnographic observations, semi-structured interviews, and institutional records on medical supply use, waste generation, and pharmaceutical administration and disposal.*

Ecological Health: Society, Ecology and Health
Advances in Medical Sociology, Volume 15, 195–215

ISSN: 1057-6290/doi:10.1108/S1057-6290(2013)0000015013

Findings – *Care provided on all three medical units had significant socioecological impacts. Cumulative impacts were greatest on the conventional unit, followed by palliative care, and lowest on the hospice unit. Variations in impacts mirrored differences in dependence on material interventions, which arose from variations in patient needs, institutional policies, and nursing cultures between the three units.*

Practical implications – *Social and economic concerns have been major drivers in reforming end-of-life medical care, and our analysis shows that ecological concerns must also be considered. Transitioning terminal patients to less materially intensive modes of care when appropriate could mitigate ecological health impacts while honoring patient preferences.*

Originality – *This chapter describes how the medicalization of dying has converged with institutional policies, practices, and actors to increase the negative consequences of medical care, and recognizes that the far-reaching impacts of clinical decisions make the provision of medical care a socioecological act.*

Keywords: Cumulative life-cycle analysis; environmental justice; ethnography; occupational health exposures; qualitative methods

INTRODUCTION

A century ago, death and dying mainly occurred at home under the care and observation of family, friends, and neighbors (Aries, 1974). Rapid advances in the science and standardization of medicine, combined with the social changes of industrialization that brought laborers from rural to urban areas, have led hospitals to become a primary site for end-of-life care (Starr, 1982). Post-World War II scientific and technological advancements further medicalized the dying process by improving chances to fight disease and prolong death (Connelly, 1997). This displacement of the dying to hospitals first met with criticism in the late 1960s with landmark studies that exposed the social isolation, anxiety, mistrust of staff, and withholding of terminal diagnoses that dying patients often experienced in hospitals (Glaser & Strauss, 1965; Sudnow, 1967). More recently, end-of-life care has undergone criticism that the high cost of care does not change the outcome of disease: patients undergoing intensive medical interventions still die in the same amount of time and with lower quality of life than those who forgo such treatments (Institute of Medicine, 1997). Furthermore, research suggests that up to 81% of all patients and up to 90% of cancer patients would prefer to die at home (Higginson & Sen-Gupta, 2000), but only 24–30% actually do (National Center for Health Statistics, 2011). These social and economic concerns regarding the medicalization of dying have been major drivers in proposals to reform end-of-life medical care (*ibid.*). Our analysis shows that ecological concerns also warrant discussion.

The aim of this paper is to argue for the inclusion of ecological concerns into health care practices, policies, and decision-making by (1) providing a comparative analysis of the socioecological consequences of care provided in three end-of-life impatient settings in the United States (conventional cancer care, palliative care, and hospice), and (2) examining the social factors that govern differential impacts in order to reveal points of intervention.

ECOLOGICAL IMPACTS OF MEDICINE

A recent health care "greening" movement has focused on reducing the impacts on humans, wildlife, and ecosystems that result from medical facilities and operations (see Jameton & Pierce, 2001). Organizations such as Health Care Without Harm have developed tools and best practices for addressing the environmental consequences of health care through environmentally preferred purchasing, energy efficiency, waste management, and similar programs (Health Care Without Harm, 2013). While these efforts have done much to mitigate the consequences of health care facilities and operations, the driving forces behind the provision of medical care – clinical practices, policies, and decision-making at the patient bedside – have not been analyzed for their socioecological impacts.

At the patient bedside, medical care practices impact local and global workers, communities, and ecosystems through life-cycle processes of commodity chains. In other words, the medical supplies and pharmaceuticals that support patient care each have an environmental legacy resulting from their extraction as natural resources through to their manufacture, distribution, consumption, and disposal as medical products (Jameton & Pierce, 2001). For example, the life cycles of petroleum-based plastics that are commonly used in medical supplies and packaging have a number of socioecological consequences. Oil extraction pollutes and degrades natural ecosystems throughout exploration, drilling, and transport, and refining phases, and causes long-term harm to wildlife including marine mammals, migratory birds, and fish populations (O'Rourke & Connolly, 2003). During manufacturing, workers who are exposed to vinyl chloride monomer (the building block of polyvinyl chloride used in intravenous (IV) bags and tubing) have an increased incidence of angiosarcoma of the liver (Creech & Johnson, 1974; Lee & Harry, 1974), while communities living near PVC manufacturing plants face health risks from chemical emissions (Markowitz & Rosner, 2002). Further downstream, both workers and communities face health risks from medical waste incineration which releases heavy metals, dioxins and furans, and persistent organic pollutants (Rowat, 1999; Sedman & Esparza, 1991). The persistent, bio-accumulative effects of dioxins released via PVC incineration have long-term impacts on both human and wildlife health (Thornton, McCally, Orris, & Weinberg, 1996). Plastic manufacturing and medical waste incineration also pose environmental justice concerns since these facilities are disproportionately sited in minority communities (Brown, 1995; Bryant & Mohai, 1992). Together, these life-cycle impacts of plastics commodity chains highlight how medical resource use at the patient bedside ties the delivery of health care to the global environment. A key factor in this study is the fact that the provision of medical care requires the simultaneous use of multiple medical supplies and pharmaceuticals, each of which has their own life-cycle impacts. Therefore, our approach examines the cumulative socioecological impacts that result from patient care.

FOCUS ON THE ECOLOGICAL IMPACTS OF END-OF-LIFE MEDICINE

Current medicalized approaches to end-of-life cancer care offer a unique setting for studying the socioecological impacts of medical care for three reasons. First, end-of-life care is an extreme example of expensive, materially intensive care: between a quarter and a third of all Medicare expenditures are spent on the 5% of patients who die each year (Barnato, McClellan, Kagay, & Garber, 2004). Second, these settings serve a large number of patients (1,500 cancer deaths each day in the United States; American Cancer Society, 2010), which suggests potentially high and far-reaching cumulative ecological impacts. Third, three existing medical models for end-of-life cancer care (conventional curative care, palliative care, and hospice) differ in their approach, thereby providing an opportunity for comparative analysis. Conventional curative care offers materially intensive treatments of surgery, radiation, and chemotherapy; the increasing effectiveness of these treatments has resulted in more patients seeking aggressive care right up to the point of death (Earle et al., 2004). Hospital-based palliative care services offer less materially intensive pain and symptom management for patients with life-limiting illness, as reflected by a 38% lower average daily cost per patient compared to conventional services (Elsayem et al., 2004). Likewise, hospice care is characterized by a comparatively low level of material interventions as evidenced by a 27% lower cost than conventional care (Emanuel & Emanuel, 1994), but typically provides end-of-life care external to the hospital. Since the three end-of-life care models differ in their reliance on material interventions, a comparative analysis of these three medical models offers an opportunity to examine the socioecological impacts resulting from common medical practices.

METHODOLOGY

We employed a multisited ethnography (Marcus, 1995) that draws on (1) institutional records from our sites to quantitatively measure impacts of resource use and waste generation, and (2) observations and in-depth interviews to qualitatively measure the factors that govern these impacts. Our approach parallels the "ethical life-cycle analysis" of Pierce and Kirby (1999) by taking into account the cradle-to-grave socioecological consequences of supply chains, and is complimented by the application of a "social metabolic" lens which recognizes that resource use is a function of social organization (Fischer-Kowalski & Hüttler, 1998).

RESEARCH SITES

This study compares common medical practices on three inpatient units: a conventional cancer ward (~30 private patient rooms), a palliative care unit (~10 private rooms), and a hospice center (~30 private rooms). The units were selected because each provided care to terminal cancer patients. The conventional and palliative care units were both located within a teaching hospital in a Midwestern city; the hospice inpatient unit was within a stand-alone facility run by a nonprofit organization in the same city. Each of these settings was situated within an ecological context (historically prairie, in this case) which provided the backdrop for the buildings, roadways, landfill, water treatment plant, and other infrastructure that supported health care delivery.

This study was reviewed and approved by the Institutional Review Board of the University of Wisconsin, and administrators at both the hospital and hospice organization granted written permission to conduct the study. Participants (physicians, nurses, housekeepers, administrative staff, and waste handlers; 73 total) provided written informed consent before entering the study. Patients provided verbal consent to allow the researcher to shadow participating staff members within their rooms. Data were collected while observing the care of 56 conventional cancer patients, 21 palliative care patients, and 56 hospice patients.

DATA COLLECTION

We used two common ethnographic methods to collect qualitative data: participant observation and semi-structured interviews. A single field researcher gathered over 255 hours of observational data from May 2008 to June 2009. Observations were made at each site in two rounds of approximately two months each; the first two months were spent on the conventional unit, the next on the palliative care unit, and the next at hospice. This approach allowed us to gather extensive information at each site, then revisit each site for further observation. We shadowed participants (i.e., physicians, nurses, housekeepers, etc.) for up to eight hours a day, and made observations on all days of the week and all shifts (day, evening, and night). The majority of observations occurred during weekday day-shifts because this is when the most physicians round and patient–physician consultations occur, and consequently when most resource-use decisions are made.

Thirty-three hours of semi-structured interviews with 36 participants (30–90 minutes long) were audio-recorded, transcribed, and coded for analysis. Interviewees were purposefully selected after at least one round at a unit to shed light on observations and to provide insight into clinical resource use decision-making. Interview questions were designed to inquire about common material resources used in each site, ecological health concerns regarding resource use, differences and similarities in end-of-life care models, and factors that govern resource use decision-making (i.e., political economy, cultural norms). A single researcher coded and analyzed all data using inductive, directed content analysis (Hsieh & Shannon, 2005; Miles & Huberman, 1994); data were entered into a digital matrix to facilitate data organization and retrieval for thematic and pattern identification. Unreferenced quotations are drawn directly from interview transcripts.

We obtained institutional records on medical supply use, waste generation, pharmaceutical administration, total patient days, and average length-of-stay from the research sites for the annual period that corresponded with observational data. This quantitative data provided insight into the scope of resource use on a per patient basis, but had a number of limitations for comparing material intensity between the sites. The hospital's conventional and palliative care units were considered a single cost unit, so medical supply data were available only for the two sites combined. Likewise, medical supply data for the hospice unit includes annual usage for both the inpatient hospice facility and hospice services provided to patients within their own homes. Similarly, waste generation data was only available for the entire hospital and hospice facilities (including inpatient units, administrative offices, laboratories, cafeterias, etc.), rather than for our specific research sites. To address this shortcoming, we have augmented our quantitative data with observational data to tease apart variations in medical supply use, waste generation, and disposal practices at each site.

FINDINGS

In this section, we first provide a quantitative comparison of the consumption and disposal of medical supplies and pharmaceuticals at each site, then describe our qualitative findings that explain the policies, practices, and decision-making which govern observed differences in the socioecological impacts of each site.

COMPARISON OF IMPACTS

Medical Supply Consumption and Waste Generation

According to institutional records, the hospice site used the lowest volume of supplies per patient and generated the least waste (Table 1) as compared to the conventional and palliative care sites. Since the hospital where the conventional and palliative care units were located considered the two wards a single cost unit, it was not possible to get separate institutional data

Table 1. Comparison of the (1) Volume of the Three Most Commonly Used Medical Supplies,[a] (2) Volume of Waste Generation,[b] and (3) Number of Doses of the Three Most Commonly Dispensed Pharmaceuticals per Patient Day[c] on the Conventional Cancer, Palliative Care, and Hospice Inpatient Units (2009).

	Conventional		Palliative Care		Hospice	
Medical supply use	Rank	Volume per patient day	Rank	Volume per patient day	Rank	Volume per patient day
Medical gloves	1	67	1	67	1	2
Wound dressings	2	5.5	2	5.5	4	0.4
Intravenous tubing	3	2.5	3	2.5	n/a	n/a
Oral swabs	35	0.01	35	0.01	2	0.8
Incontinence briefs	28	0.04	28	0.04	3	0.4
Waste generation	Rank	Average pounds of waste per patient day (percentage of total)	Rank	Average pounds of waste per patient day (percentage of total)	Rank	Average pounds of waste per patient day (percentage of total)
Total	–	30	–	30	–	14
Municipal waste	1	20 (67%)	1	20 (67%)	1	10 (71.2%)
Recycling	2	6 (20%)	2	6 (20%)	2	4 (28.5%)
Infectious	3	4 (13%)	3	4 (13%)	3	0.04 (0.3%)
Drug (purpose)	Average doses/ patient day (rank)	Average dose	Average doses/ patient day (rank)	Average dose	Average doses/ patient day (rank)	Average dose
Total	20.57	–	10.66	–	16.28	–
Acyclovir (antiviral)	1.23 (1)	519.7 mg	0.03 (63)	11.8 mg	0.01 (127)	6.8 mg
Pantoprazole (stomach acid reduction)	0.78 (2)	31.1 mg	0.32 (8)	12.7 mg	0.34 (12)	13.8 mg
Oxycodone (pain)	0.67 (3)	8.48 mg	0.41 (5)	7.69 mg	0.33 (13)	7.19 mg
Acetaminophen (pain/fever reduction)	0.64 (4)	425.6 mg	0.50 (3)	385.4 mg	0.58 (5)	364.7 mg
Morphine (pain)	0.59 (7)	16.3 mg	1.61 (1)	71.2 mg	2.33 (1)	125.3 mg
Lorazepam (sedative/muscle relaxant)	0.56 (8)	0.35 mg	0.66 (2)	0.51 mg	1.06 (2)	1.02 mg
Docusate-senna (laxative)	0.48 (12)	0.74 Tablet	0.46 (4)	0.92 Tablet	0.74 (3)	1.40 Tablet

[a]Institutional supply data were not available separately for the conventional and palliative care units because these wards operated as a single cost unit at the hospital where they were located. Likewise, data for the inpatient hospice unit were only available combined with in-home supply use.

[b]Waste generation data specific to our research sites were not available; data reflect cumulative waste generated by each facility including patient care areas, cafeterias, administrative offices, etc.

[c]"Patient day" represents the total number of days that all patients were in a medical facility (i.e., 100 patients in hospital for 1 day equals 100 patient days).

detailing the differences in supply use and waste generation for these two sites. However, our observations revealed that patient care on the conventional unit utilized larger volumes of medical supplies, and in turn generated larger volumes of waste, than the palliative care unit. For example, IV bag use per patient was highest on the conventional unit (average of 2.4 IV bags per patient per day), followed by the palliative care ward (0.7), and lowest on the hospice unit (0.07). IV bag use is indicative of supply use intensity and waste generation because each bag represents the use and disposal of a number of other materials including tubing, wound dressings, dressing change kits, medical exam gloves, and sterile packaging. IV bag use is also indicative of socioecological impacts of medical care since these PVC-containing materials are associated with the environmental, human, and wildlife health consequences described previously.

Drug Use and Administration

Pharmaceutical use is difficult to compare across the research settings for a number of reasons. The drugs that work best for managing pain and other symptoms change as diseases progress; the amount of drugs needed for symptom control increases as patients develop tolerance to them, particularly in the case of narcotics (Way, Leong, Loh, & Shen, 1969); and as patients transition from aggressive disease treatment toward symptom management, certain medications are stopped because they no longer make sense to administer (e.g., cholesterol-lowering drugs may no longer be beneficial at end-of-life). In addition, each drug has its own history and life cycle of development, manufacture, and disposal, and each patient has their own requirements for different drugs during the course of their treatment. Given these difficulties, we examine the socioecological impacts of pharmaceuticals across the three research units in general terms, by focusing on the total number of drugs administered on each unit, the average number of drugs administered per patient day (indications of occupational nursing exposures and volume of potential upstream and downstream impacts), and known concerns regarding occupational exposures to specific pharmaceuticals.

According to both institutional records and observations, the conventional unit dispensed more drugs than the other two units in terms of the largest number of drugs dispensed in total, largest average number of doses and largest volume of drugs administered per patient day. In 2009, 577 different drugs were dispensed to patients on the conventional unit, 428 on the hospice unit, and 270 on the palliative care unit. Likewise, the largest average number of medication doses per patient day was administered on the conventional unit (20.57 doses; Table 1), followed by the hospice unit (16.28 doses), and the palliative care unit (10.66 doses). In addition, observational data indicated that patients on the conventional unit received between five and six drugs on average each day in addition to chemotherapy, most of which were intended to alleviate chemotherapy side effects. Palliative care and hospice patients received an average of about two medications per patient per day. Based upon the sheer number and volume of pharmaceuticals administered, it appears that the conventional unit had the largest cumulative socioecological impacts, followed by the hospice unit and finally the palliative care setting.

Of all the drugs administered on our sites, chemotherapy agents appear to have the greatest known occupational health concerns at the point of administration. Many chemotherapy drugs are classified as hazardous substances and exposure to them has negative health impacts including increased incidence of adverse reproductive health outcomes such as miscarriage and stillbirth (Valanis, Vollmer, & Steele, 1999), as well as infertility, premature delivery, and low birth weight (Fransman et al., 2007). Nurses who are exposed to these anticancer agents may also have an increased risk of developing leukemia (Skov et al., 1992). Health concerns are greatest among pharmacists and nurses who handle chemotherapy regularly or their coworkers who come into contact with contaminated surfaces (Sessink, Boer, Scheefhals, Anzion, & Bos, 1992). Our observations confirmed that while occupational exposure to chemotherapy varied across settings, nurses on the conventional unit had the greatest occupational exposure risks to chemotherapy based on volume of drugs administered and delivery mechanism (IV administration in the conventional setting versus oral delivery in hospice). Institutional data showed that nurses on the conventional unit dispensed a total of 1,990 doses of fifteen different cancer chemotherapy drugs to their patients in 2009, or about one dose per nurse per day. No chemotherapy drugs were dispensed on the palliative care unit. Nurses on the hospice unit administered a total of 12 doses of 5 different oral chemotherapy drugs (or about one dose per nurse every three months) as palliative treatment.

Pharmaceutical Waste and Disposal

The conventional and palliative care units disposed of pharmaceutical waste via incineration, while the hospice facility relied on a combination of incineration and flushing drugs down the drain. Both methods pose ecological health risks. Pharmaceutical waste generated on the conventional and palliative care units was mingled with various other forms of hazardous waste produced at the hospital and sent to a regulated incinerator in Port Arthur, Texas – a community that has historically been about 80% African American and Latino (Cole, 1994). Toxic ash that resulted from the incineration process was then transported to the largest hazardous waste dump in the United States, located in Emelle, Alabama, a poor, rural community. This landfill was sited in 1978 when there were no minority representatives on the county industrial development board, county commission, or in the state legislature (Bullard, 2000). Though hazardous waste facilities are highly regulated by the Resource Recovery and Reclamation Act and incinerators are equipped with devices to minimize emissions (EPA, 2011), these facilities are associated with a number of public health concerns due to the release of heavy metals and persistent organic pollutants (Rowat, 1999; Sedman & Esparza, 1991). Poor and minority communities living near hazardous waste facilities bear the greatest public health burden of emissions, making medical waste disposal an issue of environmental justice, especially since the minority populations most burdened by pollution are typically underserved by the health care community (Shavers, Klein, & Fagan, 2012).

On the hospice unit, the majority of pharmaceutical waste was also disposed of via incineration using the same hauler contracted by the hospital where our other two sites were located. However, nurses on the hospice unit commonly flushed unused portions of narcotics down the drain. Little is known about the human health effects of exposure to drugs that enter the water system, but there are some concerns regarding long-term human health exposure to the cocktail of pharmaceutical residues in drinking water that treatment plants are unable to filter out (Daughton, 2003). The environmental impacts of certain types of drug waste in aquatic ecosystems have received greater attention, including reproductive problems in fish exposed to medications, particularly from birth control pills (Corcoran, Winter, & Tyler, 2010).

Surprisingly, we learned that hospice nurses who provided care to patients in their own homes flushed all pharmaceutical waste down the drain. Although this was not the standard practice on our hospice inpatient unit, it is particularly important to note since about 40% of hospice patients received in-home care (the remaining 60% received hospice care in nursing homes or other medical facilities). Only 6% of all hospice patients were ever admitted for short-term acute care to the hospice inpatient unit where unused drugs were disposed of by incineration. As a result, we presume that a large volume of unused pharmaceutical products entered the local water system, particularly since nurses reported flushing large quantities of medications that had accumulated over years of conventional care, but no one can quantify the amount of drugs that were flushed. This suggests that the cumulative impacts of down-the-drain pharmaceutical disposal from in-home hospice care are relatively large.

Cumulative Impacts of Medical Supply and Pharmaceutical Use and Disposal

The cumulative socioecological impacts that resulted from medical supply and pharmaceutical use and disposal across our three sites depend upon the number of days that patients remain within each setting. The target length-of-stay for patients on each unit was as follows: conventional care, 11 days (ranging from 3 days for oncology patients to 20 days for bone marrow transplant patients); palliative care, 3 days; hospice, 3 days. As a result, conventional care had the largest cumulative impacts of medical supply use, waste generation, and pharmaceutical use, while the differences between palliative care and hospice are less clear since resource use and waste generation were similar on these sites.

SOCIAL GOVERNANCE OF IMPACTS

Medical Supply Consumption and Waste Generation

Observed differences in supply use intensity and waste generation across our sites are not surprising since conventional patients typically undergo more medical interventions than palliative care or hospice patients. A physician on the conventional ward contrasted the settings by saying, "Conventional care means more imaging, more blood tests, chemotherapy oftentimes. And anytime you have an intervention that you do, it means everything else that's associated with it – bloods, and scans, and x-rays, and catheters, and IV's, and you know, everything else. And there are just more interventions with conventional care. So more interventions means more of everything else."

However, we identified nursing practices and infection control policies that combined to exacerbate the volume of medical supply use and waste generation on the conventional and palliative care units. Nursing practices on the conventional and palliative care units focused on emergent response to patient needs: when a patient called for help, nurses quickly selected a range of medical supplies to bring into the room in response to the situation, though typically only a small portion of the supplies were required. Infection control policies required any material that entered a patient's room to be disposed of because after crossing the threshold into the room it was considered "contaminated." Infection control was considered of utmost importance at each of our research sites, and such policies are mandated and enforced by national accreditation organizations (e.g., Joint Commission) to help reduce the transmission of hospital-acquired infections which result in medical costs ranging from $3.2 to $7.3 billion annually (Roberts et al., 2010), and nearly 100,000 deaths (Klevens et al., 2007).

Interestingly, nurses who divided their time between the palliative care and conventional units continued the more intensive habit of resource use despite palliative care's purportedly less aggressive interventional stance. Despite there being fewer "emergencies" on the palliative care unit, nurses found it hard to break the habit of bringing extra supplies with them when responding to patient calls. As a result, a large volume of unused medical supplies were disposed of on both the conventional and palliative care units. For example, following a patient's death on the palliative care unit, the room-cleaning process generated three 30-gallon bags of garbage. Two bags were filled with the waste of supplies used to treat the patient. The third bag was full of unopened supplies that had been brought into the room during an emergent situation, were not used, but were discarded because they were considered "contaminated."

In contrast, practices and policies converged to minimize supply use and waste generation at the hospice site. A nurse on the hospice unit said, "in hospice, we say there's no such thing as an emergency. Even though our patients are very ill and need very intense care, it's different from the care you would give in a hospital." As a result, hospice nurses were more likely to first assess a patient's situation and then retrieve supplies. For example, after the death of a hospice patient, a much smaller volume of waste (less than one 30-gallon garbage bag) was generated when the room was cleaned. Additionally, infection control policies on the hospice unit had evolved to designate a "clean" area in each patient room where any supply could be placed and, if unused, returned to the supply closet (a policy that was, in part, possible at the hospice facility because patient rooms were more spacious than in our other two sites).

We did, however, observe instances on the hospice unit when supplies met criteria for being "contaminated" by entering patient rooms without being placed in the "clean" area, but rather than being disposed of, they were donated to a local nonprofit agency for distribution to medical centers in developing countries. Nurses often opened bags of adult diapers in patient rooms, and stored the unused portion of diapers in the room (not in the "clean" area) for later use. Any unused diapers that remained after the patient died or was discharged were picked up by the donation organization. This supply donation program diverted nearly one ton of medical supplies from the hospice's waste stream each year, but raises an ethical dilemma concerning the trade-off between ecological impacts of waste disposal and potential health impacts of sending potentially contaminated materials to developing countries where health care systems are poorly resourced.

Drug Use and Administration

Occupational exposures to chemotherapy on the conventional unit were exacerbated by two factors: institutional policies that endorsed chemotherapy preparation at the bedside, and nurses' concerns that protective gear decreased patient satisfaction. A conventional unit pharmacist told us that, regarding IV chemotherapy delivery, "one of the most hazardous steps is actually the nurses attaching the tubing to the chemotherapy bag because you're essentially poking a sharp object into the bag. So we've pushed for having that done in the hood in the central pharmacy where it's a controlled setting and any spill would be contained and the technician always has protective gear on." Instead, the institutional policy and practice was for nurses to prepare IV bags on the inpatient unit, which had resulted in a few spills that exposed nurses, patients, and housekeepers to the toxic agents. We learned from a unit pharmacist that the push for safer chemotherapy preparation had met with resistance because of the increased technical labor required for such a protocol.

Institutional policies were in place to protect nurses when handling IV chemotherapy by requiring them to wear certain protective equipment (e.g., chemotherapy-rated gowns and exam gloves), and recommending the use of other equipment (e.g., face masks). However, some nurses reported to us that they did not wear all of the recommended personal protective equipment out of concern for patient satisfaction. One nurse stated, "… the way the drug is administered can affect the way the patient feels about what you're putting into their body…If you walk into the room with goggles and gown and mask, and they're like okay this is what you're putting in me and you can't even come into contact with it?" Nurses' concerns over patients' psychological impacts from protective equipment have been reported previously as a barrier to preventing occupational exposure to antineoplastic agents (Eisenberg, 2009), but ways to mitigate this barrier have received little attention in the literature.

Pharmaceutical Waste and Disposal

In both the inpatient and in-home hospice settings, down-the-drain pharmaceutical disposal emerged as a way for the hospice institution to remain in compliance with Drug Enforcement Agency (DEA) regulations to prevent narcotic diversion (see Office of National Drug Control Policy, 2009). When asked about the potential impacts of flushing narcotics down the drain, a hospice pharmacist said, "The DEA scares us more than the EPA. They have guns and can put you in jail. The EPA might fine you."

Cumulative Impacts and Decision-Making

This section reports the social factors that prevent patients from transitioning to palliative care and hospice when medically appropriate and patient preferred. As stated above, previous studies have illustrated that the majority of patients (cancer patients, in particular) would prefer to die at home (Higginson & Sen-Gupta, 2000), but only a small portion actually do (National Center for Health Statistics, 2011). Our ethnographic data indicates five barriers that prevent patients from shifting from conventional care to less materially intensive modes of care.

First, clinicians reported that making a terminal diagnosis is difficult because prognostication is difficult. In reference to the Medicare hospice rule that requires physicians to certify that a patient may likely die within six months, one palliative care nurse practitioner stated, "When is end-of-life? When is six months?" Second, the goals of care within the conventional model make it difficult for physicians to see the value of transitioning to palliative options unless the patient is actively dying. A conventional physician who had recently received palliative care training illustrated this point by saying, "I think most health care providers don't think of palliative care as a legitimate option for patients. It's okay if you're dying, but if you're not dying the default pathway should always be toward life prolonging treatment." As a result, participants reported that conventional physicians do not have conversations to determine patients' goals of care and whether palliative care might be a good option.

Third, conventional physicians are not trained to discuss end-of-life goals with patients, so even though a patient may be appropriate for palliative care or qualify for hospice they may not realize they are terminally ill. According to a palliative care physician, conventional doctors, "won't bother having this conversation with the patient. They'll just refer [terminally ill patients] to me and let me have the conversation … I worry that we're not training [physicians] to have these conversations." In addition, a hospice physician stated, "if you look at referrals to hospice I would guess there are pretty strong patterns of certain docs who do refer because they have more skill at having those conversations, they're more comfortable saying this [curative] treatment is really not likely to help you and it may cause your quality of life to be worse as you go through the last few months of your life." Fourth, even if a conventional physician is trained in end-of-life conversations there is a systemic lack of time for holding such discussions. An oncologist who had palliative care training stated, "If you have a patient who is actively getting chemotherapy for lung cancer, and it may be someone who is dying from their disease, there's a lot you have to do to make sure it's safe to give chemotherapy, to assess their ability to tolerate it, to dose it, to do all of that stuff. And if you have 20, 30 minutes with the patient, that's going to take up the bulk of that time. And so you don't have a lot of time to explore [their preferences]."

Finally, there is a general lack of understanding about palliative care and hospice which prevents patients from enrolling in these modes of care early enough in their disease progression to benefit from these options. According to a palliative care physician, "Patients almost universally don't know what palliative care is … I think practitioners have the same confusion – 'you're going to get my patient to forgo treatment?'… As opposed to really just having us try to provide options." Another palliative care physician said that this model of care will, "always be considered part of the death squad but the reality is that we deal with anyone who has life limiting illness…where the goal is comfort and quality of life. I have patients who probably have years to live who are getting palliative care. So they're not eligible for hospice, but they [are no longer interested in curative treatments]."

CONCLUSION

Each of the three end-of-life medical models examined here had significant socioecological consequences resulting from medical supply and pharmaceutical use that impacted local, regional, and global ecosystems through pollution and degradation from commodity chain life-cycle processes, and occupational and public health exposures to hazardous materials, many of which disproportionately affect disadvantaged social groups. However, the volume of material interventions used in each setting led to differences in cumulative socioecological impacts between the three sites. As expected, the conventional cancer care inpatient unit exhibited greater intensity of medical supply use, waste generation, and pharmaceutical administration than the palliative care and hospice units. The impacts of waste generation were higher than expected on the palliative care unit due to a combination of nursing culture and institutional infection control policies. Also surprisingly, pharmaceutical disposal practices surfaced as a problem in the in-home hospice setting where standard practice for in-home patients called for flushing all medications down-the-drain where they enter surface waters.

More importantly, our findings trace these differential impacts to the governance of institutional policies and practices that produce complex, and often contradictory, pressures to prioritize concerns about some types of adverse impacts over others. For example, institutional policies that required the disposal of unused medical supplies may have prevented the spread of infectious agents, but may also have disproportionately impacted communities located near manufacturing plants and hazardous waste facilities. Likewise, the donation program at the hospice facility decreased waste generation, but raises ethical concerns by sending potentially contaminated medical supplies to developing countries because they are unfit for use in the United States, but better than nothing for other communities. Similarly, down-the-drain pharmaceutical waste disposal allows health care facilities to comply with DEA drug diversion mandates but ignores the ecological impacts associated with such practices. Overall, our findings demonstrate the kinds of impacts resulting from complex institutional pressures, while also making institutional relationships and the assumptions behind them visible.

Policy Implications

Many of the impacts identified here are the product of institutional policies and practices and are therefore amenable to change. At the broadest level, this chapter shows how the complex and often contradictory policies and regulations that health care settings are subject to lead to negative socioecological consequences because those actors that hold the most power do not currently account for the ecological impacts of their mandates. In order to ensure more ecologically minded and broadly ethical quality clinical care, these regulatory and professional agencies need to begin accounting for the socioecological harms that result from clinical practices and decision-making, and coordinate the directives they issue to health care institutions.

In addition, our findings show that practices involved in the sourcing, use, and disposal of medical supplies and pharmaceuticals require particular concern for mitigating the socioecological impacts resulting from health care. The current invisibility of natural and human resources within the commodity chains that support health care raises the need for policies to develop appropriate measures to account for socioecological impacts.

Implications for end-of-life decision-making, in particular, revolve around developing standards and policies for transitioning patients out of materially intensive and ecologically damaging conventional medical settings, and into palliative care and hospice settings when medically and socially appropriate. The median length-of-stay that patients spend in hospice has consistently remained under one month, despite the fact that the Medicare hospice benefit is available for at least six months prior to death (National Hospice and Palliative Care Organization, 2005). Since the majority of patients would prefer to die at home, transitioning patients to less materially intensive settings when medically appropriate has the potential to honor patient preferences, decrease the economic costs of care, and decrease the cumulative ecological health impacts of care. One successful model that has the potential to support this transitioning is the concurrent provisioning of conventional curative care and palliative care, which helps identify patient preferences and gets patients into preferred settings earlier (Byock, Sheils Twohig, Merriman, & Collins, 2007). Policies that support such practices could serve to improve the patient experience of end-of-life care, lower the cumulative costs of care, and act as preventive measures for decreasing the ecological health impacts resulting from the material supply chains that support clinical care.

REFERENCES

American Cancer Society. (2010). *Cancer facts and figures*. Retrieved from http://www.cancer.org/Research/CancerFactsFigures/CancerFactsFigures/cancer-facts-and-figures-2010

Aries, P. (1974). *Western attitudes toward death: From the middle ages to the present*. Baltimore, MD: The Johns Hopkins University Press.

Barnato, A. E., McClellan, M. B., Kagay, C. R., & Garber, A. M. (2004). Trends in inpatient treatment intensity among Medicare beneficiaries at the end of life. *Health Services Research*, *39*(2), 363–375.

Brown, P. (1995). Race, class, and environmental health: A review and systematization of the literature. *Environmental Research*, *69*(1), 15–30.

Bryant, B., & Mohai, P. (1992). *Race and the incidence of environmental hazards: A time for discourse*. Boulder, CO: Westview Press.

Bullard, R. D. (2000). *Environmental racism in the Alabama blackbelt*. Environmental Justice Resource Center. Retrieved from http://www.ejrc.cau.edu/envracismalablackbelt.htm

Byock, I., Sheils Twohig, J., Merriman, M., & Collins, K. (2007). Promoting excellence in end-of-life care: A report on innovative models of palliative care. *Journal of Palliative Medicine*, *9*(1), 137–151. doi:10.1089/jpm.2006.9.137

Cole, L. W. (1994). The struggle of Kettleman City: Lessons for the movement. *Maryland Journal of Contemporary Legal Issues*, *1*, 67–80.

Connelly, R. J. (1997). The medicalization of dying: A positive turn on a new path. *Omega*, *36*(4), 331–341.

Corcoran, J., Winter, M. J., & Tyler, C. R. (2010). Pharmaceuticals in the aquatic environment: A critical review of the evidence for health effects in fish. *Critical Review of Toxicology*, *40*(4), 287–304.

Creech, J. L., Jr., & Johnson, M. N. (1974). Angiosarcoma of the liver in manufacture of polyvinyl chloride. *Journal of Occupational and Environmental Medicine*, *16*(3), 150.

Daughton, C. G. (2003). Cradle-to-cradle stewardship of drugs for minimizing their environmental disposition while promoting human health. I. Rationale for and avenues toward a green pharmacy. *Environmental Health Perspectives*, *111*(5), 757–774.

Earle, C. C., Neville, B. A., Landrum, M. B., Ayanian, J. Z., Block, S. D., & Weeks, J. C. (2004). Trends in the aggressiveness of cancer care near the end of life. *Journal of Clinical Oncology*, *22*(2), 315–321. doi:10.1200/JCO.2004.08.136

Eisenberg, S. (2009). Safe handling and administration of antineoplastic chemotherapy. *Journal of Infusion Nursing*, *32*(1), 23–32.

Elsayem, A., Swint, K., Fisch, M. J., Palmer, J. L., Reddy, S., Walker, P., … Bruera, E. (2004). Palliative care inpatient service in a comprehensive cancer center: Clinical and financial outcomes. *Journal of Clinical Oncology*, *22*, 2008–2014.

Emanuel, E. J., & Emanuel, L. C. (1994). The economics of dying: the illusion of cost savings at the end of life. *New England Journal of Medicine*, *330*(8), 540–544.

EPA (Environmental Protection Agency). (2011). *Hospital/medical/infectious waste incinerators*. Retrieved from http://www.epa.gov/ttn/atw/129/hmiwi/rihmiwi.html

Fischer-Kowalski, M., & Hüttler, W. (1998). The intellectual history of materials flow analysis, part II, 1970–1998. *Journal of Industrial Ecology*, *2*(4), 107–136.

Fransman, W., Roeleveld, N., Peelen, S., Kort, W., Kromhout, H., & Heederik, D. (2007). Nurses with dermal exposure to antineoplastic drugs: Reproductive outcomes. *Epidemiology*, *18*(1), 112–119.

Glaser, B. G., & Strauss, A. L. (1965). *Awareness of dying*. New York, NY: Aldine.

Health Care Without Harm. (2013). *About us*. Retrieved from http://www.hcwh.org/all_regions/about/

Higginson, I. J., & Sen-Gupta, G. J. A. (2000). Place of care in advanced cancer: A qualitative systematic literature review of patient preferences. *Journal of Palliative Medicine*, *3*(3), 287–300.

Hsieh, H. F., & Shannon, S. E. (2005). Three approaches to qualitative content analysis. *Qualitative Health Research*, *15*(9), 1277–1288.

Institute of Medicine. (1997). Approaching death: Improving care at the end of life. In M. J. Field & C. K. Cassel (Eds.). Washington, DC: National Academy Press.

Jameton, A., & Pierce, J. (2001). Environment and health: 8. Sustainable health care and emerging ethical responsibilities. *Canadian Medical Association Journal*, *164*(3), 365–369.

Klevens, R. M., Edwards, J. R., Richards, C. L., Horan, T. C., Gaynes, R. P., Pollock, D. A., & Cardo, D. M. (2007). Estimating health care-associated infections and deaths in U.S. hospitals, 2002. *Public Health Reports*, *122*(2), 160–166.

Lee, F. I., & Harry, D. S. (1974). Angiosarcoma of the liver in a vinyl-chloride worker. *Lancet*, *303*(7870), 1316–1318.

Marcus, G. E. (1995). Ethnography in/of the world system: The emergence of multi-sited ethnography. *Annual Review of Anthropology*, *24*, 95–117.

Markowitz, G., & Rosner, D. (2002). *Deceit and denial: The deadly politics of industrial pollution*. Berkeley, CA: University of California Press.

Miles, M. B., & Huberman, A. M. (1994). *Qualitative data Analysis: An expanded sourcebook* (2nd ed.). Thousand Oaks, CA: Sage Publications.

National Center for Health Statistics. (2011). *Health, United States, 2010: With special feature on death and dying*. Hyattsville, MD. Retrieved from http://www.cdc.gov/nchs/data/hus/hus10.pdf

National Hospice and Palliative Care Organization. (2005). *Facts and figures*. Retrieved from http://www.nhpco.org/files/public/2005-facts-and-figures.pdf

Office of National Drug Control Policy. (2009). *Proper disposal of prescription drugs: Federal guidelines*. Retrieved from http://www.whitehousedrugpolicy.gov/publications/pdf/prescrip_disposal.pdf. Accessed on June 10, 2010.

O'Rourke, D., & Connolly, S. (2003). Just oil? The distribution of environmental and social impacts of oil production and consumption. *Annual Review of Environmental Resources*, *28*, 587–617.

Pierce, J., & Kirby, C. (1999). The global ethics of latex gloves: Reflections on natural resource use in healthcare. *Cambridge Quarterly of Healthcare Ethics*, *8*, 98–107.

Roberts, R. R., Scott, R. D., Hota, B., Kampe, L. M., Abbasi, F., Schabowski, S., … Weinstein, R. A. (2010). Costs attributable to healthcare-acquired infection in hospitalized adults and a comparison of economic methods. *Medical Care*, *48*(11), 1026–1035.

Rowat, S. C. (1999). Incinerator toxic emissions: A brief summary of human health effects with a note on regulatory control. *Medical Hypotheses*, *52*(5), 389–396.

Sedman, R. M., & Esparza, J. R. (1991). Evaluation of the public health risks associated with semivolatile metal and dioxin emissions from hazardous waste incinerators. *Environmental Health Perspectives*, *94*, 181–187.

Sessink, P. J. M., Boer, K. A., Scheefhals, A. P. H., Anzion, R. B. M., & Bos, R. P. (1992). Occupational exposure to antineoplastic agents at several departments in a hospital. *International Archives of Occupational and Environmental Health*, *64*(2), 105–112.

Shavers, V. L., Klein, W. M. P., & Fagan, P. (2012). Research on race/ethnicity and health care discrimination: Where we are and where we need to go. *American Journal of Public Health*, *102*(5), 930–932.

Skov, T., Maarup, B., Olsen, J., Rorth, M., Winthereik, H., & Lynge, E. (1992). Leukaemia and reproductive outcome among nurses handling antineoplastic drugs. *British Journal of Industrial Medicine*, *49*, 855–861.

Starr, P. (1982). *The social transformation of American medicine: The rise of a sovereign profession and the making of a vast industry*. New York, NY: Basic Books.

Sudnow, D. (1967). *Passing On*. Englewood Cliffs, NJ: Prentice Hall.

Thornton, J., McCally, M., Orris, P., & Weinberg, J. (1996). Hospitals and plastics: Dioxin prevention and medical waste incineration. *Public Health Reports*, *111*(4), 298–313.

Valanis, B., Vollmer, W., & Steele, P. (1999). Occupational exposure to antineoplastic agents: Self-reported miscarriages and stillbirths among nurses and pharmacists. *Journal of Occupational and Environmental Medicine*, *41*(8), 632–638.

Way, E., Leong, E., Loh, H. H., & Shen, F. (1969). Simultaneous quantitative assessment of morphine tolerance and physical dependence. *Journal of Pharmacology and Experimental Therapeutics*, *167*(1), 1–8.

WHY IS AN INTEGRATED SOCIAL-ECOLOGICAL SYSTEMS (ISES) LENS NEEDED TO EXPLAIN CAUSES AND DETERMINANTS OF DISEASE? A CASE STUDY OF DENGUE IN DHAKA, BANGLADESH ☆

Parnali Dhar Chowdhury and C. Emdad Haque

ABSTRACT

Purpose – *The purpose of this chapter is to offer reflections on conventional theories concerning causes and determinants of diseases. It also intends to examine both theoretical and empirical bases for adopting an Integrated Social-Ecological Systems (ISES) lens as a tool for understanding complexities related to drivers, determinants and causes of diseases.*

☆ Both the authors participated equally in the planning and writing of this article.

Ecological Health: Society, Ecology and Health
Advances in Medical Sociology, Volume 15, 217–239

ISSN: 1057-6290/doi:10.1108/S1057-6290(2013)0000015014

Design/methodology/approach – *We assessed the theoretical underpinnings of a range of historical and contemporary lenses for viewing infectious disease drivers and the implications of their use when used to explain both personal (i.e. individual) and population health. We examined these issues within the empirical context of the City of Dhaka (Bangladesh) by adopting an ISES lens. Within this study an emphasis has been placed on illustrating how feedback loops and non-linearity functions in systems have a direct bearing upon various aspects of infectious disease occurrences.*

Findings – *A brief triumph over microbes during the last century stemmed in part from our improved understanding of disease causation which was built using disciplinary-specific, monocausal approaches to the study of disease emergence. Subsequently, empirical inquiries into the multi-factorial aetiology and the 'web of causation' of disease emergence have extended frameworks beyond simplistic, individualistic descriptions of disease causation. Nonetheless, much work is yet to be done to understand the roles of complex, intertwined, multi-level, social-ecological factors in affecting disease occurrence. We argue, a transdisciplinary-oriented, ISES lens is needed to explain the complexities of disease occurrence at various and interacting levels. More theoretical and empirical formulations, with evidence derived from various parts of the world, is also required to further the debate.*

Originality/value – *Our study advances the theoretical as well as empirical basis for considering an integrated human-nature systems approach to explaining disease occurrence at all levels so that factors at the individual, household/neighbourhood, local, regional and global levels are not treated in isolation.*

Keywords: Cause of disease; determinants; integrated approach; social-ecological systems; dengue; Bangladesh

Traditional thinking which treats the biological, ecological and social as discrete spheres existing in isolation and which imagines a linear path of progression of diseases is an inadequate framework for explaining disease causation in both individuals and populations. New and improved insights into the interconnections between various spheres – as they interplay across multiple scales and levels – and how these affect infectious disease emergence will enable medical and public health initiatives to more successfully prevent, control and respond to infectious disease epidemics. Rooted in this view, the objective of our study is twofold: (i) to offer a critique of conventional disciplinary and multi-disciplinary epistemological lenses which have been used to study disease causation, and (ii) to justify the claim that an integrated social-ecological systems (ISES) lens is required in order to explain the interplay between causes and determinants of disease occurrence at various levels. Such an integrated social-ecological approach is warranted, we argue, because complex patterns, processes and relationships among interconnected biological, ecological and social variables cannot be fully captured through any one disciplinary lens. In isolation, disciplinary lenses, whether conducting analyses of single or multiple dimensions (such as a 'web of disease causation'), are not able to adequately study the complexities of disease causation. We assert, therefore, that analyses of the roles of biological, ecological and social spheres must be augmented by a study of their interactions and simultaneous functions made possible by adopting a systems approach to analysis which develops more robust and nuanced explanations of disease causation.

Understanding the complexities of human disease risk has been a historic mission of modern medicine, beginning with the production of aetiologic explanations based on cause-and-effect studies of disease emergence at the individual level and moving into the formulation of epidemiologic theories of disease causality at the scale of populations. However, during the latter part of the 20th century, the dominance of the established biometrically oriented epidemiological theories, which by and large disregarded social and ecological drivers, was challenged by two major developments. First, the elaboration of theories of social epidemiology and their associated empirical substantiations (Berkman & Kawachi, 2000; Krieger, 1994, 2001) offered another way of thinking about disease causation. The analytical power of social epidemiology was bolstered by both the US National Institutes of Health (NIH) and Public Health Agency of Canada's (PHAC) recognition of the significant roles that societal phenomena and human behaviour (on both the individual and population scales) play in producing health and disease. These movements have shone an appreciative light on the social sciences and have aided their integration into biomedical and ecological knowledge domains. Second, the ISES theoretical lens (Berkes, Colding, & Folke, 2003; Liu et al., 2007; McMichael, 1999), which considers human-in-nature (i.e. as embodied component parts of nature) was developed to explain the multi-level and multi-scale complexities of disease causation and emergence. Over time, these frameworks have received wider acceptance among inter- and transdisciplinary academic domains (Parkes et al., 2005; Wilcox & Colwell, 2005). Both the integration of social epidemiology into research on the aetiologies of disease and the imperative to study health in the interplay between social and ecological systems has generated novel ways of thinking about disease causality as well as offered a new set of references for critiquing existing frameworks. Therefore, we find that 'systems thinking' under which the concept of ISES is pursued, is a potent framework to use when advancing explanations of disease emergence at multiple levels.

In this chapter, we first present a review of key biomedical frameworks that have been developed over time to describe disease causality. In the section that follows, we critique the aspiration to identify an underlying driver of disease emergences. In the three sections that follow we deepen that critique by first highlighting the contributions that the social studies of health have made to understanding disease processes and pathways. Second, we consider the contributions that have been made by moving from siloed or disciplinary thinking to interdisciplinary and transdisciplinary frameworks for studying disease causality. Third, we consider the contributions that systems thinking and integrated social-ecological frameworks can make to initiatives dedicated to understanding disease causality through a lens of complexity. In order to ground these theoretical observations materially we then apply the ISES lens in order to offer an examination of the case study of Dengue in Dhaka, Bangladesh. Dengue fever (DF) and dengue hemorrhagic fever/dengue shock syndrome (DHF/DSS) are the most prominent vector-borne viral diseases affecting humans worldwide. DF is self-limited though debilitating disease characterized by headaches, retro-orbital pain, myalgia, arthralgia, rash and in some cases hemorrhagic manifestations. DHF is characterized by hemorrhagic signs, thrombocytopenia and hemoconcentration or other evidence of vascular leakage, and can progress to shock (DSS) and death (Harris et al., 2000). They are, in short, important diseases to understand.

A JOURNEY FROM GERM THEORY TO MULTIPLE CAUSATION

Historically, the contagionists' view purported that epidemics occurred as a result of person-to-person contact that spread invisible poisons which were viewed as non-living, non-reproducing toxins (Bynam, 2008; Pelling, 1978). In turn, this perspective led to the search for specific disease agents and resulted in numerous discoveries of living micro-organisms and ultimately to the acceptance of 'germ theory'. The argument that infectious diseases are caused by a 'contagium-vivum' or a living micro-organism was brought to prominence by the seminal book by T. J. Maclagan in 1876. By the early 20th century, germ theory had become the predominant paradigm of explanation, as well as the scientific framework that was drawn upon to develop a means of intervening in infectious disease outbreaks (Krieger, 2011; Winslow, 1923).

While germ theory helped to identify the contagion, issues of causality also needed to be addressed. The notion of multi-causality in the field of aetiology began to enjoy prominence on account of numerous, significant and mutually reinforcing discoveries. These included John Snow's (1813–1858) success in establishing a link between cholera infections and water as its vector; Robert Koch's (1843–1920) isolation of Bacillus anthracis, the Tuberculosis bacillus and Vibrio cholera and further research by Patrick Manson (1878) on Lymphatic Filariasis; Ronald Ross (1987) on Malaria; and Carlos Chagas (1909) on Chagas disease. Research on all of these vector-borne diseases indicated the significance of transmission ecology and expanded the horizon of simplistic cause-and-effect theories to encompass an *agent-host-environment* framework (Parkes et al., 2005). While this work on the biology of diseases was occurring work on the importance of multi-causal explanations for disease emergence was also being developed. For example Rudolf Virchow, in his report on Typhus outbreak in Upper Silesia, Germany in 1849, argued that infectious disease outbreaks could not be solved by treating individual patients with drugs or through lifestyle modification alone. Rather, what was required were societal interventions that would improve living, working and therefore health conditions for the population at large (Brown & Fee, 2006).

Consolidating the epidemiological insights of the past century helped to refine a theory of multiple causation. For example, in the late 1950s, J. N. Morris (1957) and Taylor and Knowelden (1957) and in the 1960s Brian McMahon's and colleagues influential book *Epidemiologic Methods*, advanced the concept of 'multiple causation' of population health and disease within North American epidemiology. A key contribution MacMahon and his colleagues (1960) made was the notion of a 'web of causation' which refered to multiple causation within disease theories. Together, these scholars produced work that not only questioned single-agency germ theories of disease but also called for the development of deeper analysis (i.e. attention to poverty and politics) and more proximal (e.g. consideration of factors such as hunger or exposure to mine dusts) notions of disease determinants, akin to those that had been described by earlier scholars such as Virchow.

The need to differentiate between the determinants of disease in individuals and in populations is an even newer insight. By illustrating that the 'causes of cases' and the 'the causes of incidence' are not necessarily the same, Rose (1985) asserted that a study of these two kinds of causes needed varied approaches and lines of inquiry. For example, Rose suggested that inquiring about why some individuals suffer from hypertension is different than inquiring about why some populations suffer from hypertension, while others do not. Individual susceptibility is emphasized in the former question and interventions developed from this knowledge would be aimed at 'high risk' individuals; whereas population exposure is stressed in the latter question, and interventions would concentrate on the need for improving conditions for groups of susceptible people as a way to control and prevent disease. In effect, it is the social epidemiologic approach which has succeeded in challenging the dominant model of individual disease causation.

A SEARCH FOR THE UNDERLYING DRIVER – THE UNSEEN 'SPIDER' OF DISEASE IN POPULATION

Intrigued by notions of multi-causality, epidemiologists began to question the *a priori* nature of the premises of the 'web of causation' and to look deeper in order to understand complex phenomena through the development of more effective explanations. Rothman (1986), for example identified the concepts such as *confounding* and *synergies* or *effect modification. Confounding* takes place when an observed association is in fact distorted because the exposure is also correlated with 'another risk factor' where this additional risk factor is also associated with the outcome, but remains independent of the exposure under investigation (Hennekens & Buring, 1987). *Effect modifications* take place through interactions, which occur when the direction or magnitude of an association between two variables differs because of the effect of a third variable. This view may account for a cumulative effect of multiple risk factors which is not acting independently and therefore produces a greater or lesser effect than the sum of the effects of each factor acting on its own (Barratt & Kirwan, 2009). In addition, Rothman (1976) and Rothman and Greenland's (2005) emphasized the need for more focus on the relationships between 'necessary' and 'component' causes. They asserted that the power of a relationship between a component cause and a given result relies not only on their specific association but also on the prevalence of the other component causes needed to bring about the specific change. Krieger (1994, 2001) also critiqued the 'web of causation' after having investigated the presumption of an 'origins' of causes. In her article, 'Has Anyone Seen the Spider?' Krieger highlights that a description of origins is incompatible with a study of interactionality between multiple causes. She drew from HIV and AIDS research, and argued that the predominance of the biomedical orientation that had led to an almost exclusive concentration with in HIV research on the 'mechanisms' of transmission had consequently placed little attention on the study of social determinants of transmission. This research failure has led to a related failure in helping practitioners comprehend how social bias affects the type of questions they ask patients as well as how these views impact the ways doctors make sense of, and treat, the disease (Fee & Krieger, 1993).

A contention of this chapter is that discourses of disease causation have historically evolved without much clarity around the *context* of emergence vis-à-vis both individual and population health. By context, we are referring to the social, ecological, temporal conditions within which diseases at the scale of individuals and populations emerge. In addition, the roles of social and ecological factors have been neglected within prevailing paradigms due to a lack of distinction between *disease causation* and *disease distribution*. For example, the synonymous usage of 'cause', 'determinant' and 'drivers' is common within epidemiology. However, from the perspectives of valuing descriptions of differential scales and contexts, it is critical to distinguish between *disease causation* and *disease determinants* and to give greater descriptive *specificity* to the former rather than the latter concepts. Conforming to Rothman (1976, p. 588; also see Lucas & McMichael, 2005; McMichael, 1999; Rose, 1985), we view a 'cause' as 'an act or event or a state of nature which *initiates* (authors' emphasis) or *permits* (authors' emphasis) alone or in conjunction with other causes, a sequence of events resulting in an *effect*'. 'Disease determinants' refer to any of a group of variables, such as specific disease agents, social and environmental factors, that directly or indirectly affect the frequency and/or distribution of a disease by bringing about a change in health (Moon et al., 2000). *Causation* traditionally considers the source of disease emergence while *distribution* considers the occurrence of a disease within a specific population. The former emphasizes both individual and population scales while the later stresses how diseases are distributed among a population and why they follow certain patterns. We argue, however, that some of the problems in our understanding of diseases persist due to a lack of necessary appraisal of the limitations of each mode of description and how they have been taken up within various disciplines and used to develop methods of explanation that look for disease causation. By moving into a new era of disciplinary and methodological curiosity, traditional approaches can be enhanced to develop a suite of tools capable of studying disease emergence within the context of complex, multi-scalar social-ecological systems.

SOCIETAL FACTORS MATTER: ROLE OF PSYCHOSOCIAL VARIABLES AND SOCIAL STRUCTURE

Recalling the critiques of the 'web of causation' of disease, a further question to ask is whether investigations into the causes of diseases in a population should begin with a study of disease causation at the scale of the individual and how much can knowledge about individuals be extrapolated from to explain population disease causation. Since the 1950s, a response from within social epidemiology has been that although there are overlaps between disease causation in individuals and in populations, it is not *necessary* that the latter is preconditioned by the former nor that one explanation can be extended across the scales (see the ecological fallacy) (Smolinski, Hamburg, & Lederberg, 2003). One such example within epidemiology is the dominant concentration on individual-level biological and behavioural 'risk factors', as used within 'biomedical' and 'lifestyle' approaches to analysing disease causation. Yet, there are also calls for the development of epidemiologic theories of *disease distribution* which would incorporate the use of social factors as an important tool for explaining disease distribution among various sub-sets of populations. One of the early writers on these epidemiological issues, John Cassel (1976), lauded the contributions of psychosocial studies conducted during the 1940s and 1950s which linked vulnerability to disease to both physical and psychological stress. He built on these insights by developing the concept of the 'social environment' which is comprised of psychosocial factors generated by human interactions which modify host susceptibility by impacting neuroendocrine function rather than by way of other exogenous environmental factors. These perspectives were articulated under the rubric of a 'life course' approach which asserts that health status at any given point in life illustrates both prior living experiences and current conditions (Kuh & Ben-Shlomo, 1997). However, the origins of the stress in terms of what and who creates social exclusion and isolation, deprivation and insults; how their distribution is conditioned by societal environments; and the role of social responsibility and accountability are issues given less attention within psychosocial explanations (Krieger, 1994, 2001, 2011). This leaves considerable room for other theories to fill in the gaps, such as insights offered within 'political economic' perspectives of health which show that societal inequality is a key social determinants of health and disease (Link & Phelan, 1996; Navarro, 1986). As Krieger emphasizes, however, these explanations need to be considered as complementary because alone the political economic theoretical standpoint offers 'few principles for investigating *what* these determinants are determining' (2001, p. 671; also see Krieger, 1994). Also, the overemphasis on fundamental social causes of inequality, deprivation and exclusion suggests one particular direction for public health and policy interventions to take when addressing social factors in determining the health status is a more considered approach to the social determinants.

In response to the call to formulate multi-level frameworks for 'integrating social and biological reasoning and history' as well as integrating ecological and societal spheres, an array of multi-, inter- and transdisciplinary theoretical approaches to health and disease can be assembled. Although the above-noted developments can be applied to most health issues, in order to illustrate our arguments clearly, we concentrate mainly on emerging infectious diseases (EID) in the sections that follow.

DISCIPLINARY, MULTI-DISCIPLINARY AND INTERDISCIPLINARY LENSES: TOWARDS THE INTEGRATION OF KNOWLEDGE SPHERES

Until recently, and while some researchers are advancing an understanding of disease emergences occurring within biological, ecological and social spheres, most analysts continue to consider social and ecological factors to be separate rather than interactive within contexts of emergence (Parkes et al., 2005, p. 260). This compartmentalized disciplinary ethos has been so deeply institutionalized that the notion of interactions occurring across ecological and societal spheres is difficult to address within traditional research and practice communities. Even when interdisciplinary issues are analysed, the frameworks used tend to be disciplinarily confined – as a brief review of critiques of disciplinarily bound research into infectious diseases reveal.

During the latter part of the 20th century, serious shortcomings in disciplinary approaches to health issues were identified in North America by public health agencies. For example, the US Office of Behavioural and Social Science Research, NIH stated that:

> The disciplines concerned with health research may be separated conceptually, methodologically, and administratively, the *processes* (authors' emphasis) about which they are concerned are inextricably linked…the social, behavioural, and biological processes that affect health are interdependent. (Anderson, 1998, p. 564)

The NIH further noted that despite such interconnections, the distinctions between scientific disciplines have been reified as if the compartmentalization of the health sciences mirrors a corresponding compartmentalization of the *origins* of human illness. The latter, in fact, does not exist in this manner. In response to the call for integrative research, teams were formed but problems of communication and transformative learning remained acute. In the early years (i.e. 1970s and 1980s) of integration efforts, research teams consisted of disciplinary-oriented scientists who worked together for 'multidisciplinary' health studies, where the relation among them was 'associative' (Kessel & Rosenfield, 2008; McDonell, 2000). To address common problems or a set of problems, each team member, from a disciplinary perspective, worked independently and the results were brought together only at the end of the process (Rosenfield, 1992). As a result, a study of the interconnectedness of phenomena as it extended across disciplinary arenas was not conducted. To incorporate into analysis information about the relationships, interconnectedness and interdependencies between phenomena, inter- and transdisciplinary approaches were developed (Stokols, Hall, Taylor, & Moser, 2008). In these frameworks, the relationship between researchers was also considered to be 'relational' as it required researchers to collaborate by taking 'up some of the assumptions and worldviews and languages of the others' (McDonell, 2000, p. 27). A new era of searching for ways to transform one's own (intra) disciplinary as well as collective or interdisciplinary knowledge through interacting with and learning from other scholars had been catalysed. This was an era of transdisciplinarity where approaches to developing integrated knowledge were invited into the health sciences.

SYSTEMS THINKING AND INTEGRATED LENSES TO VIEW INTERCONNECTED, MULTI-LEVEL FACTORS

The multiplication of efforts to integrate social and ecological determinants of health and disease, along with endogenous biological factors (molecular, cellular and organ system) of diseases, has also been building bridges between biomedical proponents and social epidemiologists. The convergence of theoretical lenses in aetiologic, epidemiologic and health studies has evoked several notable integrative theoretical approaches, of which 'ecosocial theory' (Krieger, 1994, 2001), the 'social-ecological systems perspective' (McMichael, 1999) and 'Ecosystem approaches' to health (Charron, 2012; Lebel, 2003; Waltner-Toews, Key, & Lister, 2008) are principal ones. Notable and shared characteristics of these approaches are: (a) interconnectedness and interdependencies between biological, ecological and social spheres, (b) multi-level thinking and (c) a holistic framework. However, they vary in terms of structure, emphasis in analysis and in the types of intervention they produce.

When introducing 'ecosocial' theory, Nancy Krieger (1994, 2001) uses a visual metaphor of an evolving 'bush of life' intertwined at every scale (the micro to macro), with the external structure being that of society which various social groups reinforce or try to alter. The approach promotes that (a) the health of all organisms is interconnected, (b) the rigid distinction between individual and group-level analysis is unacceptable and (c) in addition to the need to pay attention to the social production of disease, it is also important to look into the processes of the social production of science. However, thus far, appreciation of this integrative model has remained primarily limited to social epidemiological literature, and evidence-based substantiations of this theory are still in their infancy (Waltner-Toews et al., 2008; Wilcox & Colwell, 2005).

The 'social-ecological systems perspective' was advocated by Anthony McMichael (1999) when in the United States, and James Kay and H. Regier (2000) and Fikret Berkes and his colleagues (2003) in Canada. They frame a disease or health problem using an ISES lens and by considering multiple biological, ecological, behavioural and societal levels and dimensions. As 'systems thinking' focuses on the relationships among these elements (Charron, 2012) within the complex whole, in their view, humans are both part of nature and largely conditioned by societal practices. A separation between the biophysical ecosystem and social system is, therefore, arbitrary and unnecessary (Berkes et al., 2003). Despite its merits, the emerging ecologically inclined multi-level frameworks have remained largely conceptual (Krieger, 2001) and much more explicit applications to aetiological inquiry and interventions would be required in order to mainstream such frameworks under the rubric of systems thinking.

The 'EcoHealth' lens focuses on the interaction between the ecological and societal dimensions of a given situation, their effect on human and animal health, as well as on how individuals and populations receive utility or impact ecosystems, including the provision of ecosystem services and sustainability. By departing from an exclusive orientation to biomedical thinking, and by adopting systems thinking, scholars championing this new arena of research and practice such as Waltner-Toews and his colleagues (2008), Jean Lebel (2003) and Dominique Charron (2012), have attempted to formulate a set of principles for 'Ecosystem approaches to health' ('EcoHealth'). This emerging field seeks to understand encapsulate the 'coupled ecological and social systems' within which human health vis-a-vis disease is generated and challenged. These principles were formulated to assist in comprehending the nested, interdependent systems in which diseases emerge, maintain themselves by self-organization and re-organization, and evolve into a multi-level system (molecular, organismal, communal, national and global) (Parkes et al., 2005; Wilcox & Colwell, 2005).

We find that in recent years wider interest in these perspectives has been generated in literature on the larger scale 'drivers' of societal and ecological changes (e.g. climate change, global poverty and inequity, human and environmental insecurity, and degradation of ecosystem services), in particular in relation to finding new ways to explain disease causation and distribution (also see, Parkes et al., 2005; Plowright, Sokolow, Gorman, Daszak, & Foley, 2008; Wilcox & Colwell, 2005). Also, a transdisciplinary approach, where 'the integrating language of relationships is taken to the extent of there being a transcendent language, a metalanguage, in which terms of all pertinent disciplines are, or can be, expressed' (McDonell, 2000, p. 27), is being recognized as a more effective way to understand and explain complex, interwoven problems concerning health and diseases. In using such a lens, representatives of various disciplines are encouraged to *transcend* their separate conceptual and methodological orientations in their efforts to develop a shared approach to a problem and its potential solutions.

SEEING DENGUE THROUGH THE INTEGRATED SOCIAL-ECOLOGICAL SYSTEMS LENS

One of the most suitable areas to study the application of an ISES lens is that of emerging and reemerging infectious diseases, such as dengue as they occur through complex interfaces between biological, ecological and societal phenomena. DF and DHF are endemic infectious diseases in the tropics today and will continue to pose a serious public health threat in the future (Guzman et al., 2010). Dengue disease in humans is caused by four closely related dengue viruses (antigenically similar but immunologically different serotypes) of genus *Flaviviruses* that are primarily transmitted by *Aedes aegypti* mosquito (Ooi, Goh, & Gubler, 2006). Here, the interplay between human, mosquito and virus contributes to the clinical spectrum and geographical distribution of dengue (Kyle & Harris, 2008). In other words, each sphere affects the others, all in the context of the dynamics of social-ecological systems.

Because there is no non-human reservoir in Bangladesh, DF can spread due to the dispersal of humans. For example, with the movement of troops during World War II, along with human population and settlement growth and the resultant destruction of forests, dengue viruses and their vectors (*A. albopictus* and *A. aegypti*) spread throughout Southeast Asia and the Western Pacific regions. During the post-World War II era, urban growth, without well-organized water and waste management systems in the tropical and sub-tropical regions, has resulted in increased populations of mosquitoes living in intimate association with crowded human populations. Specifically, unplanned and unchecked urbanization in most tropical countries, along with the plethora of water-holding containers, have supported the intense breeding of dengue vectors. Also, the emerging vector abundance has not remained confined locally as ever increasing commercial shipping and air traffic have greatly enhanced the spread of *Aedes* mosquitoes and the dissemination of dengue viruses through the rapid transit of viraemic individuals around the world. Viral adaption to the vector has also resulted in efficient transmission (Moncayo et al., 2004); nonetheless, that adaptation was not needed for the emergence of the virus from a sylvatic cycle (Vasilakis et al., 2007). In fact, *A. aegypti* species progressively adapted to rapid anthropogenic changes to the environment, primarily in transforming agricultural or fallow land into built-up urban conglomerations, which provided alternative blood sources (domestic animals and humans) and water collections for larval habitat. The agglomeration of human population to a certain threshold (in the range between 10,000 and 1,000,000 in population) has been found to be a critical factor in sustaining dengue disease transmission (Kuno, 1997).

By embracing linear thinking about disease causation and distribution, concerned public health authorities during the 1960s and 1970s opted for vector control, through environmental perturbation and pesticide application. The thrust of such government-initiated, top-down institutional programs was simplistic, postulating that because mosquito breeding precedes disease transmission controlling the vector population before the disease is detected would break the disease transmission cycle. This approach to the eradication or significant control of dengue virus within some countries of Southeast Asia (i.e. Singapore) and Latin America was visible until the 1970s. However, the consequences of neglecting the nested hierarchical relationships between numerous ecological, social, political, demographic, economic and cultural factors and of the lack of interest in understanding ISES dynamics soon became apparent. The reemergence of dengue virus, with more virulence, occurred during the 1990s and early 2000s. For example, it appeared at a several-fold higher overt attacking rate in Singapore compared to in the 1960s (Ooi et al., 2006). In Dhaka (Bangladesh), periodic outbreaks of dengue have been recorded since 1964, with the first DHF epidemic in 2000 (Karim, Munshi, Anwar, & Alam, 2012). The recent increased incidence and rapid geographical spread of dengue disease across the tropical and sub-tropical world have been profoundly influenced by some large scale phenomena including, but not limited to, population growth, urbanization, inadequate access to basic housing, climate change and air travel (Banu et al., 2012; Gubler, 2002; Wu, Guo, Lung, Lin, & Su, 2007). In fact, dengue has become the most rapidly advancing vector-borne disease with an estimated 50 million dengue infections occurring every year (Farrar et al., 2007). Consequently, world-wide, an estimated 3.61 billion people have become at risk of being infected in 124 countries (Baly et al., 2012; Beatty, Stone, & Fitzsimons, 2010).

Integrated Social-Ecological Systems Lens to Explain Dengue in Dhaka

The City of Dhaka, the capital of Bangladesh established in the 16th century, is characterized by all the attributes of a 'primate city' in the developing world. The city has a metropolitan area of 1,464 sq. km and a population of over 14 million, which is growing at a rate of 7 per cent per year. Ranked as the 9th largest city in the world (UNICEF, 2012), Dhaka is also among the most densely populated cities in the world. Approximately 300,000–400,000 migrants arrive in Dhaka every year (World Bank, 2007) in search of socio-economic opportunities, most of whom initially concentrate in marginal settlements due to poverty (CUS, 2006). More than one-third (37.4 per cent in 2007) of the total population of the city live in informal settlements, developed on private and public land (often illegally) (Hossain, 2011) – revealing a 250 per cent growth of the slum population during 1996–2006. The incidence of DF and DHF each year is now common in Dhaka, with varying degrees of admission into hospitals and of fatalities.

From 2005 to 2009, there were 3,130 admissions for DF to the 11 hospitals located in the city (Hashizume et al., 2007; Haque, Yamamoto, Malik, & Sauerborn, 2012). As stated above, although periodic outbreaks of dengue have been recorded in Bangladesh since 1964, the first dengue epidemic, with fatalities from DHF and DSS, occurred in 2000, causing 93 deaths (Karim et al., 2012). It is worth noting here that subsequent infection with DF strains is likely to cause DSS. The Directorate General of Health Services (DGHS) registered 23,872 cases of DF and DHF, with 233 fatal cases, between 2000 and 2009 (Banu et al., 2012). Considering that there is no effective vaccine or specific treatment of the disease per se yet, reducing or eliminating larval habitats of the principal vector (i.e. *Aedes Aegypti* mosquitoes) has been considered an important component of suitable vector control programs by the Dhaka City Corporation authority (Aldstadt et al., 2011). This notion triggered the development of vertically structured (i.e. top-down) governmental programs for larval control in domestic water storage and collection containers, as well as in relation to the use of insecticides. However, the increasing level of resistance by mosquitoes, householders' rejection of the treatment of their drinking water and difficulties in attaining high and regular levels of population coverage by insecticide spray have resulted in a general failure of vector and dengue disease control programs (Guzman et al., 2010). Such a failure has been mirrored in a 2–3 year cycle of dengue reemergence in Dhaka during 2000–2008 (Karim et al., 2012).

In contrast to these more traditional approaches to dengue control, ISES thinking underscores the importance of looking for complex patterns through considering the interplay between social, biological or ecological systems. However, using an ISES lens also called a 'coupled social-natural systems' explanation (Berkes et al., 2003; Gunderson & Holling, 2002; Liu et al., 2007) to show dengue to be a unique type of EID, albeit one in its infancy (Waltner-Toews et al., 2008; Wilcox & Colwell, 2005). Only a handful of analyses thus far have attempted to provide a rationale and demonstrate the merits of such efforts (Wilcox, & Colwell, 2005). As Liu and his colleagues (2007) have determined, the outcomes of ISES analysis are *nonlinear dynamics with thresholds, reciprocal feedback loops, time lags, resilience, heterogeneity* and *surprises.* It is beyond the scope of this chapter to elaborate on each of these aspects; however, we would like to present some relevant key observations we derived from our study using an ISES lens to study a case of dengue in Dhaka.

As stated above, the City of Dhaka has undergone unprecedented demographic, socio-economic and physical changes as a result of the effects of globalization since the mid-1990s. Specifically, it has been through capitalizing on the low wages of female workers that the export-oriented knitting and garment industry has grown exponentially. The propensity and aspirations to earn higher wages in Dhaka more than in any other part of the country has been a *driver* of the continuous influx of rural populations into this city. In the face of these mounting demographic pressures, many government sectors and the supply of basic utilities has failed. One of the most critical sectors that has been severely affected is evidenced by the city's inability to provide a sustained supply of electricity to the households, as priorities have been given to critical industries (such as the garment industry in order to sustain the country's export levels) and seasonal irrigation for agriculture used to address food insecurity issues. Locally known as the 'loadshedding' of electricity, the intermittent interruptions of power supplies have been a common phenomenon for the last few decades. It is extremely difficult to predict both when loadshedding will occur and for how long any individual period of loadshedding will last (minutes, hours, occasionally days). This, in turn, has resulted in serious disruptions in water supply to households via municipal pipelines. Consequently, the need to store water (for drinking, taking baths, doing household chores and other necessary work) in containers of various sizes within the houses has created a perfect environment for mosquito breeding, especially during monsoon seasons. The prevalence of multiple serotypes of dengue virus along with the multiplication of mosquito breeding containers in one of the most densely settled cities of the world has produced the necessary conditions for the emergence of DF, DHF and DSS in Dhaka. These multi-level socio-ecological positive feedback loops which cause dengue occurrence in Dhaka are depicted in Fig. 1.

Dengue disease affects various types of phenomena negatively at multiple levels. At the micro-level, it impacts individuals' health and wellbeing through a loss of income. At the community level, social order and cohesion are hampered and the sub-national economy suffers due to labour shortages. At the macro-level, national income vis-à-vis the economy declines due to loss of labour productivity and rises in public health costs. These negative feedback loops add further complexities to EID in low-income countries like Bangladesh (Fig. 1). As stated above, more than one-third of the population of the City of Dhaka is poor and lives in dense squatter settlements, and their livelihood is based on daily wage earning. Hospital reporting by this segment on this population is fragmented, and often the infected person travels across the city for work as well as to other towns and rural areas to visit extended family members. The rapidly

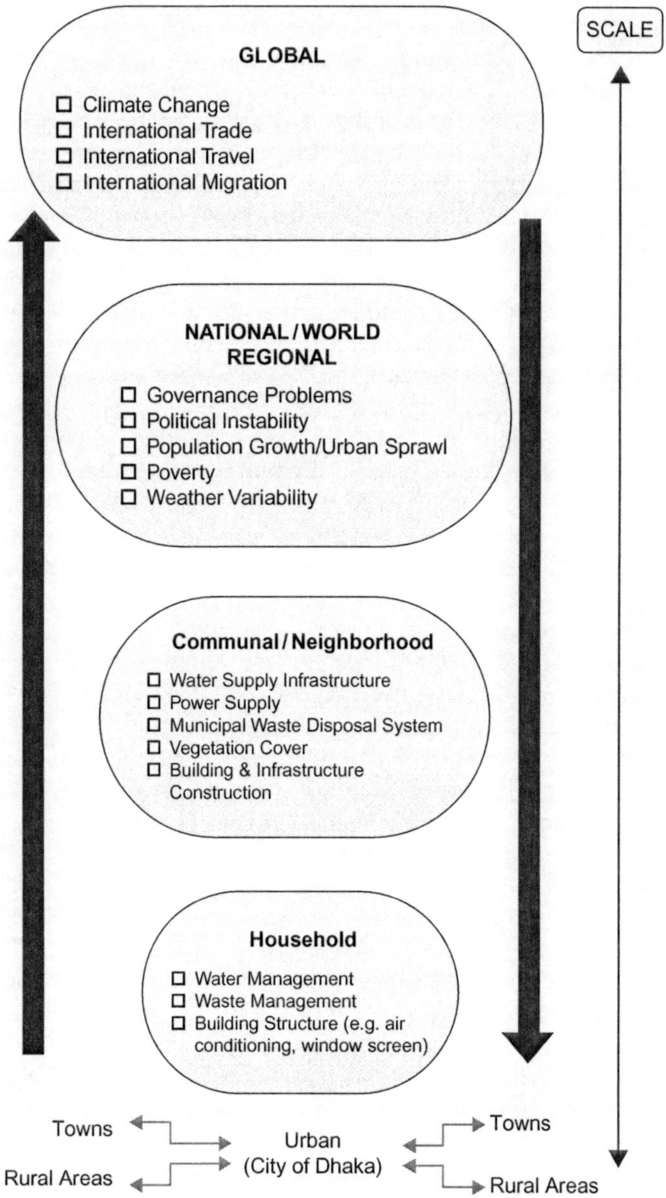

Fig. 1. Multi-Level Socio-Ecological Systems within which Dengue Occurs in Dhaka.

changing urban complex, characterized by high population growth, increased mobility and migration, income inequality, inadequate water and waste management infrastructure and limited awareness of the *Aedes* mosquito habitat among the city dwellers; these changes are enabling the sustenance of the two to three year dengue cycle. Therefore, the risk of exposure to dengue virus among the population and the potential for a dengue epidemic in the future remain a major public health concern.

The case of the City of Dhaka reveals that the biomedical and 'individually oriented' theories of disease causation, in which population risk is treated as the sum of individuals' risks, are grossly inadequate for the task of explaining the complex interplay of systems driving dengue disease emergence and their population effects within the context of the rapidly changing social-ecological systems of Bangladesh. Underscoring the dynamics of the integrated social and natural systems within which the people of Bangladesh and the natural components of the Brahmaputra-Ganges deltaic systems are interacting is important here. The risks for burden of diseases are being materialized within the conjoined social-ecological systems and affected groups within a population such as specific social classes, genders, ages, ethnicities and groups are configured in relation to socio-political relations of power and authority.

In the context of Dhaka, it is important to underscore that the complexity of 'coupled systems' has not yet been well understood, largely for two reasons: (i) the traditional separation of ecological and social sciences, and (ii) most work has been theoretical rather than empirical. Therefore, much more theoretical and empirical work needs to be carried out, recognizing that the 'ISES' lens is only one of many *necessary* frameworks needed to comprehend the complexity of disease occurrence in human society (Liu et al., 2007). Accepting that 'epistemological pluralism' (Miller et al., 2008), whereby integration across knowledge perspectives ('horizontal integration') and among different types of knowledge users ('vertical integration') (Parkes et al., 2005) can be realized is fundamental to the study of infectious disease emergence. As describe above, the 'political economy' and 'political ecology' lenses, which offer a 'social production of disease' framework, can bring out valuable complementary insights into social class and power structures, as well as their manifestation as micronutrient deficiencies, underweightness and addictive behaviour such as, smoking and alcohol use. Complementarily, biomedical lenses built from multiple causation frameworks can also be combined with ISES lenses to advance knowledge about the microbial and genetic patterns and processes within the human body and in nature.

CONCLUSION

In conclusion, we suggest that an ISES lens is needed to explain the causes and determinants of diseases since many pivotal aspects driving an emergence cannot be accounted for by using separate biomedical and ecological lenses. Emerging theoretical frameworks, such as ISES, provide new means to comprehend and systematize our thoughts on the integral interfaces between humans and nature and the simultaneous working of these facets in determining the state of population health and disease distribution across various scales – ranging from local to global. As well, these systems-oriented frameworks encourage researchers to transcend disciplinary knowledge frameworks through 'viewing' a common problem or agenda using multiple lenses. As the spirit of transdisciplinarity is neither to nullify nor to undermine disciplinary knowledge and practice; rather, it is intended to help build on and complement them. In these contexts, the mission of constructing bridges between disciplines should be pursued. Attempts to adopt an 'integrated' Social-Ecological Systems approach represent a step forward in this direction.

We presented some preliminary observations from our research on our case study of Dhaka which investigated how the interplay between ecological and societal factors helps drive the emergence and sustenance of dengue disease. Particular attention was paid to how the dimensions of non-linearity, multi-level feedback loops, surprises, heterogeneity, time lags and resilience can help explain the disease phenomena better. Our understanding of these complexities is still tentative and much deeper analysis is required for the development of more meaningful theoretical and action plan formulations. Nonetheless, our template can be applied to many other integrated systems at local, sub-national, national and global levels and used to develop further empirical evidence to support the ISES lens. We conclude that there is an urgent need to transcend the disciplinary 'silos' (e.g. biomedical or sociological) and to consider the interconnectedness of biological, ecological and social spheres for advancing our knowledge of disease causation and distribution. Relatedly, there is also an urgent need for *knowledge translation* achieved through building strong and sustained links between knowledge producers and knowledge users. Knowledge without action would bring about limited change, but knowledge with application to real world problems could make significant changes to how disease occurrences are understood and managed and ultimately to the status of health and wellbeing experienced by populations, such as vulnerable groups in Dhaka.

ACKNOWLEDGMENTS

This research was supported by the International Development Research Centre (IDRC), Ottawa, Canada funded project, entitled "Climatic Variability, Social-Ecological Changes and Dengue Disease in Bangladesh" (Grant 106040-001). Financial assistance from the Manitoba Health Research Council (MHRC) through the Graduate Studentships Award and from IDRC (Ottawa) through the IDRC Doctoral Research Awards (IDRA) to the first author of this study to conduct the required field research is gratefully acknowledged. We thank Mr. Cameron Zywina for his editorial assistance.

REFERENCES

Aldstadt, J., Koenraadt, C. J. M., Fansiri, T., Kijchalao, U., Richardson, J., Jones, J. W., & Scott, T. W. (2011). Ecological modeling of Aedes aegypti (L.) pupal production in rural Kamphaeng Phet, Thailand. *PLoS Neglected Tropical Diseases*, *5*(1), e940.

Anderson, N. B. (1998). Levels of analysis in health science: A framework for integrating sociobehavioural and biomedical research. *Annals of the New York Academy of Sciences*, *840*(1), 563–576.

Baly, A., Toledo, M. E., Rodriguez, K., Benitez, J. R., Rodriguez, M., Boelaert, M., & van der Stuyft, P. (2012). Costs of dengue prevention and incremental cost of dengue outbreak control in Guantanamo, Cuba. *Tropical Medicine and International Health*, *17*(1), 123–132.

Banu, S., Hu, W., Hurst, C., Guo, Y., Islam, M. Z., & Tong, S. (2012). Space-time clusters of dengue fever in Bangladesh. *Tropical Medicine & International Health*, *17*(9), 1086–1091.

Barratt, H., & Kirwan, M. (2009). *Confounding, interactions, methods for assessment of effect modification*. Retrieved from http://www.healthknowledge.org.uk/public-health-textbook/research-methods/1a-epidemiology/confounding-interactions-methods

Beatty, M. E., Stone, A., & Fitzsimons, D. W. (2010). Best practices in dengue surveillance: A report from the Asia-Pacific and Americas Dengue Prevention Boards. *PLoS Neglected Tropical Diseases*, *4*, e890.

Berkman, L., & Kawachi, I. (Eds.). (2000). *Social epidemiology*. Oxford: Oxford University Press.

Berkes, F., Colding, J., & Folke, C. (2003). *Navigating through social-ecological systems: Building resilience for complexity and change*. Cambridge, UK: Cambridge University Press.

Brown, T. M., & Fee, E. (2006). Rudolf carl virchow. *American Journal of Public Health*, *96*(12), 2104–2105.

Bynam, W. (2008). *The history of medicine: A very short introduction*. Oxford, UK: Oxford University Press.

Cassel, J. M. (1976). Potentialities and limitations of epidemiology. In A. Katz & J. Felton (Eds.), *Health and the community: Readings in the philosophy and sciences of public health* (pp. 432–435). New York, NY: The Free Press.

Centre for Urban Studies (CUS). (2006). *Slums of urban Bangladesh: Mapping and census, 2005*. Dhaka: Centre for Urban Studies, University of Dhaka.

Charron, D. F. (Ed.). (2012). *Ecohealth research in practice: Innovative applications of an ecosystem approach to health*. Ottawa: International Development Research Centre.

Farrar, J., Focks, D., Gubler, D., Barrera, R., Guzman, M. G., Simmons, C., Kalayanarooz, S., et al. (2007). Editorial: Towards a global dengue research agenda. *Tropical Medicine and International Health, 12*(6), 695–699.

Fee, E., & Krieger, N. (1993). Understanding AIDS: Historical interpretations and the limits of biomedical individualism. *American Journal of Public Health, 83*(10), 1477–1486.

Gubler, D. J. (2002). Epidemic dengue/dengue hemorrhagic fever as a public health, social and economic problem in the 21st century. *Trends Microbiol, 2*, 100–103.

Gunderson, L. H., & Holling, C. S. (2002). *Understanding transformations in human and natural systems*. Washington, DC: Island Press.

Guzman, M. G., Halstead, S. B., Artsob, H., Buchy, P., Farrar, J, Gubler, D. J., & … Peeling, R. W. (2010). Dengue: A continuing global threat. *Nature reviews: Microbiology, 8*(12), S7–S16.

Harris, E., Videa, E., Perez, L., Sandoval, E., Tellez, Y., Perez, M. D. L., … Balmaseda, A. (2000). Clinical, epidemiological, and virological features of dengue in 1998 epidemic in Nicaragua. *American Journal of Tropical Medicine and Hygiene, 63*(1,2), 5–11.

Haque, M. A., Yamamoto, S. S., Malik, A. A., & Sauerborn, R. (2012). Households' perception of climate change and human health risks: A community perspective. *Environmental Health, 11*(1), 1–12.

Hashizume, M., Armstrong, B., Hajat, S., Wagatsuma, Y., Faruque, A. S. G., Hayashi., T., & Sack, D. A. (2007). Association between climate variability and hospital visits for non-cholera diarrhoea in Bangladesh: Effects and vulnerable groups. *International Journal of Epidemiology, 36*(5), 1030–1037.

Hennekens, C. H., & Buring, J. E. (1987). *Epidemiology in medicine*. Philadelphia, PA: Lippincott Williams & Wilkins.

Hossain, S. (2011). Informal dynamics of a public utility: Rationality of the scene behind a screen. *Habitat International, 35*, 275–285.

Karim, M. N., Munshi, S. U., Anwar, N., & Alam, M. S. (2012). Climatic factors influencing dengue cases in Dhaka city: A model for dengue prediction. *The Indian Journal of Medical Research, 136*(1), 32–39.

Kay, J., & Regier, H. (2000). Uncertainty, complexity, and ecological integrity: Insights from an ecosystem approach. In P. Crabbe, A. Holland, L. Ryszkowski & L. Westra (Eds.), *Implementing ecological integrity: restoring regional and global environmental and human health. NATO Science Series, Environmental Security*. Dordrecht: Kluwer Academic Publishers.

Kessel, F., & Rosenfield, P. L. (2008). Toward transdisciplinarity research: Historical and contemporary perspectives. *American Journal of Preventive Medicine, 35*(2S), S225–S234.

Krieger, N. (1994). Epidemiology and the web of causation: Has anyone seen the spider? *Social Science and Medicine, 39*(7), 887–903.

Krieger, N. (2001). Theories for social epidemiology in the 21st century: An ecosocial perspective. *International Journal of Epidemiology*, *30*(4), 668–677.

Krieger, N. (2011). *Epidemiology and the people's health: Theory and context*. Oxford, UK: Oxford University Press.

Kuh, D., & Ben-Shlomo, Y. (Eds.). (1997). *A Life course approach to chronic disease epidemiology*. Oxford, UK: Oxford University Press.

Kuno, G. (1997). Factors influencing the transmission of dengue viruses. In D. J. Gubler & G. Kuno (Eds.), *Dengue and dengue hemorrhagic fever* (pp. 61–88). Wallingford, UK: CAB International.

Kyle, J. L., & Harris, E. (2008). Global spread and persistence of dengue. *Annual Review of Microbiology*, *62*, 71–92.

Lebel, J. (2003). *Health: An ecosystem approach*. Ottawa, Canada: International Development Research Centre. Retrieved from http://www.idrc.ca/in_focus_health/

Link, B. G., & Phelan, J. C. (1996). Understanding sociodemographic differences in health: The role of fundamental social causes. *American Journal of Public Health*, *86*(4), 471–473.

Liu, B. M., Abebe, Y., McHugh, O. V., Collick, A. S., Gebrekidan, B., & Steenhuis, T. S. (2007). Overcoming limited information through participatory watershed management: Case study in Amhara, Ethiopia. *Physics and Chemistry of the Earth*, *33*, 13–21.

Lucas, R. M., & McMichael, A. J. (2005). Association or causation: Evaluating links between environment and disease. *Bulletin of the World Health Organization*, *83*(10), 792–795.

MacMahon, B., Pugh, T. F., & Ipsen, J. (1960). *Epidemiologic methods*. Boston, MA: Little, Brown and Company.

McDonell, G. (2000). Disciplines as cultures: Towards reflection and understanding transdisciplinarity. In M. A. Somerville & D. Rapport (Eds.), *Recreating integrated knowledge* (pp. 25–38). Oxford, UK: EOLSS Publishers.

McMichael, A. J. (1999). Prisoners of the proximate: Loosening the constraints on epidemiology in an age of change. *American Journal of Epidemiology*, *149*(10), 887–897.

Miller, T. R., Baird, T. D., Littlefield, C. M., Kofinas, G., Chapin III, F. S., & Redman, C. L. (2008). Epistemological pluralism: Reorganizing interdisciplinary research. *Ecology and Society*, *13*(2). Retrieved from http://www.ecologyandsociety.org/vol13/1ss2/art46/

Moncayo, A. C., Fernandez, Z., Ortiz, D., Diallo, M., Sall, A., Hartman, S., … Weaver, S. C. (2004). Dengue emergence and adaptation to peridomestic mosquitoes. *Emerging Infectious Diseases*, *10*(10), 1790–1796.

Moon, G., Gould, M., et al. (2000). *Epidemiology: An introduction*. Buckingham, UK: Open University Press.

Morris, J. N. (1957). *Uses of epidemiology*. Edinburgh, UK: E&S Livingston Ltd.

Navarro, V. (1986). *Crisis, health and medicine: A social critique*. New York, NY: Tavistock.

Ooi, E., Goh, K., & Gubler, D. J. (2006). Dengue prevention and 35 years of vector control in Singapore. *Emerging Infectious Diseases*, *12*(6), 887–893.

Parkes, M. W., Bienen, L., Breilh, J., Hsu, L., McDonald, M., Patz, J. A., … Yassi, A. (2005). All hands on deck: Transdisciplinary approaches to emerging infectious disease. *EcoHealth*, *2*(4), 258–272.

Pelling, M. (1978). *Fever and english medicine, 1825–1865*. Oxford, UK: Oxford University Press.

Plowright, R. K., Sokolow, S. H., Gorman, M. E., Daszak, P., & Foley, J. E. (2008). Causal inference in disease ecology: Investigating ecological drivers of disease emergence. *Frontiers in Ecology*, *6*(8), 420–429.

Rose, G. (1985). Sick individuals and sick populations. *International Journal of Epidemiology*, *14*, 32–38.

Rosenfield, P. L. (1992). The potential of transdisciplinary research for sustaining and extending linkages between the health and social sciences. *Social Science and Medicine*, *35*, 1343–1357.

Rothman, K. J. (1976). Causes. *American Journal of Epidemiology*, *104*, 587–592.

Rothman, K. J. (1986). *Modern epidemiology*. Boston, MA: Little, Brown and Company.

Rothman, K. J., & Greenland, S. (2005). Causation and causal inference in epidemilogy. *American Journal of Public Health*, *95*(Suppl. 1), 144–150.

Smolinski, M. S., Hamburg, M. A., & Lederberg, J. (Eds.). (2003). *Microbial threats to health: Emergence, detection, and response*. Washington, DC: National Academies Press.

Stokols, D., Hall, K. L., Taylor, B. K., & Moser, R. P. (2008). The science of team science. Overview of the field and introduction to the supplement. *American Journal of Preventive Medicine*, *35*(Suppl. 2), S77–S89.

Taylor, I., & Knowelden, J. (1957). *Principles of epidemiology*. Boston, MA: Little, Brown and Company.

UNICEF. (2012). *State of the world's children 2012: Children in an urban world*. New York, NY: United Nations Children's Fund.

Vasilakis, N., Shell, E. J., Fokam, E. B., Mason, P. W., Hanley, K. A., Estes, D. M., & Weaver, S. C. (2007). Potential of ancestral sylvatic dengue-2 viruses to re-emerge. *Virology*, *358*(2), 402–412.

Waltner-Toews, D., Key, J. J., & Lister, I. E. (Eds.). (2008). *The ecosystem approach: Complexity, uncertainty, and managing for sustainability*. New York, NY: Columbia University Press.

Wilcox, B. A., & Colwell, R. R. (2005). Emerging and reemerging infectious diseases: Biocomplexity as an interdisciplinary paradigm. *EcoHealth*, *2*(4), 244–257.

Winslow, C. E. A. (1923). *The evolution and significance of the modern public health campaign*. New Haven, CT: Yale University Press.

World Bank. (2007). *Dhaka: Improving living conditions for the urban poor*. Dhaka: The World Bank Office.

Wu, P. C., Guo, H. R., Lung, S. C., Lin, C. Y., & Su, H. J. (2007). Weather as an effective predictor for occurrence of dengue fever in Taiwan. *Acta Tropica*, *103*, 50–57.

PERPETUATING A REDUCTIONIST MEDICAL WORLDVIEW: THE ABSENCE OF ENVIRONMENTAL MEDICINE IN THE AMERICAN ADHD CLINICAL PRACTICE GUIDELINES

Manuel Vallée

ABSTRACT

Purpose – *Since the beginning of the 20th century environmental health researchers have known about the association between toxicant exposure and disease. However, that knoweldge has not been well integrated into mainstream medicine. Shedding light on why is the focus of this chapter.*

Methodology/approach – *To shed light on this issue I analyze the 2011 American Academy of Pediatrics' clinical practice guidelines for Attention Deficit/Hyperactivity Disorder (ADHD), focusing specifically on the omission of environmental health research pertaining to ADHD symptoms and exposures, such as lead and mercury.*

Ecological Health: Society, Ecology and Health
Advances in Medical Sociology, Volume 15, 241–264

ISSN: 1057-6290/doi:10.1108/S1057-6290(2013)0000015015

Findings – *I found that while environmental researchers have been documenting the link between lead and ADHD for over forty years, the American Academy of Pediatrics has completely omitted this research from its 2012 clinical practice guidelines. Moreover, I argue this omission can be traced to competitive pressures to protect medical jurisdiction, and a reductionist worldview that emphasizes treatment over prevention.*

Originality/value of paper – *This is the first attempt to analyze the way clinical practice guidelines help reinforce and perpetuate dominant medical perspectives. Moreover, to shed explanatory light, this chapter offers a synthetic explanation that combines materialist and ideological factors.*

Research implications – *Beyond the specific case of ADHD, this chapter has implications for understanding how and why environmental health research is omitted from other materials produced by mainstream medicine, such as materials found in the medical school curriculum, continuing medical education, medical journals, and on the medical association web sites.*

Keywords: Clinical practice guidelines; ADHD; dominant epidemiological perspectives; American Academy of Pediatrics; lead poisoning; environmental medicine

INTRODUCTION

In *An Invitation to Environmental Sociology* Michael Bell (2004) introduces the concept of "environment" to make the point that the environment isn't something that is external to us, but rather that we are inextricably intertwined with it through the water we drink, the food we eat, and the air we breathe. In turn, this deep connection means our treatment of the external environment affects human health. To demonstrate this point Bell uses a series of examples (including pesticide use, waste incineration, river pollution) that illustrate the way industrial pollution has led to increased human morbidity and mortality.

The exceptionally close relationship between health and environment is one that has been acknowledged in the medical literature since, at least, the beginning of the 20th century (Davis, 2007; Markowitz & Rosner, 2002).

However, this understanding has often been absent from mainstream medicine's conceptions of disease and their treatment protocols. To be clear, while some sectors of the medical profession (such as environmental health researchers) have been instrumental in identifying and explicating the pollutant/disease relationship, other sectors have actively suppressed the dissemination of this information (Davis, 2007). To shed light on this matter this chapter examines the role played by clinical practice guidelines, with a particular focus on the practice guidelines developed for Attention Deficit/Hyperactivity Disorder (ADHD), a neurobehavioral disorder characterized by difficulties of inattention, hyperactivity and/or impulsiveness, and which is estimated to affect 5% of children (Winterstein et al., 2008).

In trying to understand the reproduction of dominant disease conceptions, clinical practice guidelines are of strategic analytical interest because they are the Medical profession's systematic attempt at shaping the clinicians' diagnostic and treatment practices, as well as their understandings of disease. Relatedly, if clinical practice guidelines obfuscate the relationship between disease and pollutants, it will have profound consequences for diagnostic, treatment, and prevention activities. Moreover, the ADHD practice guidelines are of particular interest for two reasons, with one being the fact that treating clinicians are aware of, and integrate, the practice guidelines into their practice (Rushton, Fant, & Clark, 2004). Second, ADHD is a condition whose association with toxicant exposure has been well documented in recent decades (Schettler, Stein, Reich, Valenti, & Wallinga, 2000).

In what follows I articulate the conceptual framework guiding this analysis. Then I summarize the voluminous literature linking ADHD to toxicants, and analyse mainstream medicine's failure to incorporate those findings into ADHD clinical practice guidelines. Apart from demonstrating mainstream medicine's failure to acknowledge environmental medicine, I also examine the social consequences of that failure, which includes identifying who benefits from it, and who does not. The penultimate section considers the causes of the omission and provides a synthetic explanation that considers materialist and ideological factors. I conclude the chapter by highlighting implications of this analysis and making suggestions for future research. Beyond the ADHD case, this analysis provides insight for understanding how and why developers of clinical practice guidelines omit vital information about the relationship between pollution, disease, and health effects.

CONCEPTUAL BACKGROUND

The way we conceptualize a medical condition matters profoundly for how we treat/address that condition. For example, if asthma is primarily conceptualized as an individualistic, nonenvironmental, gene-based disorder, then physicians will emphasize symptom suppression, most probably through medications. On the other hand, if asthma is conceptualized as being brought on by environmental conditions (such as a moldy living environment or a maladapted diet), clinicians will steer patients to environment-changing interventions, such as mold remediation and/or diet alterations. In turn, the conceptualization of disease has significant economic implications for treatment manufacturers (such as pharmaceutical manufacturers) and medical subdisciplines, whose fortunes rise and fall according to the way a disease is conceptualized.

For these reasons many social scientists have researched the battles over the conceptualization of disease. In this regard the work of Brown and colleagues (2001) has been quite useful, as they recognize that competing etiological perspectives can exist for a particular condition (in their case, breast cancer), with some occupying more dominant positions than others. They argue that the position of each etiological perspective[1] is mediated by the activities of government agencies, the medical profession, the private sector (such as the pharmaceutical industry), and patients. While much research has illuminated the ways etiological paradigms are shaped by medical professions (Abbott, 1988; Pawluch, 1983; Smith, 1994), there is a dearth of research on the way professions influence etiological battles through clinical practice guidelines.

Clinical practice guidelines are documents that attempt to articulate high level medical care, and which aim to standardize the way clinicians diagnose and/or treat specific ailments (Nigam, 2012). The guidelines reflect the post-WW II emphasis on standardizing medicine, and the development of these guidelines has proliferated over the last two decades, with the National Guideline Clearinghouse currently listing more than two thousand guidelines on its website (Weisz et al., 2007). Clinical practice guidelines are particularly important for shaping dominant etiological perspectives as they are intended to systematically influence the clinical practices of all physicians addressing the condition in question. Moreover, while research has shown that the guidelines are only mildly effective for some conditions, they have been particularly effective for ADHD, as 91% of pediatricians are familiar with the ADHD guidelines, with 78% of them reading and incorporating them into their practice (Rushton et al., 2004).

To elucidate the role played by clinical practice guidelines I examine the case of ADHD, which is a neurobehavioral condition that mostly affects children. The dominant etiological perspective for ADHD is an individualistic one, which implicitly, and sometimes explicitly, attributes the disorder to a gene-based brain defect. This simplistic conceptualization is at odds with the most current genetic research, which views genetic expression as being influenced by both genes and the environment (Nigg, 2012; Schettler et al., 2000). It is also at odds with the environmental medicine research, which, since the early 1970s, has been producing a vast and burgeoning literature on the relationship between toxicant exposure and ADHD symptoms. In particular, this literature has highlighted the significant roles played by lead, cadmium, mercury, PCBs, dioxin, tobacco, and organophosphate pesticides (Grandjean & Landrigan, 2006; Hubbs-Tait, Nation, Krebs, & Bellinger, 2005; Schettler et al., 2000).

From a public health perspective the environmental medicine literature is particularly important for highlighting factors that are harming children which, in turn, provides knowledge with which society can develop and implement preventive measures. It is also important from a clinical perspective as it helps clinicians identify potential causes of disorders that otherwise go unacknowledged, and take measures towards mitigating these sources. Despite these benefits, however, mainstream medicine has done a particularly poor job of integrating environmental medicine research into ADHD clinical practice guidelines, thereby perpetuating a decontextualizing individualistic understanding of ADHD. In the next two sections I summarize the environmental medicine literature pertaining to ADHD and toxicants, and then demonstrate how practice guidelines systematically fail to acknowledge, let alone integrate, the findings of environmental medicine.

THE ENVIRONMENTAL MEDICINE LITERATURE ON ADHD

Why Focus on Chronic Lead Poisoning?

Over the last forty years environmental health researchers have found a growing number of toxicants to be associated with ADHD symptoms, including lead, mercury, manganese, cadmium, PCBs, dioxins, tobacco smoke, solvents, and organophosphate pesticides (Boucher et al., 2012;

Schettler et al., 2000). While each of these has a burgeoning literature about its association with ADHD and, thus, deserves detailed analysis, time and space limitations dictate that I restrict my focus to lead.

One reason for focusing on lead is that it was the first neurotoxicant linked with symptoms now associated with ADHD (Byers & Lord, 1943). A second is that lead is one of the two toxicants (mercury being the other) whose impact on human cognition is best documented (Hubbs-Tait et al., 2005). Third, lead exposure is seen as a significant public health issue: in 1999 one million American children (3.96%) had Blood Lead Levels (BLL) over the Centers for Disease Control's (CDC) then threshold level of concern (i.e., 10 μg/dL), which is the level at which cognitive and behavior effects were to be expected, and the level at which clinicians were expected to pursue intervention strategies (Schettler et al., 2000). In 2012 the CDC further lowered the reference value to 5 μg/dL, and now encourages clinicians to intervene at lower BLLs (CDC, 2012a). The lowered reference value means many more children are now estimated to suffer from lead-related problems than were previously believed. Indeed, the latest CDC data finds that 184,177 American children (i.e., 0.6%) six years or younger have BLLs between 5 and 9 μg/dL (CDC, 2012). Moreover, the most recent research suggests the even the CDC's new reference value is not low enough, as cognitive and behavioral effects can be found at blood-lead levels as low as 2 μg/dL (Braun, Kahn, Froehlich, & Auinger, 2006; Boucher et al., 2012; Canfield et al., 2003), and many researchers have concluded there are no "safe" levels of lead exposure for children. While the CDC does not track BLL under 5 μg/dL, if it did, there would be another dramatic increase in the number of children recognized as being lead-poisoned.

Another reason for focusing on lead are the impacts it has had on the nation's intellectual capacity and, in turn, economic productivity. Specifically, the lead exposure occurring between 1960 and 1980 reduced by 50% the children with far above average intelligence (IQ scores above 130 points), and increased the number of children with below 70 IQ scores (Needleman, Leviton, & Bellinger, 1982). Grosse, Matte, Schwartz, and Jackson (2002) estimate the ensuing lost productivity resulted in economic losses of between $110 and $319 billion USD for each cohort born during that period. Furthermore, Landrigan and colleagues (2002) estimate lead poisoning costs each contemporary birth cohort $43 billion. With regard to ADHD in particular, it is estimated that 21.1% of ADHD children (379,800) have BLLs over 2 μg/dL (Braun et al., 2006), which is conservatively estimated to cost families and society (through healthcare costs and lost productivity) $267 million USD annually (Gould, 2009).

Beyond children who are already exposed, there are also concerns for lead poisoning of future children. The Agency for Toxic Substances and Disease Registry estimates that 3–4 million children live within one mile of at least one hazardous-waste site (Hubbs-Tait et al., 2005), and the 2011 CERCLA National Priorities list identifies lead, along with arsenic and mercury, as one of the three hazardous substances of greatest concern at these sites (US ATSDR, 2011). As well, recent research shows that lead has contaminated our food stream. In a recent study, the Environmental Law Foundation (2010) (a nonprofit environmental group) evaluated 145 foods marketed to children, and found that 125 had lead levels that exceeded California's maximum allowable daily limit (Ashline, 2010)

Sources of Lead Exposure

Lead is a soft and malleable metal that has had myriad industrial uses, including building materials, bullets, fishing weights, pewter, toys, lead-acid batteries, radiation shields, pesticides (e.g., lead-arsenate), and as a gasoline additive. Over the last three hundred years there have been two sharp increases in global lead pollution, with one occurring in the 18th century (courtesy of the Industrial Revolution), and the other in the 20th century (courtesy of its use in plumbing fixtures, in paint pigments, as a gasoline additive, and as a pesticide (lead-arsenate) on fruit orchards) (Hubbs-Tait et al., 2005; Markowitz & Rosner, 2002). The latter increase was particularly deleterious to human health, with Hubbs-Tait and colleagues (2005) arguing:

> if one set out to maximize population exposures to a known neurotoxicant, it would be difficult to improve on the strategy of lining the walls of residences with it and insuring that it is emitted, submicron-sized and highly breathable, by a mobile source (the automobile) with a virtually unlimited reach. (p. 62)

Unsurprisingly, researchers have found that lead levels in contemporary humans far exceeds those found in premetallurgical humans. Whereas Flegal and Smith (1992) estimate the natural BLL of humans to be 0.016 μg/dL, researchers found the mean blood-lead level for American preschool children (1976–1980) to be 15 μg/dL, which is nearly 1,000 times higher (Hubbs-Tait et al., 2005), and 200% higher than the current reference level of 5 μg/dL.

Over the 20th century there were remarkable lead-related public health achievements, as farmers ceased using lead-arsenate as a pesticide, and lead was banned from paint in 1978. As well, in 1980 Japan banned lead from its gasoline, and most industrialized countries followed suit in the mid-1990s, with the United States phasing it out in 1996 (OECD, 1999). At the same time, however, many countries have failed to implement this important public-health measure, which means automobile engines in many countries are still pumping lead into the atmosphere. Moreover, industrial lead-emissions actually increased in some areas, as was the case in Massachusetts during 1991–1997, when lead emissions increased 77% (Schettler et al., 2000) and further poisoned the air inhaled by the citizens of that and neighboring states.

Additionally, while farmers stopped using lead-arsenate as a pesticide, no systemic remediation was undertaken to remove the lead that was deposited (either by pesticide, industry emissions, or gasoline emissions) over the previous century, which means it is still persisting in soils and is contaminating our food supply (Environmental Law Foundation, 2010; Peryea & Creger, 1994).

Another insidious source of lead includes residential tap water that is drawn from lead-based pipes, or from leaded solder that connects drinking water pipes (CDC, 2013). Another is toys. Over the last fifteen years the US Consumer Products Safety Commission (CPSC, n.d.) has ordered lead-based recalls on hundreds of children's toys, including wooden horses, action figures, crayons, art sets, children's jewellery, dolls, entertainment centers, and books.

The Impact of Lead on Learning

While severe lead poisoning (i.e., blood-lead levels of over 60 μg/dL) is associated with brain swelling, convulsions, and death, researchers have also found that less severe exposures are detrimental to children's neurological development. The first evidence of this came in 1943, when Byers & Lord reported that 19 out of 20 children exposed to lead had poor school performance, impulsive behavior, short attention span, and restlessness. Moreover, over the last three decades, researchers have established that mild lead poisoning correlates with numerous child neurodevelopmental problems, including lowered IQ (Bellinger, 2008; Lanphear et al., 2005), attention deficits (Chiodo, Jacobson, & Jacobson, 2004; Grandjean & Landrigan, 2006; Surkan et al., 2007), reduced inhibition (Chiodo et al., 2004; Surkan et al., 2007), and impulsiveness (Boucher et al., 2012; Grandjean & Landrigan, 2006; Schettler et al., 2000).

With regard to ADHD in particular, in their summary of the scientific literature Schettler et al. (2000) report that children with higher levels of lead exposure are more "distractible … impulsive, easily frustrated, not persistent, and unable to follow directions" (p. 60), each of which compromises a symptom in the DSM-IV diagnosis for ADHD. Moreover, in an editorial for the *Journal of Attention Deficit Disorder*, Bustos & Goldstein (2008) write "the significant correlation between blood and dentine lead levels on the one hand, and intelligence, behavior, and learning, on the other, suggest that lead may be a significant contributor to symptoms typically associated with ADHD in some children" (p. 425).

Lead levels are typically assessed through one of three mechanisms: hair analysis, teeth analysis, and blood analysis, with the latter being the most commonly used. With regard to ADHD, every study that has investigated the relationship between BLL and ADHD has found an association between the two, with only one study failing to reach the traditional $p \leq 0.05$ threshold for statistical significance (Eubig, Aguiar, & Schantz, 2010). The relationship was even present among children whose BLLs were *below* the CDC's recently lowered threshold of 5 μg/dL (Boucher et al., 2012; Canfield et al., 2003; Chiodo et al., 2004). For example, Boucher et al. (2012) observed the lead-ADHD association among children whose BLLs were between 1.6 and 2.7 μg/dL.

Reinforcing these points, several studies have found a positive relationship between BLL and ADHD diagnoses (Braun et al., 2006; Froehlich et al., 2009; Kim et al., 2010). In particular, Froehlich et al. (2009) found that children with BLL between 1.3–5 μg/dL had a more than twofold greater risk of meeting criteria for DSM-IV-defined ADHD than those with lower levels of lead (i.e., nondetectable to 0.8 μg/dL). Froehlich and colleagues (2009) estimate that for ADHD cases among 8 to 15 year old children, 25.4% of the cases (corresponding to 598,000 cases) could be attributable to lead exposures (BLLs of 1.3 to 5 μg/dL).

Recommendations from the Literature

Because of the weight of the evidence, numerous authorities (Bellinger, 2008; Bustos & Goldstein, 2008; CDC, 2012a; Jones, 2009; Landrigan, 2000) have concluded there are no "safe" levels of lead exposures, and many have used this as a call for systematically screening all children for lead exposure.

For example, in an editorial in the *Journal of Attention Disorders*, Bustos & Goldstein (2008) argue:

> Because a child's chronic exposure to low lead levels can cause developmental neurological anomalies that may be problematic to detect through routine physical examination, blood lead screenings become essential. All children are considered at risk and must be screened for lead poisoning as part of required prevention services offered through the Early and Periodic Screening, Diagnosis and Treatment Program (EPSDT). (p. 425)

Moreover, the US CDC (2007) argues "preventive care for every child should include obtaining an environmental history and identifying occupational lead exposure of family members" (p. 5), a recommendation that, ironically, is also supported by the American Academy of Pediatrics (2005).[2] As well, the CDC (2007) recommends clinicians "assess all children for developmental and behavior status and seek further evaluation and therapy to reduce developmental or behavioral problems, as necessary" and "consider the potential influences of lead when conducting developmental screening" (p. 10). Furthermore, CDC guidelines suggest performing "a diagnostic blood test on all children suspected of having lead exposure or an elevated BLL and institute the recommended management guidelines if a child's BLL increases to ≥10 μg/dL" (*ibid.*).

Beyond screening, numerous authors have argued that the effects at low BLLs confirm the need to conduct interventions that would reduce exposure (Boucher et al., 2012; Markowitz, 2000). Typically, the advocated interventions for lead poisoning can be broken down into four types of actions. First, there is the identification and elimination of all source of lead exposure (Dignam et al., 2008), which includes educating parents about sources and effects of lead (CDC, 2007; Gracia & Snodgrass, 2007; Markowitz, Bijur, Ruff, Balbi, & Rosen, 1996), and assessing home environments and other environments where the child spends much time. If the home environment is a significant source of lead, it is recommended to remove the child from the source of lead, either by abatement of the home or by placement in alternative lead-free housing, such as a relative's home (Guidotti & Ragain, 2007; Markowitz et al., 1996).

The second intervention strategy is behavior modification, which includes encouraging children to wash their hands regularly before eating and "stopping behaviors that bring lead-containing materials to the mouth, such as pica, nail biting," or licking toys and other household goods (Markowitz, 2000, p. 67). Behavior modification also includes "encouraging parents to remove loose paint, to wash surfaces regularly, to reduce surface dust, to observe their children closely, and to inhibit their inappropriate hand-to-mouth behavior" (*ibid.*). The third intervention is nutritional counselling, which consists, essentially, of ensuring that children are eating sufficient quantities of calcium and iron, as deficiencies in both have been associated with a lowered ability to excrete lead, higher blood-lead levels, and an exacerbation of lead effects (Konofal et al., 2007; Markowitz, 2000).

The fourth intervention is extracting lead from the child's body through the use of chemical chelators, such as CaNa2EDTA, succimer (DMSA), dimercaprol (BAL), and D-penicillamine, (Gracia & Snodgrass, 2007; Markowitz, 2000). There is a general consensus that children with BLLs over 45 μg/dL should be treated with these drugs (CDC, 1991; Markowitz, 2000). However, no consensus exists for children with BLLs under 45 μg/dL, a group for which there are three schools of thought. The first argues that, based on the established toxicity of lead in the 20 to 45 μg/dL range and given the relative safety of the available chelators, every attempt should be made to reduce lead levels in children, and that DMSA should be used (Markowitz, 2000). The second group argues against chelation, based on the lack of research showing the efficacy of chelation for BLLs under 45 μg/dL (*ibid.*). The third group acknowledges that lead is toxic below 45 μg/dL and emphasizes that there is at least a short-term benefit of reducing BLLs through chelation (*ibid.*).

ANALYZING THE CLINICAL PRACTICE GUIDELINES

Despite the vast and burgeoning environmental medicine literature on ADHD, the American Academy of Pediatrics (2001, 2011) has failed to acknowledge it in two of its three ADHD clinical practice guidelines, and only devote one paragraph to it (to argue against using lead screening to diagnosis ADHD) in the 2000 guidelines. While all three practice guidelines deserve closer scrutiny, this chapter focuses squarely on the 2011 practice guidelines, as they are the most current and, thus, had the greatest chance of incorporating the environmental medicine literature. The 2011 practice guidelines make no mention of lead, nor, for that matter, any other toxicant, which has significant implications for both diagnosis and treatment.

Regarding therapy, the 2011 practice guidelines emphasize behavior therapy and medications (depending on the age group), and completely fail to discuss remedial strategies for toxicant exposure, such as assessing the child's environments for toxicant exposures, remediation of the home environment, educating the parents about sources of potential toxicant exposures, behavior modification, and/or chelation therapies.

While the environmental medicine literature has conclusively demonstrated the association between lead exposures and learning problems in general, and ADHD symptoms in particular, it is somewhat understandable the AAP subcommittee chose to avoid recommending chelation therapies. For starters, while there is a consensus that children with BLLs over 45 μg/dL should be chelated, there is no clear consensus for children with BLLs under 45 μg/dL, which represents the bulk of children with lead poisoning (CDC, 2012). Second, research on treatment outcomes has been limited to a handful of studies, and the findings have been equivocal. On the one hand, two studies (Ruff, Bijur, Markowitz, Ma, & Rosen, 1993; Tong, Baghurst, Sawyer, Burns, & McMichael, 1998) found an association between drops in BLLs and IQ improvement. Additionally, a rodent study found that reducing brain-lead levels can lessen lead-induced impairments in learning ability, attention, and the regulation of arousal and/or emotion (Stangle et al., 2007). On the other hand, a large randomized human study found that chelation only slightly improved cognitive outcomes for children with BLLs between 20 and 44 μg/dL (Dietrich et al., 2004; Rogan et al., 2001). While Stangle et al. (2007) hypothesize the latter findings could be due to the fact chelation was not done long enough to sufficiently reduce brain-lead, no one has tested this hypothesis. Third, beyond the lack of evidence about chelation's effectiveness on cognitive functioning, there may also be safety concerns with the intervention, as medical errors involving chelation led to the deaths of two children (CDC, 2006).[3]

Apart from avoiding chelation therapy, the subcommittee failed to recommend any other lead management strategy recommended in the environmental medicine literature. While it would be unreasonable to expect clinicians to initiate, let alone manage, the assessment and remediation of the child's home environment, it is not unreasonable to expect clinicians to, at a minimum, alert parents to this potential problem, provide them with the educational resources to help them reduce exposures, and provide them with contact information for the public health agencies that could help assess the child's environment. Failing to do so is failing to help the estimated 21.1% of children who are estimated to have BLL associated with learning problems (i.e., BLLs > 2 μg/dL) (Braun et al., 2006).

Beyond therapy recommendations, the subcommittee also made diagnostic recommendations, with one of their six action statements being that primary clinicians should "rule out any alternative cause" for ADHD

(AAP, 2011, p. 1012). Despite that recommendation, however, the committee made no mention of having children evaluated for lead or any other toxicant. The omission means clinicians are not alerted to a possible factor that could be contributing to the symptoms in question, which means they are more likely to settle on an ADHD diagnosis, and contribute to an overdiagnosis of the disorder.

This finding is curious in light of: (1) the explicit action statement to "rule out any alternative cause," (2) the robust environmental medicine literature showing an association between lead exposure and ADHD symptoms, and (3) the strong recommendations by the CDC and other authorities to screen *all children* with developmental issues (Bustos & Goldstein, 2008; CDC, 2007). If the guidelines were the only item read about ADHD, one would get the distinct impression that toxicant exposures are in no way connected to ADHD. Moreover, it begs the question why the subcommittee would choose to give this impression. In the penultimate section I provide a preliminary explanation for this discrepancy. Before that, however, I discuss the social consequences of the subcommitte's decision.

THE SOCIAL CONSEQUENCES

The AAP's decision to ignore environmental medicine has many social consequences as many are harmed by it, while others actually benefit from the approach. Those harmed include the 21.1% (Braun et al., 2006) of children with ADHD who are estimated to have BLLs over 2 μg/dL, and who are not being evaluated for lead exposure. The clinicians' failure to consider lead exposure means that, a central source of the symptoms is being overlooked, which condemns children to a resource-sapping dependence on physicians for symptom-suppressing psychiatric medication. Adding insult to injury, many of these children end up suffering from side effects associated with these medications (including suppressed appetite, growth impairment, restlessness, tremors, insomnia, hypertension and cardiac arrest (Rappley, 1997)), thereby compounding their problems. Apart from the children, their parents and relatives are also psychologically, emotionally, and financially impacted by having to care for a child with ADHD symptoms.

Beyond the family, schools and school districts are increasingly taxed with providing special education resources to a growing number of children (Chambers, Shkolnik, & Pérez, 2003; Parrish, Harr, Anthony, Merickel, & Esra, 2003). This is partially salient in these "austere" times, associated with cutbacks in education funding. Then there are taxpayers, who fund much of the healthcare and special education costs for these harmed children. Finally, failing to highlight and address lead poisoning means the problem will continue to exist outside both the public consciousness and political sphere, thereby condemning many more children to future lead exposure, and to potentially developing the ADHD symptoms that will tax them, their families, their schools, and the tax-paying base.

Besides those who are harmed by ignoring environmental medicine, numerous entities actually benefit from it, including drug manufacturers, the medical profession, and industries. As long as American pediatrics continues to ignore the presence and effect of toxicants, they will favor symptom-suppressing treatment-based paradigms over preventive ones. Moreover, while there are a number of treatments offered for ADHD (including cognitive therapy, homeopathy, acupuncture, dietary intervention, and parental training), pharmaceutical medications are currently the most favored intervention in America, which means pharmaceutical manufacturers benefit most from the problem.

American pediatrics is, arguably, another group that benefits from ignoring environmental medicine, as failing to mitigate or prevent toxicant poisoning guarantees children will continue to develop ADHD symptoms, thereby ensuring a flow of patients. This situation sits in contrast to the 1950s (Pawluch, 1983), when improving public health drastically reduced the number of sick children, and reduced demand for pediatricians.

A third group benefitting from the problem are what John Mckinlay (1994) calls the "manufacturers of disease," which, in this case, includes businesses who create products with high levels of lead and/or use production processes with high lead emission rates (Markowitz & Rosner, 2002). As long as mainstream medicine continues to ignore the relationship between lead exposure and ADHD symptoms, attention will be cast away from the manufacturers and industries who contribute to ADHD's proliferation. While many people are familiar with lead exposure from leaded gasoline and lead paint, lead exposure occurs through a number of other consumer products, including window blinds, toys, plastic products, clothing, and foods, such as juices and baby foods (which are consumed in elevated amounts, by bodyweight, by infants and children, at ages critical to neurodevelopment). While many manufacturers use lead to lower production costs, the subsequent poisoning effectively transfers the burden of production costs to the harmed children, their families, and the larger society. In a sense, one could argue society is subsidizing industry's poisonous production practices, and the medical profession, wittingly or not, is helping to perpetuate this situation.

ACCOUNTING FOR THE OMISSION

In trying to account for the practice guideline's omission of neurotoxicant information, I turn to two sociological explanations: (1) professional struggles over jurisdiction, and (2) mainstream medicine's adherence to a worldview that dismisses preventive efforts.

Many sociologists have analyzed the way the medical profession and its subdisciplines continuously seek to expand, or at least maintain, their professional jurisdiction (Abbott, 1988; Pawluch, 1983; Zola, 1972). As well, sociologists have identified both how this drive for jurisdiction expansion invariably leads to turf battles between medical disciplines, and the tactics medical disciplines deploy in these struggles (Abbott, 1988; Conrad & Schneider, 2009; Winnick, 2005).

Of particular pertinence is Andrew Abbott's work, which argues that medical disciplines will shape their medical approach and treatment practices according to that which gives them the best chance of expanding, or at least protecting, their jurisdiction. As an example, Abbott highlights the case of psychiatry, which shifted from a psycho-social approach to a biological approach, as a result of growing competition from other mental health disciplines. Specifically, while psychiatry ruled the mental health world in the post-WWII decades, in the late 1960s and early 1970s it began to face growing competition from other mental health providers (such as psychologists and social workers), which prompted the discipline to emphasize pharmacological treatments over psychotherapeutic ones, "virtually ceding the psychotherapeutic world to its competitors" (pp. 312–313).

With regard to pediatrics, Pawluch (1983) elucidated how a drop in infant mortality during the 1940s prompted the pediatrics discipline to expand their jurisdiction, in the 1950s, to include child behavior problems, which included addressing the behavioral symptoms we now associate with ADHD. In the 1960s their hold over the field was strengthened with the emergence of methylphenidate (also known as Ritalin), an amphetamine-based drug that gave them a demonstratable way to suppress symptoms, and, thus, increased their "scientific credibility" (Epstein, 1996) to pronounce authoritatively about the condition.

At the core of this ADHD treatment is an etiological perspective that ignores any potential environmental causes of the symptoms, and implicitly, when not explicitly, attributes the condition to a brain defect. Whether intentional or not, upholding this etiological perspective is in the pediatric discipline's financial interests, as it ensures that underlying environmental sources of ADHD symptoms will go unaddressed, thus ensuring a steady stream of patients. From a materialist perspective it makes sense that the American Academy of Pediatrics would avoid any mention of environmental contributors to ADHD, for doing otherwise could undermine the discipline's explanatory model for ADHD, their rational for administering psychostimulants, their scientific credibility to speak authoritatively about this condition, and eventually their control over the market for ADHD patients. On the other hand, while a materialist analysis identifies objective economic interests, such an analysis needs to be complemented with an analysis of ideology.

A Reductionist Worldview

Beyond market competition, the guidelines' omission of environmental medicine can be attributed to a worldview that is averse to preventive perspectives. Brandt and Gardner (2000) argue that throughout the 20th century the American medical profession has been antagonistic towards American public health and hostile to preventive initiatives. For example, in the 1920s organized medicine (including the AMA and its lobbyists) derailed proposals for neighborhood public health centers, due to concerns that the centers would provide "free care" to potential paying patients (Brandt & Gardner, 2000). As well, for similar reasons, in 1921 the medical profession defeated public health attempts to provide prenatal and postnatal care for mothers and infants (*ibid.*). Likewise, while American public health championed universal health care for Americans throughout the 20th century, American medicine consistently opposed it (Quadagno, 2009).

While mainstream American medicine has aggressively protected its economic interests against public health, Brandt and Gardner (2000) caution against attributing this hostility to mere self-interest. While the profession has demonstrated plenty of self-interested behavior, Brandt and Gardner argue the hostility can also be traced to medicine's deep ideological conviction in the biomedical paradigm and its disease-focused reductionism. This approach, gave physicians an understanding of disease and its cure that enabled them to decouple disease from its social roots, and made them think public health's broad social and environmental agenda was unnecessary. The physicians' attraction to the biomedical paradigm was due to its "science-based objectivity and technique," which "contrasted with the tumultuous world of public health" (p. 711). Physicians were deeply skeptical about public health as they believed efforts to change disease-causing social conditions were difficult to begin with, and rendered even more difficult by the fact these efforts are tainted by "politics, advocacy, individual noncompliance, and social diversity" (*ibid.*). In the physicians' worldview, "medicine could not solve the problems of poverty, illiteracy, and inequity–but it could, at least potentially, cure the diseases that these social forces produced" (*ibid.*). Moreover, in many instances medical advocates argue that the improved health and longevity brought by medical interventions would ultimately "reduce poverty and other social inequities" (*ibid.*). For these reasons, physicians believed "public health was best served by the medical care and treatment offered to individual patients by individual doctors" (*ibid.*).

Importantly, this professional attitude has had a strong influence on medical training and, consequently, medical practice, orienting both toward disease-focused reductionism, and away from the community and disease-causing social forces (Brandt & Gardner, 2000). While many physicians of the 1920s recognized the importance of prevention-oriented research and teaching, these goals were obscured by the medical schools' focus on intensive scientific and clinical training (*ibid.*). Moreover, while reformers attempted to introduce preventive medicine at many schools in the 1930s and 1940s, these attempts failed to alter the dominance of the disease-focused reductionist paradigm (*ibid.*).

The issue is particularly germane to contemporary pediatrics, as medical schools continue to under-emphasize prevention-oriented environmental medicine. Kilpatrick and colleagues (2002) found that one in four medical schools offer no instruction in environmental medicine, and those that do average less than 10 hours of instruction over four years. Similarly, over two thirds of medical school deans reported their school's emphasis on environment is "minimal" (Graber, Musham, Bellack, & Holmes, 1995), with a similar pattern prevailing in residence training (Kilpatrick et al., 2002). Moreover, in her survey of pediatricians, Kilpatrick et al. (2002) found that only 14.9% had received specific training in environmental history-taking, and less than 22% own a copy of the *Handbook of Pediatric Environmental Health*. In turn, this had an impact on medical practice as pediatricians "reported low self-efficacy regarding environmental history-taking, discussing environmental exposures with parents, finding diagnosis and treatment related to environmental exposures" (*ibid.*, p. 826). Thus, while the pediatrics profession has a financial interest in producing practice guidelines that promote an individualistic understanding of ADHD, it seems the absence of toxicant information is also part of a historical tradition of ignoring environmental medicine.

CONCLUSION

This chapter began by discussing Michael Bell's (2004) concept of the "environment," which captures the notion that our health is inextricably intertwined with the health of the environment. Acknowledging and integrating this biological principle would help mainstream medicine be more cognizant of environmental pollutants and their role in producing disease, as well as help patients both undo the damage of chemical exposures and avoid health-damaging exposures in the future. Instead, however, mainstream medicine perpetuates individualizing understandings of disease that obscure the environmental causes of disease and obfuscate industry's role in the production of disease, both of which disempower individuals, families and communities.

To illustrate these points I analyzed the AAP's 2011 Clinical Practice Guidelines for ADHD, and demonstrated that despite the wealth of environmental medicine research showing a link between lead and ADHD the AAP summarily ignored that research in its ADHD guidelines. Beyond tracing the social consequences of the omission, I argued the omission could be traced to two factors: (1) the competitive pressures that drive medical professions to pursue jurisdiction-protecting behavior, and (2) the medical profession's adherence to a reductionist worldview that is dismissive, if not hostile, to preventive approaches.

While this chapter focused on the exclusion of lead research in the ADHD clinical practice guidelines, similar cases could be made for the exclusion of other toxicants associated with ADHD, such as mercury, manganese, cadmium, dioxins, PCBs, organophosphate pesticides, and tobacco (Boucher et al., 2012, Schettler et al., 2000). This is particularly true of mercury, due to the many sources of exposures, including dental amalgams, fish consumption, atmospheric deposition from coal-fired power plants, and exposures to broken compact fluorescent light bulbs (Clifton, 2007). Moreover, cadmium and manganese are a growing concern, due to the growing amounts industry is releasing into the atmosphere (Grandjean & Landrigan, 2006).

While this study focused on the ADHD practice guidelines, sociologists are likely to find similar omissions in most practice guidelines produced by the American Academy of Pediatrics, as well as, for that matter, American mainstream medicine more generally. Still, sociologists should identify the existing variation, as such variation can provide analytical leverage with which to identify the economic, political, professional, cultural and other social forces that shape dominant etiological perspectives and clinical practices.

Analytical leverage can also be achieved by analyzing differences between the clinical practice guidelines of different countries, as such differences often reflect the way medicine is influenced by significant social forces. For instance, in her study of France, Germany, the United Kingdom and the United States, Lynn Payer (1996) found significant differences between the medical practices of each country and, in many cases, the variation could be traced to cultural differences. Moreover, Vallée (2009) found that significant French–American differences in psychostimulant usage could be traced to significant differences in the organization of the ADHD clinical practice field, with pediatricians (with their reductionist conceptualization of ADHD) dominating the American field, and child psychiatrists (with their psycho-affective conceptualization) exerting a strong influence over the French field.

Beyond focusing on clinical practice guidelines, sociologists need to investigate how the guidelines square with the medical profession's other ideology-shaping activities, such as its policy statements, the industry-sponsored seminars it organizes, the medical education it oversees, and the continuing medical education conferences it sponsors. Divergences between these other activities and the guidelines can also illuminate the powerful forces shaping medicine. For example, while the AAP summarily ignores lead in its ADHD guidelines, in 2005 it published a policy statement on preventing lead exposure in children. Such a discrepancy might be explained by the conflicts-of-interest possessed by the developers of the guidelines, as Lisa Cosgrove's (2011) research has found in other cases. An alternative explanation might be found in the fact that ADHD is a behavioral disorder, and that perhaps the AAP feels its grasp over this field is more tenous. In any event, studying these types of discrepancies can shed important light on the social forces that are affecting the development of clinical practice guidelines, and, by extension, medical practice.

In trying to understand how mainstream medicine omits environmental medicine, sociologists also need to study the role played by mainstream medical journals (such as *JAMA* and *The New England Journal of Medicine*), which includes investigating the presence of environmental perspectives and whether environmental medicine research faces access barriers. Furthermore, we need to better understand the processes through which medical associations influence the larger public, such as through press releases, shaping entertainment content, and/or other marketing activities.

Better understanding of these issues can help short-circuit social processes that currently marginalize and, in some instances, exclude environmental medicine research. Consequently, liberating that research would help society change behaviors that are currently damaging the environment and, in turn, human health.

NOTES

1. Where Brown and colleagues use the more encompassing term "dominant etiological paradigm," I use "dominant etiological perspective" because I'm focusing strictly on etiological perspectives.

2. The AAP recommendation for taking an environmental history of all children is at odds with the the 2011 AAP clinical practice guidelines for ADHD, which make absolutely no mention of lead.

3. At the same time, it bears mentioning that "medical errors" are due to errors by the administrator of the medication, as opposed to the inherent danger of the medication. Moreover, deaths due to medical errors are not uncommon. For example, between 2000 and 2004 the American Association of Poison Control Center received 238 reports of children severely injured by medical errors, with 21 cases resulting in death (Tzimenatos, Bond, & Pediatric Therapeutic Error Study Group, 2009).

REFERENCES

Abbott, A. (1988). *Systems of professions: An essay on the division of expert labor*. Chicago: University of Chicago Press.

American Academy of Pediatrics. (2000). Clinical practice guideline: Diagnosis and evaluation of the child with attention-deficit/hyperactivity disorder. *Pediatrics*, *105*, 1158–1170.

American Academy of Pediatrics. (2001). Clinical practice guideline: Treatment of the school-aged child with attention-deficit/hyperactivity disorder. *Pediatrics*, *108*(4), 1033–1044.

American Academy of Pediatrics. (2005). Lead exposure in children: Prevention, detection, and management. *Pediatrics*, *116*, 1036–1046.

American Academy of Pediatrics. (2011). ADHD: Clinical practice guideline for the diagnosis, evaluation, and treatment of attention-deficit/hyperactivity disorder in children and adolescents. *Pediatrics*, *128*(5), 1007–1022.

Ashline, J. (2010, June 18). Lead in baby food? Here's the real story. *Orange County Register*. Retrieved from http://themomblog.blog.ocregister.com/2010/06/18/lead-in-baby-food-heres-the-real-story/56983/

Bell, M. (2004). *An invitation to environmental sociology* (2nd ed.). Thousand Oaks, CA: Pine Forge Press.

Bellinger, D. (2008). Neurological and behavioral consequences of childhood lead exposure. *Public Library of Science – Medicine*, *5*(5), e115.

Boucher, O., Jacobson, S., Plusquellec, P., Dewailly, E., Ayotte, P., Forget-Dubois, N., ... Muckle, G. (2012). Prenatal methylmercury, postnatal lead exposure, and evidence fo attention deficit/hyperactivity disorder among inuit children in arctic québec. *Environmental Health Perspectives*, *120*(10), 1456–1461.

Brandt, A., & Gardner, M. (2000). Antagonism and accommodation: Interpreting the relationship between public health and medicine in the United States during the 20th century. *American Journal of Public Health*, *90*, 707–715.

Braun, J. M., Kahn, R. S., Froehlich, T., & Auinger, P. (2006). Exposures to environmental toxicants and attention deficit hyperactivity disorder in U.S. children. *Environmental Health Perspectives*, *114*, 1904–1909.

Brown, P., Zavestoski, S., McCormick, S., Mandelbaum, J., & Luebke, T. (2001). Print media coverage of environmental causation of breast cancer. *Sociology of Health and Social Behavior*, *35*, 34–52.

Bustos, R., & Goldstein, S. (2008). Including blood lead levels of all immigrant children when evaluating for ADHD. *Journal of Attention Disorders*, *11*(4), 425–426.

Byers, R., & Lord, E. (1943). Late effects of lead poisoning on mental development. *American Journal of Diseases in Children*, *66*, 471–494.

Canfield, R., Henderson, C., Cory-Slechta, D., Cox, C., Jusko, T., & Lanphear, B. (2003). Intellectual impairment in children with blood lead concentrations below 10 μg per decilitre. *New England Journal of Medicine*, *348*, 1517–1529.

Centers for Disease Control (CDC). (1991). *Preventing lead poisoning in young children: A statement by the Centers for Disease Control*. Atlanta: CDC.

Centers for Disease Control (CDC). (2006). Deaths associated with hypocalcemia from chelation therapy — Texas, Pennsylvania, and Oregon, 2003–2005. *MMWR*, *55*(08), 204–207.

Centers for Disease Control (CDC). (2007). Interpreting and managing blood lead levels <10 μg/dL in children and reducing childhood exposures to lead. *MMWR*, *56*(RR08), 1–14.

Centers for Disease Control (CDC). (2013). *About lead in drinking water*. Retrieved from http://www.cdc.gov/nceh/lead/leadinwater/

Centers for Disease Control (CDC). (2012). *CDC's National Surveillance Data (1997–2011)*. Retrieved from http://www.cdc.gov/nceh/lead/data/StateConfirmedByYear1997-2011.htm

Centers for Disease Control (CDC). (2012a). *Low level lead exposure harms children: A renewed call for primary prevention*. Retrieved from http://www.cdc.gov/nceh/lead/ACCLPP/Final_Document_030712.pdf

Chambers, J., Shkolnik, J., & Pérez, M. (2003). *Total expenditures for students with disabilities, 1999–2000: spending variation by disability*. Retrieved from http://csef.air.org/publications/seep/national/final_seep_report_5.pdf

Chiodo, L. M., Jacobson, S. W., & Jacobson, J. L. (2004). Neuro-developmental effects of postnatal lead exposure at very low levels. *Neurotoxicology and Teratology*, *26*, 359–371.

Clifton, J. (2007). Mercury exposure and public health. *Pediatric Clinics of North America*, *54*, 237–269.

Conrad, P., & Schneider, J. (2009). Professionalization, monopoly, and the structure of medical practice. In P. Conrad (Ed.), *The sociology of health & illness: Critical perspectives* (pp. 194–200). New York, NY: Worth Publishers.

Consumer Product Safety Commission (CPSC). (n.d.). Retrieved from http://cs.cpsc.gov/ConceptDemo/SearchCPSC.aspx?fromDate=01%2f01%2f1998&toDate=05%2f08%2f2013&query=lead+recalls

Cosgrove, L. (2011). The DSM, big pharma, and clinical practice guidelines: Protecting patient autonomy and informed consent. *The International Journal of Feminist Approaches to Bioethics*, *4*, 11–25.

Davis, D. (2007). *The secret history of the war on cancer*. New York, NY: Basic Books.

Dietrich, K., Ware, J., Salganik, M., Radcliffe, J., Rogan, W., Rhoads, G., ... Jones, R. (2004). Treatment of lead-exposed children clinical trial group: Effect of chelation therapy on neuropsychological and behavioral development of lead-exposed childre after school entry. *Pediatrics*, *114*, 19–26.

Dignam, T., Logo, J., Meyer, P., Norman, E., Sayre, A., & Flanders, W. (2008). Reduction of elevated blood lead levels in children in North Carolina and Vermont, 1996–1999. *Environmental Health Perspectives*, *116*(7), 981–985.

Environmental Law Foundation. (2010). *Lead in children's foods*. Retrieved from http://envirolaw.org/

Epstein, S. (1996). *Impure science: AIDS, activism, and politics of knowledge*. Berkeley, CA: University of California Press.

Eubig, P., Aguiar, A., & Schantz, S. (2010). Lead and PCBs as risk factors for attention deficit/hyperactivity disorder. *Environmental Health Perspectives*, *118*(12), 1654–1667.

Flegal, A. R., & Smith, D. R. (1992). Lead levels in preindustrial humans. *New England Journal of Medicine*, *326*, 1293–1294.

Froehlich, T. E., Lanphear, B. P., Auinger, P., Hornung, R., Epstein, J. N., Braun, J., & Kahn, R. (2009). Association of tobacco and lead exposures with attention-deficit/hyperactivity disorder. *Pediatrics*, *124*, e1054–e1063.

Gould, E. (2009). Childhood lead poisoning: Conservative estimates of the social and economic benefits of lead hazard control. *Environmental Health Perspectives*, *117*(7), 1162–1167.

Graber, D. R., Musham, C., Bellack, J. P., & Holmes, D. (1995). Environmental health in medical school curricula: Views of academic deans. *Journal of Occupational and Environmental Medicine*, *37*(7), 807–811.

Gracia, R., & Snodgrass, W. (2007). Lead toxicity and chelation therapy. *American Journal of Health-System Pharmacy*, *64*, 45–53.

Grandjean, P., & Landrigan, P. (2006). Developmental neurotoxicity of industrial chemicals. *Lancet*, *368*, 2167–2178.

Grosse, S., Matte, T., Schwartz, J., & Jackson, R. (2002). Economic gains resulting from the reduction in children's exposure to lead in the United States. *Environmental Health Perspectives*, *110*, 563–569.

Guidotti, T., & Ragain, L. (2007). Protecting children from toxic exposure: Three strategies. *Pediatric Clinics of North America*, *54*, 227–235.

Hubbs-Tait, L., Nation, J., Krebs, N., & Bellinger, D. (2005). Neurotoxicants, micronutrients, and social environments: Individual and combined effects on children's development. *Psychological Science in the Public Interest*, *6*(3), 57–121.

Jones, A. (2009). Emerging aspects of assessing lead poisoning in childhood. *Emerging Health Threats Journal*, *2*, e3.

Kilpatrick, N., Frumkin, H., Trowbridge, J., Escoffrey, C., Geller, R., Rubin, L., ... Nodvin, J. (2002). The environmental history in pediatric practice: A study of pediatricians' attitudes, beliefs, and practices. *Environmental Health Perspectives*, *110*(8), 823–827.

Kim, Y., Cho, S., Kim, B., Hong, Y., Shin, M., Yoo, H., ... Bhang, S. (2010). Association between blood lead levels (< 5 μg/dL) and inattention-hyperactivity and neurocognitive profiles in school-aged Korean children. *Science of the Total Environment*, *408*, 5737–5743.

Konofal, E., Cortese, S., Marchand, M., Mouren, M. C., Arnulf, I., & Lecendreux, M. (2007). Impact of restless legs syndrome and iron deficiency on attention deficit/hyperactivity disorder in children. *Sleep Medicine*, *8*, 711–715.

Landrigan, P. (2000). Pediatric lead poisoning: Is there a threshold? *Public Health Reports*, *115*, 530–531.

Landrigan, P., Schecter, C., Lipton, J., Fahs, M., & Schwartz, J. (2002). Environmental pollutants and disease in American children: Estimates of morbidity, mortality, and costs for lead poisoning, asthma, cancer, and developmental disabilities. *Environmental Health Perspectives*, *110*, 721–728.

Lanphear, B., Hornung, R., Khoury, J., Yolton, K., Baghurst, P., Bellinger, D., ... Roberts, R. (2005). Low-level environmental lead exposure and children's intellectual function: An international pooled analysis. *Environmental Health Perspectives*, *113*(7), 894–899.

Markowitz, G., & Rosner, D. (2002). *Deceit and denial: The deadly politics of industrial pollution*. Berkeley, CA: University of California Press.

Markowitz, M. (2000). Lead poisoning: A disease for the next millenium. *Current Problems in Pediatrics*, *30*(3), 62–70.

Markowitz, M., Bijur, P., Ruff, H., Balbi, K., & Rosen, J. (1996). Moderate lead poisoning: Trends in blood lead levels in unchelated children. *Environmental Health Perspectives*, *104*(9), 968–972.

Mckinlay, J. (1994). A case of refocusing upstream: The political economy of illness. In P. Conrad (Ed.), *The sociology of health and illness: A critical approach* (pp. 551–561). New York, NY: Palgrave Macmillan.

Needleman, H., Leviton, A., & Bellinger, D. (1982). Lead-associated intellectual deficit (letter). *New England Journal of Medicine*, *306*, 367.

Nigg, J. (2012). Future directions in ADHD etiology research. *Journal of Clinical Child & Adolescent Psychology*, *41*(4), 524–533.

Nigam, Amit. (2012). Changing health care quality paradigms: The rise of clinical guidelines and quality measures in American medicine. *Social Science & Medicine*, *75*, 1933–1937.

OECD. (1999). *Phasing lead out of gasoline: An examination of policy approaches in different countries*. Paris: OECD.

Parrish, T., Harr, J., Anthony, J., Merickel, A., & Esra, P. (2003). *State special education finance systems, 1999–2000*. Retrieved from http://csef.air.org/publications/csef/state/statpart1.pdf

Pawluch, D. (1983). Transitions in pediatrics: A segmental analysis. *Social Problems*, *30*(4), 449–465.

Payer, L. (1996). *Medicine and culture: Varieties of treatment in the United States, England, West Germany, and France*. New York, NY: Henry Holt and Company.

Peryea, F., & Creger, T. (1994). Vertical distribution of lead and arsenic in soil contaminated with lead arsenate pesticide residues. *Water, Air and Soil Pollution*, *78*, 297–306.

Quadagno, J. (2009). Why the United States has no national health insurance: Stakeholder mobilization against the welfare state, 1945–1996. In Peter Conrad (Ed.), *The Sociology of Health & Illness: Critical Perspectives* (pp. 301–320). New York, NY: Worth Publishers.

Rappley, M. (1997). Safety issues in the use of methylphenidate: An American Perspective. *Drug Safety*, *17*(3), 143–148.

Rogan, W., Dietrich, K., Ware, J., Dockery, D., Salganik, M., Radcliffe, J., … Rhoads, G. (2001). The effect of chelation therapy with succimer on neuropsychological development in children exposed to lead. *New England Journal of Medicine*, *344*, 1421–1426.

Ruff, H., Bijur, P., Markowitz, M., Ma, Y., & Rosen, J. (1993). Declining blood lead levels and cognitive changes in moderately lead-poisoned children. *JAMA*, *13*, 1641–1646.

Rushton, J., Fant, K., & Clark, S. (2004). Practice guidelines in the primary care of children with attention-deficit/hyperactivity disorder. *Pediatrics*, *114*(1), e23–e28.

Schettler, T., Stein, J., Reich, F., Valenti, M., & Wallinga, D. (2000). *In harm's way: Toxic threats to child development*. Retrieved from www.igc.org/psr

Smith, B. (1994). Black lung: The social production of disease. In P. Conrad (Ed.), *The sociology of health and illness: A critical approach* (pp. 62–75). New York, NY: Palgrave MacMillan.

Stangle, D., Smith, D., Beaudin, S., Strawderman, M., Levitsky, D., & Strupp, B. (2007). Succimer chelation improves learning, attention, and arousal regulation in lead-exposed rates but produces lasting cognitive impairment in the absence of lead exposure. *Environmental Health Perspectives*, *115*(2), 201–209.

Surkan, P., Zhang, A., Trachtenberg, F., Daniel, D., McKinlay, S., & Bellinger, D. (2007). Neuropsychological function in children with blood lead levels <10 μg/dL. *Neurotoxicology*, *28*, 1170–1177.

Tong, S., Baghurst, P., Sawyer, M., Burns, J., & McMichael, A. (1998). Declining blood lead levels and changes in cognitive function during childhood: The Port Pirie Cohort Study. *JAMA*, *280*, 1915–1919.

Tzimenatos, L., Bond, G., & Pediatric Therapeutic Error Study Group. (2009). Severe injury or death in young children from therapeutic errors: A summary of 238 cases from the American Association of Poison Control Center. *Clinical Toxicology*, *47*, 348–354.

United States Agency for Toxic Substances and Disease Registry (US ATSDR). (2011). *The ATSDR 2011 substance priority list*. Retrieved from http://www.atsdr.cdc.gov/SPL/index.html

Vallée, M. (2009). *Deconstructing America's Ritalin epidemic: Contrasting US–France Ritalin usage*. Doctoral dissertation. Retrieved from http://udini.proquest.com/view/deconstructing-americas-ritalin-goid:506688338/

Weisz, G., Cambrosio, A., Keating, P., Knaapen, L., Schlich, T., & Tournay, V. (2007). The emergence of clinical practice guidelines. *The Milbank Quarterly*, *85*(4), 691–727.

Winnick, T. (2005). From quackery to 'complementary' medicine: The American profession confronts alternative therapies. *Social Problems*, *52*(1), 38–61.

Winterstein, A., Gerhard, T., Shuster, J., Zito, J., Johnson, M., Liu, H., & Saidi, A. (2008). Utilization of pharmacologic treatment in youths with attention deficit/hyperactivity disorder in medicaid database. *The Annals of Pharmacotherapy*, *42*(1), 24–31.

Zola, I. (1972). Medicine as an insitution of social control. *Sociological Review*, *20*, 487–504.

PART IV
SOCIAL-ECOLOGICAL APPROACHES TO STUDYING INSTITUTIONAL CHANGE FOR HEALTH

A SUSTAINABLE DEVELOPMENT AGENDA FOR THE UK NATIONAL HEALTH SERVICE (NHS): AN ORGANIZATIONAL LEARNING MODEL FOR DEFINING AND SUPPORTING GOALS

Claire Marsh

ABSTRACT

Purpose – *This chapter presents an organizational learning approach to understanding the objectives and challenges of an National Health Service (NHS) Sustainable Development agenda, which involves the integration of social, ecological and economic concerns into organizational functioning, for example the construction and management of buildings, design and delivery of services, and employment of staff.*

Methodology – *The approach is used to frame an empirical analysis of 11 Projects conducted by NHS organizations aiming to advance this agenda during a particularly active phase in the mid to late 2000s. The approach helps identify the assumptions of organizational purpose, strategy and*

Ecological Health: Society, Ecology and Health
Advances in Medical Sociology, Volume 15, 267–288

ISSN: 1057-6290/doi:10.1108/S1057-6290(2013)0000015016

practice inherent in proposals for Sustainable Development and expose the challenges these are likely to pose. This framing helps articulate a vision and identify the actor groups, and their guiding assumptions, which need to be engaged if progress is to be made.

Findings – *The vision of Sustainable Development being promoted was predominantly one of an NHS contributing to the economic and social determinants of health through its procurement, service development and employment activities. Contributions to environmental determinants only took place where financial gains to the NHS organizations themselves made activities, such as investment in renewable energy, viable in the short-term. Within most Projects strategic tools able to predict and measure benefits had to be developed on-the-job and most received help from external agencies to do this work.*

Social implications – *Rather than expecting individual NHS organizations to progress this agenda alone, others involved in the shaping of collective assumptions of the NHS' purpose and strategies for growth must be engaged in what can be viewed as a social process of learning.*

Keywords: Organizational learning; sustainable development; NHS change; public health

INTRODUCTION

This chapter presents an organizational learning model as a guide to understanding the priorities and challenges associated with proposals for integrating principles of Sustainable Development across the UK National Health Service (NHS). Key components of this agenda are first identified as progressing from efficient management of resources (e.g. waste and carbon dioxide emissions associated with buildings) to the proactive integration of the principles of Sustainable Development into all its activities in the name of health promotion. The organizational learning model is used to explore why such proposals significantly challenge dominant assumptions of organizational purpose and strategy, as it is built up on socio-ecological assumptions of health where alternative biomedical assumptions dominate. The value of this approach to understanding how progress can be made is explained with reference to empirical analyses of 11 cases of NHS organizations working to advance some aspect of Sustainable Development principles. The findings from this analysis lead to the conclusion that progress is dependent on the development of collective assumptions, held across different parts of society, about the purpose and functioning of the NHS. The actors who hold these assumptions must also be brought into the debate if progress is to be made and the processes whereby this can happen warrant as much attention as the development of specific objectives of Sustainable Development for the NHS.

SUSTAINABLE DEVELOPMENT IN THE UK NHS

The NHS in the United Kingdom comprises the four individually managed health care systems of England, Northern Ireland, Scotland and Wales. Funded primarily through taxation, and therefore largely free at the point of use to residents of the United Kingdom, these services provide comprehensive healthcare ranging from consultations with general practitioners to community-based services such as midwifery, mental health support, nursing care in the home, and all aspects of hospital treatment. Such an extensive health-care system clearly utilizes significant resources. Statistics provided by the NHS Confederation (2013) refer to a total NHS expenditure, of £104.333bn in 2011/12, on services which provide for 1m patients every 36 hours. In 2010, the NHS employed 1.35 m people.

Since the publication of the United Kingdom's first two national Sustainable Development Strategies (DEFRA, 1999, 2005), there has been a political agenda to address and manage the environmental, social and economic impacts of the NHS, along with all other publicly funded services in the United Kingdom. The latter of these Strategies outlines five principles (living within environmental limits; ensuring a strong, healthy and just society; achieving a sustainable economy; promoting good governance; and, using sound science responsibly), and advocated that publicly funded sectors should lead these by example. At present Sustainable Development, as an agenda for the NHS, is steered by an advisory body named the NHS Sustainable Development Unit (SDU) which initially directed its messages to NHS organizations on the reduction of carbon dioxide equivalent (CO_2e) emissions attributable to activities such as travel, buildings, and the supply chain (SDU, 2009). However, they now advocate the integration of the principles of resource reduction and management of environmental, social and economic impacts, into the heart of decision-making regarding service design and delivery. Their 'Route map for Sustainable Health' (SDU, 2011) calls on everyone with an interest in the health of the public, to help shift the NHS from its current emphasis on treatment of individual diseases, where environmental, financial, and social costs are secondary considerations, towards a service which directs its resources at enhancing the conditions which lead to good public health. An example given in this Route map of a service that such an NHS would provide is the provision of integrated health funds to vulnerable people to enable them to insulate their homes better. Such a service has the potential to reduce ill health during winter, reduce hospital emissions and potentially reduce the amount of money required for hospital services. It also has the potential to enable savings on fuel bills for the people who receive the fund, therefore increasing their standard of living.

The SDU refer to the concept of 'virtuous circle thinking' to help articulate this shift. Originally included over a decade ago in a report by the independent think-tank 'The King's Fund', the report entitled 'Claiming the Health Dividend' (Coote, 2002) described a 'virtuous circle' of NHS resources where social, economic and environmental gains could result from good practices such as investment into local economic and environmental conditions. These interventions were intended to positively influence the collective health of the local population and were based on the premise that that this was a necessary part of controlling NHS expenditure and improving health services for the future. This concept is illustrated in Fig. 1.

Such an approach has implications for how health budgets are set and utilized. Instead of allocating funds by disease group, and setting targets for disease-specific clinicians to work in isolation on these, at least some funds need to be attributed across departments and organizations to enable people to work together on shared objectives related to the determinants of public health. The potential for health budgets and supporting structures to be arranged in this way within the NHS, and in other countries' health systems, has been a topic of interest for some parts of the public health community for some time. Writing from this perspective, a lack of incentive structures for the NHS to engage with social, economic and environmental determinants of health has been described (Hunter, 2001, 2003; Hunter & Marks, 2005). The argument presented is that the NHS is currently incentivized to tackle disease treatment mainly through acute hospital settings. There is a tendency for health services in general, not only the NHS, to maintain this situation, and to grow in their own interests, particularly serving the interests of the medical profession which is often the most powerful group within this context. This process occurs at the expense of a focus on upstream determinants of public health, where the interest groups have less power and voice. Whilst the SDU do not directly identify these

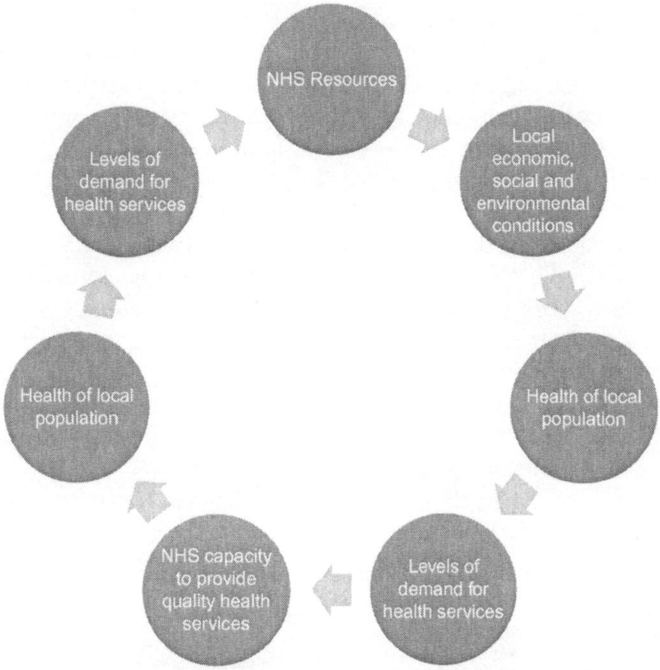

Fig. 1. The virtuous circle of NHS resources (Adapted from A. Coote 2002, p. 2).

already-existing insights about the priorities of the NHS, they are clearly relevant to their calls for engagement by the NHS in the determinants of health under the guise of a Sustainable Development agenda. The support necessary to help these people to challenge dominant interests, change current structures, and progress different priorities to those currently and likely to be reproduced requires much attention.

One way of articulating the scale and nature of the task is through the theoretical framing of organizational learning. This framing helps to make explicit what is being advocated, the challenges it makes to how things are currently done, as well as raises questions about who needs to be engaged if changes are to be achieved. It is important to identify those groups who can help advance alternative ideas as well as to acknowledge those who maintain the status quo. In this chapter, a generic framework of organizational learning for Sustainable Development provided by Sterling (2001, 2003), particularly relevant for use in UK public policy contexts, is used to help reflect upon the NHS Sustainable Development agenda and the challenges posed to its progress by the dominance of a particular model of health service delivery which currently maintains prominence across the NHS. The value of understanding the challenges in this way is illustrated with reference to the empirical analysis of 11 Projects which sought to integrate some elements of the principles of Sustainable Development into NHS practice, during a particularly active phase of the history of this agenda, the mid to late 2000s.

This chapter is aimed at those with some influence in supporting NHS organizations in this remit, including national bodies such as the SDU, NHS England (and corresponding bodies in Scotland, Wales, and Northern Ireland), the Department of Health, senior members of more localized NHS bodies such as the Clinical Commissioning Groups, and even individual Trusts. It is also aimed at those working 'on the ground' in the hope that it helps them to understand and articulate the challenges and opportunities they face as they forge useful alliances for sustainable change. Finally it should be of interest to those working in other countries' health systems particularly those where, like in the NHS, acute disease treatment is favoured over models of intervention based on a range of determinants of health.

ORGANIZATIONAL LEARNING AND SUSTAINABLE DEVELOPMENT

From an organizational learning perspective, often attributed to Argyris and Schön (1978), how an organization behaves is determined by particular assumptions regarding its purpose and functioning and any change in behaviour relies on the exposure and revision of these assumptions. Individuals working within an organization will find it difficult to change assumptions to any significant degree and will tend to achieve better results by learning how to enact the same assumptions in more efficient ways (doing the same thing better), rather than attempting to revise the parameters upon which these are based in the first place (doing better things). This is because of the infrastructure, shared values and collective ideas and practices which will have evolved to maintain current behaviours within institutions. These ideas about the barriers to change which are likely to exist within organisations such as the NHS, have, already been applied to the context of Sustainable Development in order to build insights into why organizational progress in this area can be slow. Doppelt (2010) and Senge and Carstedt (2001) have outlined how most organizations conform to particular assumptions of purpose, and of cause and effect, which focus on 'producing products and services and delivering them to the customer in the fastest and cheapest way possible' (Doppelt, 2010, p. 34). In the case of the NHS the service is the delivery of health care. The strategy organizations use to achieve such goals is linear and resource intensive as it involves the extraction of resources, polluting methods within product and service production and the generation of toxic waste in the air, water and soil. Overtime, most organizational structures and practices, and therefore those of most NHS organizations, will have evolved to support the development of a sustainable approach to service delivery.

Traditional assumptions of purpose and ideas about how to deliver products and services contrast with those inherent in new proposals for Sustainable Development in which it is advocated that organizations acknowledge at least some value in enhancing conditions outside of their traditional remits and jurisdictions. Proposals for Sustainable Development also require alternative strategies for product and service delivery based on more circular models of cause and effect which have been termed 'ecological design principles' (e.g. Van der Ryn & Cowan, 2007). In ecological design, resources are not extracted and utilized regardless of wider and longer-term impacts; rather the hope is that the negative impacts are minimized. These models are described as circular because efforts are made to turn the waste products from one part of production or service delivery (e.g. material waste) into resources that accrue system-wide benefits (e.g. using waste products as an energy source). Achieving such an approach requires better knowledge of all the components of the related sub-systems (social, ecological and economic) which comprise the whole and of the relationships which exist between them, than is currently likely to be in use. Table 1 below, helps to illustrate the contrasts likely to be found within Western public policy contexts between the outward-looking, circular assumptions proposed

Table 1. The challenge of organizational learning for Sustainable Development (adapted from S. Sterling, 2001, 2003).

Types of organizational assumptions	The nature of assumptions likely to dominate at present	The nature of assumptions proposed by advocates of Sustainable Development
Purpose	Objectivity	Participation
Strategy	Disconnection	Connection
Practice	Manipulative	Transformative

for organizations by advocates of Sustainable Development and those organizational assumptions which currently dominate. This generic model can also be used to help identify and further understand the contrasts likely to exist in a particular context (Sterling, 2001, 2003). The table provides a generalized indication of the nature of these contrasts at three levels: purpose, strategy and working practices. Whilst it is of course a simplified version of reality, this table posits that the dominant perception of an organization's *purpose*, within any Western public policy context, is likely to be based on a vision of itself as objective and separate from its external environment and that this stems from a mechanistic view of the world which has shaped collective thinking in the industrialized World since Descartes and Newton. This has been described as a 'Cartesian Worldview' (Heron & Reason, 1997). Within the United Kingdom as elsewhere, this worldview has resulted in the development of *strategic* approaches to managing organizations which complement this mechanistic view of how things work. Such strategies break systems down into disconnected parts and establish prescriptive goals for each individual part in order to control the outcomes of interventions from the top-down. Such cosmologies lead to *working practices* which manipulate resources external to the organization – whether they be, social, economic or environmental – in order to achieve discrete goals. Within this approach, little emphasis is placed on the impacts of these resource use patterns across the rest of the system. With reference to education policy in the United Kingdom specifically, but citing wider applicability to other parts of the public sector, Sterling describes this approach to strategy as the 'new managerialism'. In building the essence of his model, Sterling cites Capra (1996) as being central to a movement which seeks to articulate the dominance of such mechanistic thinking, and crucially, that the assumptions which comprise it, are not infallible truths; they are theories which are open to challenge, and indeed they are being challenged across science and society.

The organizational learning model (Table 1) identifies the characteristics of alternative theories inherent in proposals for Sustainable Development principles which are insights that need to be integrated into any organizations' functioning if progress towards a sustainable development agenda is to be made. Integral to making these conceptual and practical shifts is the demand that an organization be perceived not as separate from but as embedded within its wider context and therefore that it develops a strategy that does not focus only on delivering prescribed targets but also on the important task of strengthening relationships between its activities and other parts of its system. Strategic organizational approaches for Sustainable Development focus on developing connections amongst departments and other agencies and on cultivating shared objectives, shaped by local context, where specific results for any one organization are not necessarily easy to organize from the top down nor in isolation from one another. Sterling (2001, 2003) suggests such models of organizational functioning are based on complexity science and serve as a noteworthy challenge to conventional management theory (Stacey, 2007). Important demarcations between the approaches are that when using a complexity-based strategy, organizational members need to work not on achieving isolated objectives but on contributing to system-wide benefit through collaboration and partnership for emergent, not-necessarily controllable, gains. Furthermore, individuals should not work to manipulate systems' resources to achieve specialist goals but should rather strengthen and transform the social, economic and environmental resource base for the good of the whole system.

Utilization of this model of organizational learning as a format for presenting the challenges of Sustainable Development in the NHS is useful because it helps identify the assumptions and theoretical constructs being debated. With respect to purpose, the model states that a sustainable organization would acknowledge its role in supporting the functioning of the social, ecological and economic systems of which it is a part. For the NHS, such a purpose is evident in the proposals made by SDU's 'Route map to Sustainable Health' in which they refer to a virtuous circle of healthcare resourcing, where health interventions are designed to enhance social, economic and environmental conditions with a view to realising health benefits in the wider population. There is much theoretical insight already available within the international public health community (e.g. Barton & Grant, 2006; Baum, 2008; Brown, Grootjans, Ritchie, Townsend, & Verrinder, 2005) to support this vision for health services. Acknowledging this theoretical base allows one to identify appropriate strategies of intervention for health, which support the enhancement of the complex determinants of health which are social, economic and environmental. The Healthy Settings framework (e.g. Dooris, Dowding, Thompson, & Wynne, 1998), and its related Health Promoting Hospitals framework (e.g. Whitehead, 2004) are examples of approaches suited to this perspective on health system intervention. These Frameworks promote the formation of multi-disciplinary and multi-organizational partnerships tasked with developing a shared understanding of health-risk factors and working on collaborative strategies to address these issues intersectorally.

The model also reminds us, however, that these ideas challenge dominant assumptions that are based on models of organizations operating as isolated entities within their social, economic and ecological systems, and of them organizing their product and service delivery around discrete, not complex, objectives. With respect to the NHS and health services more generally, critiques of the prevalence of objective and atomized medical cosmologies abound. As introduced above, health services evolve to serve their own interests primarily interest which cohere with, growing their ability to treat individual diseases through specialist treatments. They base their intervention strategies on evidence obtained through randomized controlled trials not suited to intervening within a constellation of health determinants identified in socio-ecological models of health. The influence of such dominant assumptions on progress vis-a-vis the planning and delivery of sustainable health systems must be acknowledged. The fact that these theoretical debates have not been identified explicitly within current guidance for NHS Sustainable Development signals a missed opportunity. The naming of key ideas evident within these debates helps to identify the actor groups, and their respective interests, responsible for the maintenance of dominant forms of NHS functioning, as well those associated with the progression of alternatives.

Finally, this model of organizational learning for Sustainable Development is built on a perspective of organizational knowledge informed by a 'participative paradigm' (Heron & Reason, 1997), and 'pragmatism' (Elkjaer, 2003). This implies that particular approaches are most appropriate for achieving progress towards a stated vision, in this case Sustainable Development. From this perspective, the assumptions which guide behaviour in organizations are not cognitive outcomes of the individuals involved, but rather arise from the relationships and power dynamics of everyday work life ideas and practices. Individuals are therefore not able to significantly revise these assumptions unless opportunities arise within these institutional contexts to enable them to catalyze change. Such a perspective draws attention to the importance of context in determining whether or not progress can be made and the need for individuals to develop capacity to understand and capitalize on levers for change which may arise with in the wider social system.

USING THE MODEL TO DEFINE AND INTERPRET PROGRESS

Context

The political pressure to develop an NHS Sustainable Development agenda, expressed in the form of national targets and guidance, has varied since it was first evident in the late 1990s. With the arrival of the 2nd UK Government Sustainable Development Strategy (2005), came some momentum for a role for the NHS in contributing to social, economic and environmental determinants of health. For the NHS, this led to the development of a 'Good Corporate Citizenship Toolkit' (SDC, 2010) aimed at supporting all organizations under its remit to manage the environmental, social, and economic impacts attributable to its service delivery. Demands that the NHS should take this agenda seriously also came from outside of its boundaries. As has been introduced above, the King's Fund produced a widely-cited report called 'Claiming the Health Dividend' (Coote, 2002), which championed the concept of 'virtuous circle thinking', attracting support in this area from other organizations with an interest in public health. By the late 2000s, the Good Corporate Citizenship agenda had been somewhat overshadowed by the drive to reduce CO_2e emissions, which at least initially, shifted the focus onto internal organizational resource use and emissions from such activities as buildings and travel, over which NHS organizations had direct responsibility. However, before this shift, the mid to late 2000s saw many NHS organizations actively engage with the broader notion of Good Corporate Citizenship.

There also existed at the time a Government Watchdog called the Sustainable Development Commission (SDC) which was responsible for monitoring progress across the public sector. The SDC strongly supported the NHS in its corporate citizenship endeavour, publicizing examples of good practice on their website, and including contact details of the people who led them, in efforts to encourage others to learn from their progress. In 2006, these people were contacted to take part in an interview-based study which sought to define what they were trying to do and the challenges and opportunities they faced, in a format informed by the model of organizational learning for Sustainable Development described above. Now that there appears to be a renewed emphasis on the idea that the NHS has a broader role to play in enhancing social, economic and environmental conditions, as upstream determinants of health, it is timely to reflect on what this empirical research tells us about the over-arching vision of NHS Sustainable Development, and the challenges and opportunities which can and do influence its progress.

Projects Included in the Study

For this study, people from every Project listed on the website in 2006 were contacted, and 11 agreed to take part in semi-structured interviews.

Table 2. A Summary of the Initiatives Included in this Study.

Type 1 (led by the NHS)	Type 2 (led by external agencies)
1.1 Renewable energy Wind turbine and energy management	*2.1 Food procurement* Developing a sustainable (local) supply chain
1.2 Renewable energy Biofuels	*2.2a and 2.2b Procurement* Developing a sustainable supply chain
1.3 Service re-development (e.g. A) Sustainable build and community regeneration	*2.3 Recruitment* Developing NHS employment opportunities
1.4 Service re-development (e.g. B) Sustainable build & community regeneration	*2.4 Meat procurement* Developing a sustainable (local) supply chain for meat
1.5 Strategy development Linking Public Health and Environment	
1.6 Strategy development Integrated Sustainability Strategy with a focus on energy, waste and travel.	
1.7 Food procurement (e.g. A) Developing a sustainable (local) supply chain	

These people are referred to as 'Project leaders' from here on. Table 2 summarizes the nature of their initiatives, which included the: management of environmental resource use (1.1, 1.2, 1.6); integration of principles of Sustainable Development into large-scale service re-design and new build (1.3, 1.4); integration of these principles into supply chains (1.7, 2.1, 2.2, 2.4); recruitment of local people to NHS jobs (2.3); and the development of public health strategy to achieve environmental gains (1.5). Two distinguishable types of projects were immediately evident: there were those initiatives led by people working within the NHS (categorized here as Type 1) and those led by people working in organizations external to the NHS (categorized here as Type 2).

Research Design

The aim of the research was twofold. First the study sought to identify the rationales behind each initiative and to define these according to the parameters included in the organizational learning model outlined above, which were the Project's *purpose*, *strategic approach* and *working practices*. Semi-structured interview guides, therefore, included questions about each of these three areas. In order to ensure that the respondents' definitions enabled development of a detailed vision of what a sustainable health service might entail, prompts were also included to ensure coverage of key aspects at each level where tensions between what was being attempted and what existed as the status quo were likely. With regard to strategy, there were prompts to ensure coverage of approaches to prioritization, evaluation and types of working relationships, as these are areas where complexity-based models of management are likely to challenge dominant models. Second, interviewees were asked about the gaps which may or may not have existed between what they would have liked to achieve and what they did achieve, in order to understand the constraining and enabling factors they experienced. No assumptions were made about how dominant theories of purpose, strategy and practice may be influencing the progress of their Projects so no further prompts were included at this point. This was to enable the interviewees free-reign in describing what has not been explored significantly in this way before.

Box 1. Structure for interviews and analysis

1. Project Definition
    - Purpose
        - Direct aims
        - Indirect aims
    - Strategy
        - Prioritization and planning
        - Evaluation
    - Practice
        - Internal relationships
        - External relationships
        - Relationships with community
2. Challenges and opportunities
    - Constraints
    - Opportunities
    - Skills and knowledge gaps
    - Ideas for the future

Interviews were transcribed and the interview guides provided a broad coding frame for thematic analysis of the transcripts. The coding structure was quite prescriptive regarding parameters of a Project definition, for which responses were required (i.e. purpose, strategy and practices), and the sub-categories for each of these, to which a response was also desired if possible. However, there was flexibility in the themes which were actually collated and additional aspects of the definitions were also recorded if they arose. In line with the more open questioning regarding the gaps, constraints and opportunities they experienced, the initial coding structure for this part was very broad. Box 1 above summarizes the structure used for the interviews and thematic analysis.

RESULTS

Table 3 summarizes the results from these interviews, and its contents presented in conjunction with quotes from Project Leads which are presented after the table.

The analysis revealed two broad categories of Projects. In the first Category (1), are Projects concerned with a relatively narrow interpretation

Table 3. Summary of Results.

	Category 1	Category 2
	1.1 (renewable energy – wind turbine), 1.2 (renewable energy – biofuels), 1.6 (energy, waste, transport)	1.3 (service re-development – A), 1.4 (service re-development – B), 1.5 (public health and environment), 1.7 (food procurement – A), 2.1 (food procurement – B), 2.2 (procurement), 2.3 (recruitment), 2.4 (meat procurement)
Purpose	Cost saving Risk management Compliance	Socio-economic gain Minimising environmental impact Improving health of community
Strategy	Cost Benefit Analysis	Context-specific priorities Measuring processes, as well as outputs Developing measures of outputs through trial and error
Practice	Initiatives carried out by one department, mainly in isolation	Partnership building Making connections

of Sustainable Development, based mainly on the efficient use of internal organizational resources. These are Projects 1.1, 1.2 and 1.6 which all focus on cost saving, risk management, and compliance, using a cost-benefit-analysis approach to gaining organizational investment.

> if you're saving money, that's really the key aim (1.1 renewable energy – wind turbine)

> last year we saw the highest gas prices we had ever seen…at which point I was given permission and some money to write a specification for an alternative heating method (1.2 renewable energy – biofuels)

Such initiatives were carried out by non-clinical, estates management staff, with little input from other departments.

> there wasn't really a lot of involvement from anyone else in the Trust…. (1.1 renewable energy – wind turbine)

In the second Category (2) are those initiatives which are concerned with contributing to the social, economic or environmental conditions of their communities. Four Projects led by NHS staff are included in this Category (1.3, 1.4, 1.5, 1.7), as well as all the Projects led by external organizations (2.1, 2.2, 2.3, 2.4). The focus of all Projects in Category 2 was to have some impact on the determinants of health, with economic determinants cited as key objectives in 1.3, 1.4 and 1.7.

> obviously (the hospital development) is in a particularly poor part of the UK, with … very high levels of deprivation. (It) is the worse borough in the country in terms of the index of deprivation … if you're investing that amount of money in health infrastructure the project should have a wider social economic impact (1.3 service re-development)

> I think we were talking about something like £90m or £100m so you know if there was investment of that nature happening in the City, and it is quite a deprived area, well there are obviously mixed areas of quite high deprivation and some more prosperous areas of the city, we knew we should be able to do something additional through that investment (1.4 service re-development)

> the GDP of (this County) is less than 75% of the European average … so we could have a real impact if we took our 1.5m budget and plugged it straight into the (County's) economy (1.7)

Economic objectives were also key in 2.1, 2.2, 2.3 and 2.4. For example, Project 2.2 sought to increase the proportion of NHS spend in its geographical region, and Projects 2.1 and 2.4 had similar objectives with specific relation to food procurement. Project 2.3 sought to promote local employment by the NHS, and set targets for training provided to potential employees, to make them more able to apply for NHS jobs. Project 2.1 was unique within the Projects studied, being the only one to have explicit environmental objectives in addition to economic ones. This Project sought not only to increase the proportion of local food suppliers, but also to improve the environmental credentials of the food supply chain.

All Type 2 Projects (those led by people external to the NHS) also cited additional objectives concerning the establishment of new working relationships between the NHS and its suppliers or potential employees. Projects 2.1, 2.2 and 2.4 each aimed to link the NHS with a wider supplier base than had previously existed, building the necessary infrastructure (e.g. websites), and creating spaces for improved communication to take place (e.g. 'meet the buyer' events).

> Kind of negotiating between the two (NHS and suppliers), and even so much as just putting them together, finding the contacts, acting as a sort of go-between really, that was sort of a large part of the role (2.1 food procurement)

> We came into this very clearly thinking right, we're gonna link SMEs (Small – Medium Enterprises) with NHS spend and you know, we're gonna get them exchanging contracts and talking together (2.2 procurement)

> Engaging the local Trusts...in terms of their specific requirements and how those can be influenced, and engaging with the supplier base to identify areas where perhaps we haven't identified suppliers in the past (2.4 meat procurement)

This took place through training and support aimed at potential suppliers from the local area.

> (We) developed a (procurement) workshop and we've rolled 10 of those. I think we've had about ... probably about 200 or more, 300 attendees (2.2 procurement)

> Developing the training & guidance and providing my own expertise and knowledge to any suppliers (2.2 procurement)

> So the actual going out & finding suitable producers, helping them to get into the NHS contracts, so that might have been ... on Health and Safety, and auditing, having the necessary knowledge to get involved ... (2.1 food procurement)

It also took place through training and support aimed at potential employees from the local area;

> encouraging NHS employers ... to increase the proportion of people from deprived backgrounds who got that employment ... and making sure people were employed and trained, recognising that these people were perhaps not so immediately job ready (2.3 employment)

In summary, goals for Type 2 Projects aimed to impact the determinants of health, and these were mainly economic ones. They also hoped to develop long-term relationships between the NHS and its suppliers and employees.

Projects 2.1, 2.2 and 2.4 assessed success against measures including 'number of local contracts secured' and 'number of small businesses receiving training in selling to the NHS'. Similarly Project 2.3 measured for training provided to potential employees. The nature of all Projects in Category 2, whether led by people in the NHS, or by people outside, was collaborative, and partnership focused requiring the development of cross-department, and cross-organizational relationships.

Challenges and Opportunities

Category 1 Projects faced relatively few challenges as they were able to make the case for organizational support in terms of cash savings. They were relatively simple Projects carried out by staff linked to estates management. As long as they could make the financial case, they got the support of their organization to develop the Project. The financial case did depend on the availability of grants for initial investment and on commodity prices within the energy sector. Finances were very restrained at least for 1.2, and the dependency of the case on changing fuel prices was evident. Of the 3 Projects in this Category, Project 1.6 was the broadest. This occurred in Wales where there was a Country-wide commitment for every NHS organization to achieve Environmental Management Standards accreditation. Therefore individual trusts did not have to make a financial case for every part of their environmental management strategy, as the need for continual improvement was taken as a given.

For Category 2 Projects, the factors allowing Projects to progress were largely located outside the NHS. A broadly accepted idea that the areas were economically deprived appeared particularly influential in 1.3, 1.4, 1.7 and 2.3. In these areas there was an acknowledgement that the NHS was one of the biggest organizations and employers and should therefore contribute to the economic growth of the region. Measurable outputs from these Projects were required, particularly where financial input had been high. To this end, it was possible for people to measure what they did (e.g. number of people employed as a result of the Project), but not the health benefits which may arise from the initiatives; however many did express a desire to attribute gains to specific activities, even if this was deemed much more difficult. In Projects 1.3 and 1.4, consultants were employed to help them strategically plan for health outcomes, although actual measurement of these outcomes was not part of this process.

> They (the consultants) basically tutored us on what you could do and what you might expect (1.4 service re-development)

> We did some work with (consultants) and the London Health Commission to sort of scope out the kind of non-health outcomes of the whole project...when I say non-health, that's probably the wrong way of putting it but the non-clinical outcomes of the project in terms of potential for local employment, transport, reducing emissions, using more modern technology, forming closer links with the local community and so on, a very big agenda (1.3 service re-development)

Similarly, the Local Multiplier 3 tool known as LM3 (Sacks, 2002) was used in Projects 1.7 and 2.2 to estimate the local economic gains to be achieved from local NHS spend, but as the approaches above, does not attribute health gains to this spend.

DISCUSSION

Application of the model of organizational learning to the NHS context in order to analyse these Projects raises several points of discussion which are summarized in Table 4. Overall, this analysis shows that the progress these Projects have made have been achieved within the current tensions playing out in everyday practice. On the one hand there are advocates of an NHS which would like to direct significant resources towards the enhancement of social, economic and environmental conditions as determinants of health. On the otherhand, there are institutionalized health system goals which prioritize health treatments over working on these upstream health determinants. Those Projects which promise measurable cost-savings to NHS organizations (1.1, 1.2) are therefore relatively easy to progress because they do not challenge dominant organizational goals. They could even support these goals by providing additional funds for investment in disease treatment. However, there were a number of Projects analysed which promise contributions to determinants of health that succeeded to some degree, and the levers which enabled their success should be of interest. In this research, economic determinants featured most highly in Projects' purposes. This can be linked to the existence of regional economic growth strategies which were located outside the NHS at the time and, which encouraged support from the public sector. It is useful to speculate as to whether or not regional economic strategies currently exist to support such goals, as there is currently much less emphasis on public sector funding within the prevailing political climate.

With respect to other social and environmental goals, it is worth speculating about the kinds of levers which could exist or be developed to facilitate NHS engagement with these tools for organizational change.

Table 4. Organizational Learning for Sustainable Development in the NHS.

	Dominant model of health system	Most significant aspects of real-life Projects (2006)	Advocated in sustainable models	How can vision be progressed further?
Purpose	Deliver disease treatment to public	— Save money — Contribute to economic determinants	Enhance conditions which determine health	Identify levers that exist to promote all determinants (e.g. Wanless, 2002; Marmot, 2010)
Strategy	Linear, top-down, discrete goal setting	— CBA arguments easy to develop — some 'complex' outcomes intended, but measurement difficult	Non-linear, multi-level priority setting around broad principles	Develop tools and models for complex public health interventions (e.g. Healthy Settings, Health Promoting Hospitals)
Practice	Manipulate resources to achieve discrete goals	— Most Projects reliant on partnerships and shared agendas — A no. of Projects focused on these relationships as primary goal	Transform resource base to enhance determinants of health	Continue to provide resources to develop these forms of working.

There have been two high-profile UK Public Health reviews on factors which lead to inequalities in health (Marmot, 2010; Wanless, 2002). The Marmot Report made explicit reference to ecological factors as well as economic; whereas, the Wanless Report focussed on the importance of such determinants for the controlling demand for NHS services in the long-term. The strength of such reports to influence policy and practice across government and broader society will depend on the prevailing political climate. The way such ideas are progressed in the newly formed structures of the NHS following the White Paper (Great Britain, 2010) is worthy of further investigation.

The Projects' strategic approaches can also be understood in terms of the tensions of competing assumptions highlighted by Table 4. Projects 1.1 and 1.2 did not challenge dominant forms of strategy, with traditional cost-benefit-analysis used to justify investment in renewables, with clearly predictable returns. Most other Projects were trialling forms of strategy appropriate to supporting more complex interventions, and indeed reported some dissatisfaction in their ability to do so. The organizational learning model helps identify potential approaches informed by complexity science, such as 'healthy settings' to aid this process, but the use of such approaches is under-researched. The development of appropriate tools and techniques, and also the improvement of their credibility within an NHS were evidence-based policy and prediction of impact is pervasive, should be supported.

The potential for the organizational learning model to be influential at multiple levels within the health system also warrants further attention. As has been briefly described, this model is built on the perspective that assumptions guiding behaviour arise from social processes. The fact that progress took place in many of these Projects through collaboration and the formation of alliances which sought to develop mutual interests and to challenge dominant interests, suggests that this view may warrant further articulation. It has implications for action-based organizational learning initiatives which bring together different parties to define mutual goals, set actions and aid critical reflections upon progress. Existing theoretical models identified as relevant to a vision of Sustainable Development within the NHS can be used as guides in this analytical process. Action research methods, such as action learning and appreciative inquiry (Reason & Bradbury, 2005), are well developed for the context of organizational change, specifically with respect to Sustainable Development (Reason et al., 2009) and their potential for supporting learning of the kind identified by the model could also be further explored. The support and resources required for people to engage in such learning processes must also be the subject of further research.

CONCLUSION

In this chapter, the challenges of integrating Sustainable Development principles into the day-to-day functioning of the NHS have been presented as a challenge of organizational learning. This approach has been used to illustrate the key theoretical debates and actor groups which must be engaged with if any significant progress is to be made. Table 4 summarizes a 'bird's eye' view of the levels at which the agenda must be approached. Some parties will be better placed to influence debates at the level of organizational purpose. In the NHS these include the SDU and the

Department of Health, as well as senior managers in more localized parts of the NHS responsible for priority setting across geographical areas. Some parties will be better placed to influence debates at the level of strategic approaches to service delivery. Academics, policymakers and strategists working on tools and models of complex interventions for health must engage in debates with those who currently promote alternative strategic models of health intervention based largely on linear, top-down, prescriptive planning. The fact that such approaches are currently viewed as the gold standard to developing the evidence base for health priorities is problematic and must be challenged. Those tasked with progressing Sustainable Development initiatives in individual organizations, of the kinds analysed for this chapter, will be best suited to challenge daily working practices. The people who attempt to address sustainability through such initiatives as procurement and building design can seek to develop innovative partnerships and cultivate shared objectives with other parties across the health system. However, these three levels of activity cannot be viewed in isolation. As this organizational learning perspective helps to illustrate, developing understanding across the different levels is also vital and the processes through which this analysis is supported require more attention than they have been given to date.

ACKNOWLEDGEMENTS

All those interviewed for the research documented here, particularly Helen Ross and other members of the SHARP network (Sustainable Health through Action Research in Practice), William Young, and the Sustainability Research Institute, University of Leeds.

REFERENCES

Argyris, C., & Schön, D. (1978). *Organizational learning in action: A theory in action perspective.* Boston, MA: Addison-Wesley.

Barton, H., & Grant, M. (2006). A health map for the local human habitat. *The Journal for the Royal Society for the Promotion of Health, 126*(6), 252–253.

Baum, F. (2008). *The new public health* (3rd ed.). Melbourne: Oxford University Press.

Brown, V. A., Grootjans, J., Ritchie, J., Townsend, M., & Verrinder, G. (2005). *Sustainability and health. Supporting global integrity in public health.* London: Earthscan.

Capra, F. (1996). *The web of life: A new synthesis of mind and matter.* London: Flamingo.

Coote, A. (2002). *Claiming the health dividend.* London: Kings Fund.

DEFRA. (1999). *The UK government sustainable development strategy*. London: HMSO.

DEFRA. (2005). *Securing the future*. London: HMSO.

Doppelt, B. (2010). *Leading change toward sustainability: A change management guide for business, government and civil society*. Sheffield: Greenleaf Publishing Ltd.

Dooris, M., Dowding, G., Thompson, J., & Wynne, C. (1998). The settings-based approach to health promotion. In A. Tsouros, G. Dowding, J. Thompson & M. Dooris (Eds.), *Health promoting universities: Concept, experience and framework for action*. Copenhagen: WHO Regional Office for Europe.

Elkjaer, B. (2003). Social learning theory: Learning as participation in social processes. In M. Easterby-smith & M. A. Lyle (Eds.), *The blackwell handbook of organizational learning and knowledge management* (pp. 38–54). Oxford: Blackwell Publishing.

Great Britain. (2010). *Equity and excellence: Liberating the NHS* (Cm 7881). London: HMSO, Department of Health.

Heron, J., & Reason, P. (1997). A participatory inquiry paradigm. *Qualitative Inquiry*, *3*(3), 274–294.

Hunter, D. (2001). The new NHS plan. A new direction for english public health? *Critical Public Health*, *11*(1), 75–81.

Hunter, D. (2003). *Public health policy*. Oxford: Blackwell.

Hunter, D., & Marks, L. (2005). *Managing for health: What incentives exist for NHS managers to focus on wider health issues?* London: King's Fund.

Marmot, M. (2010). *Fair society, healthy lives*. London: UCL.

NHS Confederation. (2013). *Key statistics website*. Retrieved from http://www.nhsconfed.org/priorities/political-engagement/Pages/NHS-statistics.aspx. Accessed on 20 April 2013.

Reason, P., & Bradbury, H. (Eds.). (2005). *The handbook of action research: Participative inquiry in practice* (2nd ed.). London: Sage.

Reason, P., Coleman, G., Ballard, D., Williams, M., Gearty, M., Bond, C., … Maughn Mcclachlan, E. (2009). *Insider voices: Human dimensions of low carbon technology*. Bath: Centre for Action Research in Professional Practice.

Sacks, J. (2002). *The money trail: Measuring your impact on the local economy using LM3*. London: New Economics Foundation and the Countryside Agency.

Senge, P., & Carstedt, M. G. (2001). *Innovating our way to the next Industrial Revolution. MIT Sloan Management Review*, *42*(2), 24–38.

Stacey, R. (2007). *Strategic management and organizational dynamics: The challenge of complexity (to ways of thinking about organizations)*. Harlow: Pearson Education.

Sterling, S. (2001). *Sustainable education: Revisioning learning & change*. Dartington: Green Books.

Sterling, S. (2003). *Doctoral thesis – Whole systems thinking as a basis for paradigm change in education: Explorations in the context of sustainability*. Retrieved from http://www.bath.ac.uk/cree/sterling.htm. 10 Feburary 2006.

Sustainable Development Commission (SDC). (2010, 30 April). *NHS good corporate citizenship web-based assessment tool*. Retrieved from http://www.corporatecitizen.nhs.uk

Sustainable Development Unit (SDU). (2009). *Saving carbon, improving health*. Cambridge: SDU.

Sustainable Development Unit (SDU). (2011). *Routemap for sustainable health*. Cambridge: SDU.

Van der Ryn, S., & Cowan, S. (2007). *Ecological design, tenth anniversary addition*. Washington, DC: Island Press.

Wanless, D. (2002). *Securing our future health: Taking a long term view*. London: HM Treasury.

Whitehead, D. (2004). The European Health Promoting Hospitals (HPH) project: How far on? *Health Promotion International*, *19*(2), 259.

ENVIRONMENTAL HEALTH RISK GOVERNANCE IN PRACTICE: LESSONS LEARNED FROM A FLEMISH CASE STUDY APPROACH

Kristien Stassen and Pieter Leroy

ABSTRACT

Purpose – *Risk governance is being successfully inserted into scientific and political agendas as a way to understand and address complex problems, such as health problems that have both social and ecological dimensions. However, to date, the debate on risk governance has remained rather conceptual. This chapter addresses these lacunae through describing research that was conducted on the Flemish environmental health governance arrangement and draws on this empirical case study to suggest ways that risk governance can be put into practice.*

Methodology – *The case study analysis was conducted through a mixed methods study that combined document analysis and data gathered through 22 in-depth interviews with environmental and health scientists as well as policymakers.*

Findings – *This research shows that the Flemish environmental health risk governance arrangement has succeeded in increasing the exchange of*

Ecological Health: Society, Ecology and Health
Advances in Medical Sociology, Volume 15, 289–309

ISSN: 1057-6290/doi:10.1108/S1057-6290(2013)0000015017

information between: (1) governments at a variety of levels; (2) scientists, policymakers and the general public and (3) environmental and public health civil servants. The analysis also provides insights into some shortcomings and makes recommendations for ameliorating this arrangement: (1) the integration of environmental health objectives into all relevant policy domains, (2) the need for additional research into environmental health indicators and (3) the facilitation of the co-production of knowledge and multi-actor governance.

Originality/value of paper – *Empirical contributions and analysis about risk governance and policy formation processes are not often conducted. The added value of this Flemish case study is that it presents an example of good practice from which lessons for future risk governance arrangements can be drawn.*

Keywords: Risk governance; discursive institutionalism; environment and health

INTRODUCTION

Environmental health, ecosystem-based approaches to health and human ecology are complementary disciplines which have been converging over the last two decades (Parkes, Panelli, & Weinstein, 2003). All three disciplines take into account a holistic, transdisciplinary and integrated perspective of human health as inextricably linked to the health of the biophysical, social, cultural, political and economic environments in which people live (Crescenti & Dakubo, 2010). Parkes et al. (2003) emphasize, however, the need for the further conceptual and methodological integration of knowledge and frameworks in order to ensure the development of comprehensive insights into complex biophysical, social and public health systems. These diverse knowledge fields also have to be integrated into policy strategies and intervention programs in order to improve public health outcomes. With these goals in mind the work of Renn (2005, 2008), and in particular his concept of risk governance, promise to be useful.

Over time risk governance has been effectively inserted into scientific and political agendas as a way to accurately deal with complex issues, such as health problems, that have both social and ecological dimensions. However, the debate on risk governance has remained rather conceptual; studies on 'how to actually do it' and 'how to bring risk governance into practice' are still rare (Keune, 2012; Runhaar, Driessen, Van Bree, & Van der Sluijs,

2010). The main goal of this paper, therefore, is to empirically contribute to a further elaboration of the risk governance concept by conducting research that actually translates the principles of stakeholder participation, transdisciplinarity and (policy) integration into practice. To do this, a case study of the Flemish environmental health risk governance arrangement is discussed.

The Flemish case study sought to inform two debates: (1) the epistemological debate about complexity and risks and their impact on knowledge production and decision-making, so far underscored in Ecohealth and Human Ecology literature, and (2) the debate on the laborious ways integrative concepts such as risk governance gradually stabilize and institutionalize, thereby altering pre-existing institutional patterns. This research was guided by two empirical questions:

1. What impact does the new discourse on complexity as a feature of environmental health issues have on the institutional dynamics of the Flemish environmental health risk governance arrangement?
2. How does this Flemish environmental health risk governance arrangement perform?

The value of this case study approach is that it presents good practices, identifies lessons for future risk governance arrangements and increases mutual learning on how knowledge production and policy-making can be put into practice when dealing with complex health problems which can be linked to the environment and even to specific ecosystems. The theoretical contribution of this chapter to risk governance research comes predominantly from practical lessons learned which will be valuable to those studying health at the interface between social and ecological systems. In particular, this research will be meaningful to those working in new policy domains that are starting to be informed by ecological health principles, such as work being developed in relation to initiatives of the World Health Organization and the International Development and Research Centre.

In the section that follows, the background and ambitions of the risk governance concept will be introduced and its aims to deal with complex environmental health problems discussed. In the third section, our methodological approach and the evaluation framework which was developed to assess the Flemish environmental health arrangement is described. The fourth section, then, empirically elucidates the institutional dynamics of the Flemish environmental health risk governance arrangement, discusses the research findings, and suggests some intriguing questions for further research. The chapter concludes with a summary of the main thoughts emerging from this work and highlights some of the implications it has for social health studies and the newly emerging field of ecosystem based approaches to health research.

RISK GOVERNANCE FOR DEALING WITH COMPLEX ENVIRONMENTAL HEALTH PROBLEMS

Environmental and ecological health problems are intrinsically complex. This complexity is attributable to two distinguishable aspects. First, there is natural scientific complexity, such as: (1) the multiple, interlinked yet non-linear cause–effect relationships that are often caused by cumulative exposures to multiple (low dose) agents and influenced by a wide range of intervening variables; (2) the long delay periods between exposure to multiple causes of disease and their manifestation as multiple health effects which may vary from annoyance due to loss of well-being through to mortality; and, (3) the unrestrictive nature of time (e.g. impacts on future generations), and/or geographic scale (e.g. environmental health drivers are not necessarily bound by national borders) (Briggs, 2008; Knol, Briggs, & Lebret, 2010). As a consequence, knowledge about environmental health risks is clouded with uncertainties and imperfect understanding (Van der Sluijs, 2007). Another dimension of the complexity is that environmental health problems are embedded within broader social, economic, cultural and political contexts; thus, they are interwoven with – often conflicting – moral, financial, economic, environmental, socio-cultural and socio-political norms, values, interests and perceptions. In sum, environmental health issues are socially, scientifically and politically complex. Scientific and societal complexity creates challenges for: (1) scientists working to produce robust, definitive knowledge, (2) politicians endeavoring to manage this complexity appropriately and (3) most notably, for those working at the science–policy interface (Leroy, Driessen, & Van Vierssen, 2010a).

Cognizant of these science–policy challenges, when introducing the concept of risk governance, Renn (2005) sought to present an analytical system that could be used to improve the study and management of complex risks. As an inclusive and integrative framework, risk governance involves the application of the substance and core principles of good governance within the context of complex risks and risk-related decision-making. In contrast to more traditional approaches to risk, risk governance, as a double concept, reflects quite substantial changes in: (1) the epistemology of risks and complexity, encompassing shifts in knowledge ambitions, knowledge production and knowledge organization, and (2) the overall governance debate, implying changing relations and responsibilities between scientists, governments, markets and civil society. What follows is a brief literature review which discusses both aspects in order to engender a better understanding of risk governance and its value for the governance of natural spaces and social activities within them in relation to issues of public health and well-being.

Epistemological Debates on Risks

Epistemological debates on risk are closely connected with the historical evolution of philosophical paradigms, most notably the move from positivism to constructivism. Positivism is characterized by the interest-free search of universal truth because it claims a full knowability of things when using linear and reductionist approaches based on a technocratic separation of facts and values. The positivistic paradigm, therefore, regards risk as an objective, real and measurable threat or danger, which is predominantly defined as the probability of any adverse event (likelihood), multiplied by the impact of that event (seriousness) (Rosa, 1998). Importantly, by developing a dichotomy between facts and values one can make a clear distinction between risk assessment (facts) and risk management (value-laden decisions).

Despite its strengths, the positivistic paradigm has been increasingly criticized. For instance, Douglas and Wildavsky (1982) were one of the first to challenge the status of 'risk' as an objective measure and took a weak constructionist view to argue that dangers are mediated by cultural processes. More recent sociological work has adopted an interpretive approach to study the relationship between lay and expert knowledge about risk (Giddens, 1990) and social responses to risk, such as Beck's work on risk society (1992). Lupton (1999), on the other hand, has adopted a strong constructionist position arguing that risks are solely a product of discourses, strategies, practices and institutions (Gabe, Bury, & Elston, 2004). Overall, constructionists emphasize the value-laden character of risks and acknowledge their social, political and cultural embeddedness. Recognizing the best features and shortcomings of each paradigm, Rosa defines risk as, 'a situation or event where something of human value has been put at stake and where the outcome is uncertain' (1998, p. 19). This conception of risk acknowledges that certain states of the world can be objectively defined as risks, yet it also takes into account the issue of scientific uncertainty and makes clear that conceiving of and handling risks always implies an (e)valuation and, hence, is a value-laden practice.

This epistemological shift is neither linear nor without controversy. Nevertheless it has undoubtedly influenced both scientific methods and the organization of knowledge production. As to the latter, across scientific fields and political domains, organizations are engaging more in participatory processes of knowledge co-production which includes developing strategies of mutual learning between all relevant stakeholders, such as representatives from medical, environmental and social sciences, NGOs, and industrial and citizen groups. This type of knowledge production is also referred to as transdisciplinary knowledge production (Thompson-Klein et al., 2001), the co-production of knowledge (Jasanoff, 2004) and Mode 2 Science (Nowotny, Scott, & Gibbons, 2003). These epistemological shifts are also being increasingly reflected methodologically as researchers argue that 'risk assessment must become less reductionist and less focused on obtaining complete information on all aspects of individual hazards … [and that] risk perception should receive increased attention' (Grandjean, 2005, p. 656). Integrated risk assessment is one such response as it aims to present all relevant knowledge and values in a balanced, integrated, holistic, and transparent way in order to better support the decision-making process. Within this more holistic approach to risk assessment, the meaning of integration has three dimensions as it requires: (1) taking into account all relevant aspects of the cause–effect chain simultaneously, (2) studying this chain in an interdisciplinary way and (3) involving stakeholders in a participatory process (Briggs, 2008).

To summarize, the key features of the epistemological, organizational and methodological challenges of knowledge production processes vis-à-vis risk are 'societal participation, mutual learning, and opening up pre-existing organizational and institutional boundaries'; furthermore, these must be addressed in order to ensure a more responsible, legitimate, and effectively jointly produced knowledge outcome (Leroy, Driessen, & Van Vierssen, 2010b). However, despite some large American and European-funded research projects (Briggs, 2008; National Research Council, 2009) integrated risk assessment as applied to environmental health and governance is still in its infancy.

Governance

These epistemological shifts in the scientific domain are paralleled by shifts in the political domain. Driven by technological change, internationalization,

Europeanization, and modernization – which are themselves in part responses to complex social life risks – a new range of practices have emerged (Van Kersbergen & Van Waarden, 2004). Reasons for this emergence include a decreased ability of central governments to regulate society and an increase in common concern about accountability, responsibility and legitimacy (Van Kersbergen & Van Waarden, 2004). This shift is often referred to as the move from (traditional, sectoral, state-based) government towards governance (in which more societal actors take their part in responsibility and accountability). The general shift from government to governance encompasses three important evolutions: (1) an increased demand for co-operation between, and the integration of, different policy fields as complex problems that often transcend traditional policy borders (multi-sector governance); (2) a similar demand for exchange and co-operation between different policy levels (multi-level governance) and (3) an increased appeal for participatory approaches which take into account the voices of all relevant stakeholders in private, semi-private and public spheres in the decision-making process (multi-actor governance). In sum, governance refers to the increasing blurring of the boundaries between policy fields, policy levels and the science–policy–society interface.

Risk Governance

From the above discussion it may be clear that the concept of risk governance is being developed at the cross roads of these epistemological and political shifts. According to Renn (2008), risk governance applies the principles of good governance to the identification, assessment, management and communication of risks. In contrast to the single, disciplinary focus of traditional governmental approaches, risk governance encourages openness and transparency and takes into account multiple causes, pathways and health effects, as put forward by the multiple stakeholders involved (Briggs, 2008). Through stakeholder participation, social and cultural aspects (like perception, values, and emotions) become embedded within knowledge production and decision-making processes with the outcome of cultivating co-responsibility for preventing and reducing complex risks. As a result, risk governance has become a logical framework to use when the goal is to research and manage complex issues in an integrated way. Based upon the concept of risk governance, Renn (2005) developed a conceptual framework informed by a set of key principles intended to develop sound risk governance. Using a cyclical, iterative, and interlinked process, Renn set out to develop a method in which all relevant scientific, economic, social and cultural aspects and stakeholders are taken into account.

The principles and priorities that led to an interest in risk governance parallel principles underlying ecosystem approaches to health as the latter is in effect a constellation of methods which 'encourages the investigation of environmental health problems from an inter-sectoral, trans-disciplinary and multi-stakeholder perspective to allow for a better understanding of the determinants of health and to develop appropriate interventions' (Crescenti & Dakubo, 2010, p. 39). Risk governance has, however, tackled some of the theoretical issues of policy formation that ecohealth is grappling with and may offer some valuable insights in how to work effectively on policy initiatives relevant to the governance of complex social-ecological health issues. However, despite concerted efforts, to date there is a dearth of practical applications and empirical examples of risk governance not only within the arena of environmental health but also within the field of ecohealth and environmental health (Keune, 2012; Runhaar et al., 2010). This chapter, therefore, aims to help address this need to translate knowledge into action by conducting empirical research on the emergence of the Flemish environmental health risk governance arrangement.

ANALYTICAL PERSPECTIVE AND METHODOLOGICAL APPROACH

The case study of the Flemish environmental health arrangement has been conducted using a discursive institutional approach. Discursive institutionalism is a theoretical perspective within the social sciences employed to study institutions and which exists as an alternative to rational choice or historical and sociological institutionalism. The specific features of discursive institutionalism include its valuing of: (1) the important role of discourses in influencing actors' preferences, interests and behaviour (Hajer, 1995) and (2) the role of these discourses in assuring institutional stability, while simultaneously triggering and legitimizing institutional change (Scott, 2001). In other words, discourses are considered to be the driving force behind institutional preservation and change and entail processes of institutional renewal. As an approach, discursive institutionalism tends: (1) to pay more attention to the cognitive frames of agencies rather than to their material interests; (2) be a more dynamic, agent-centred approach rather than follow static and path-dependent patterns; and, (3) promote the dynamics of valuation and norm setting rather than using static approaches (Schmidt, 2008).

Informed by discursive institutionalism, we assume that an increased awareness of complexity and the emergence of new and changing discourses related to issues of complexity, have influenced the institutional dynamics of the Flemish environmental health risk governance arrangement. Our hypothesis is that the changing discourses regarding complexity have the potential not only to find their way into new concepts and new methodological tools for research but also to gradually inform the development of novel organizational facilities, policy programmes, budgets, rules and as other outcomes of decision-making processes.

To investigate the impact of changing discourses vis-à-vis complexity within the Flemish environmental health risk governance arrangement, we conducted a systematic discourse analysis of documents and transcripts of 22 in-depth interviews in order to get a detailed and balanced picture of the institutional dynamics at play. The document analysis gave some initial indications of the key historical developments, social events, discourses, actors and formal rules of the game involved. This information was then used to inform the development of the research instrument used to conduct the in-depth interviews.

A valid discourse analysis asks for a clear methodology for how to select documents and interviewees. As to the former, mainly primary sources were studied including original research papers, policy documents, legislation, annual reports and advisory reports, etc. As to the latter, an actor map was built, distinguishing governmental, private, societal and scientific agencies which also highlighted those particularly responsible for the interaction between actors (e.g. advisory boards). While the actor map is merely a heuristic instrument that does not pretend exhaustiveness, it did plot key agencies and the primary interaction forums relevant to the study.

The findings of the interviews were analysed against the information generated through the document analysis in order to confirm findings as well as to gain additional information as documents tend to only describe the compromises reached and rarely offer nuanced information about the discussions behind them or the informal rules of the game in play. As such, document analysis and in-depth interviews were complementary methods which was necessary in order to increase the credibility and validity of the research results. The 22 respondents were selected based on their role and position, ensuring a well-balanced number of scientists and policy-makers, on the one hand, and respondents from environment and health perspectives on the other. Figure 1 presents an overview of the interviewees, either located in the scientific or the political sphere, and engaged either in the environment or the health domain – with a few of them crossing

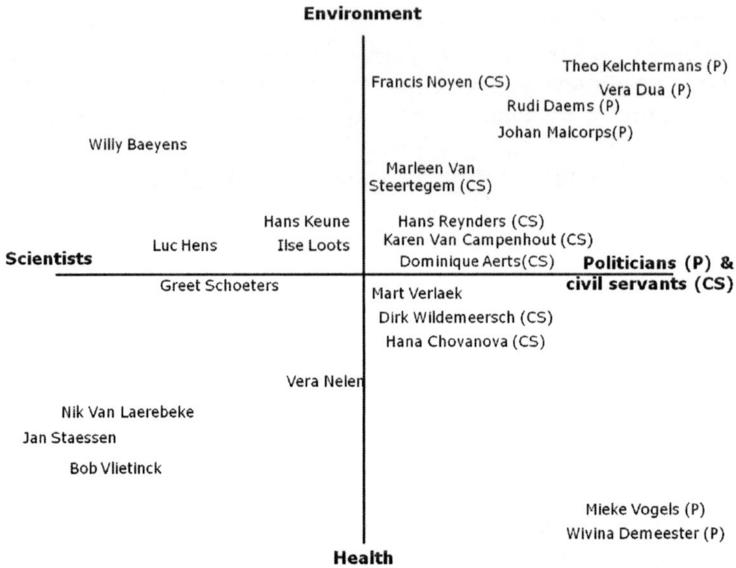

Fig. 1. Graphic Overview of the Positions and Backgrounds of the Interviewees.

the boundaries between these fields. All interviews were semi-structured. The questions related to the changing discourses about complexity and uncertainty, the perceived interaction between science and policy, etc. The duration time of the interviews varied between one and two and a half hours. All interviews were taped, transcribed in Word 2007 and imported into NVivo Software. This software was only used to gather all information derived by the interviews related to a specific subject efficiently and to deduce overviews quickly based on the NVivo coding.

Importantly, the work described in this chapter is one aspect of a larger research project in which the dynamic emergence of the Flemish environmental health policy arrangement has been studied over a period of forty years. In this broader study, the gradual, long-term developments were analysed against the background of the international and European context as well as in relation to the impact of four local events that produced serious environmental health concerns and which ultimately generated discursive shifts and dynamic institutional responses in Flanders/Belgium. These incidents were related to: (1) lead pollution due to metallurgic activities; (2) cadmium pollution due to zinc smelters; and (3) dioxin deposition by two waste incinerators, shortly followed by (4) dioxin contamination of the food chain. For the impact of the European environmental health discourse on national public health governance, we refer to Stassen, Gislason, and Leroy (2010). For more details on how these local events generated a gradual rethinking of complexity (discursive shifts) and institutional renewal, we refer to Stassen, Smolders, & Leroy (2013).

In order to assess the Flemish environmental health risk governance arrangement an evaluation framework was developed (Table 1) based on the literature review presented above, a checklist of criteria for the evaluation of environmental policy integration (European Environmental Agency, 2005), and the analytical framework for evaluating environmental health risk governance regimes (Runhaar, Driessen & Van der Sluijs, 2009, 2010).

Table 1. A Set of Criteria and Indicators to Evaluate the Performance of an Environmental Health Arrangement.

Criteria	Indicators
Political commitment, vision and leadership	• Overarching EH-strategy • Political leadership for EH so that the public and administrations are continually encouraged to deepen their EH thinking • Legal texts or public statements • Long and medium-term EH objectives
Policy integration at the governmental level	• A strategic department to co-ordinate EH across sectors • Integration of EH objectives in related policies • Mechanisms for co-operation with higher or lower levels of governance (international and European network) • Adequate resources (budget and staff) • Policy integration expressed in legislation and decision-frameworks
Knowledge development for decision-making	• Outcome of knowledge production appreciated by the scientific world? • Outcome of knowledge production useful for and used to inform policymakers? • Adequate uncertainty management and integration of plural, legitimate perspectives? • *Ex ante* environmental health assessment for proposed policies?
Science–Policy interaction	• Organizations and tools which facilitate participation, create a common language and stimulate co-production of knowledge?

Table 1. (*Continued*)

Criteria	Indicators
	• Persistent, stable, durable, flexible and adaptive organizations at the science–policy interface? • Scientifically valid and policy relevant tools at the science–policy?
Participation of stakeholders	• Involvement of stakeholders in the knowledge production process • Involvement of stakeholders in the decision-making process • Participation characterized by ○ direct or indirect? ○ formal or informal? ○ output (decision-oriented, reaching consensus) or input (deliberative, mapping out diversity) oriented?
Outcome of the arrangement and monitoring the outcome	• Does the arrangement succeed in early detecting EH problems and setting them on the political and scientific agenda? • Does the arrangement succeed in reducing EH-risks to levels that are acceptable to all stakeholders? • Does the arrangement succeed in increasing social basis of EH policy? • Is the arrangement legitimated by all stakeholders? • Is the arrangement a source of inspiration for other regions and countries? • Is there a systematic process to monitor and evaluate the EH objectives and targets (e.g. EH indicators)?

RESULTS AND DISCUSSION

As indicated by Stassen et al. (2013), a series of local environmental health events that occurred within a relatively short time period gradually shifted the discourses of Flemish politicians, scientists, and the population in general about the complexity of environmental health issues. Related to (scientific) knowledge production, it was determined that there is an increased need to: (1) manage uncertainties appropriately; (2) extend the research team to include several scientific disciplines (social scientists, communication experts, medicine, and the environmental sciences) as well as to include non-scientific forms of expertise; and, (3) shift the research focus from mortality and severe health effects caused by a short-term, high-dose exposure to the study of moderate health effects and the impact of long-term, low-dose exposure of mixed pollutants on well-being generally and specifically on the unborn child and women's fertility. Yet, these discourses require a more integrated approach to the knowledge-production process both at the organizational (inter-disciplinary and transdisciplinary research teams) as well as at the methodological (integrated risk assessment) levels. A hallmark of this integrated approach is the practice of taking into account both environmental and health factors but also related social, cultural and economic issues.

Within Flanders, changes observed in key discursive governmental activities reflect the need to: (1) differentiate between various target groups when establishing environmental quality standards, as some societal groups are more vulnerable than others; (2) better co-ordinate and even integrate environmental and public health policy; (3) increase stakeholders' participation in order to take into account all types of knowledge, perceptions, values, etc. in the decision-making process and (4) develop efficient and effective communication strategies in response to scientific uncertainty. Overall, making a shift to multi-actor and multi-sector governance was recognized as important.

Under the influence of the participation of the Green Party in the Flemish Government from 1999 until 2004, each discursive shift has transformed (to a greater or lesser extent) the Flemish environmental health arrangement by giving opportunities to new agencies and organizational structures to enter the arena (such as the Flemish Environmental Health Network). These opportunities have been created through establishing new rules inducing altered power relations (such as the formation of the decision-framework for uncertain risks and the Flemish Decree on Preventive Health Policy), and developing tools to ameliorate challenges arising at the science–policy interface (such as phased action plans and the use of environmental health indicators). Together, the actions taken have helped to effectively and efficiently manage complex environmental health risks and to prevent health problems due to environmental pollution in the near future. A schematic overview is presented in Fig. 2.

As indicated above, in order to facilitate the decision-making process on complex environmental health risks and the science–policy interface, the Phased Action Plan and a decision-making framework for dealing with uncertain environmental health risks have been developed. The Phased Action Plan was established to ensure that the scientific data generated by the Flemish Centre of Expertise were being translated into concrete policy measures. The Action Plan must assess how significant the data are in

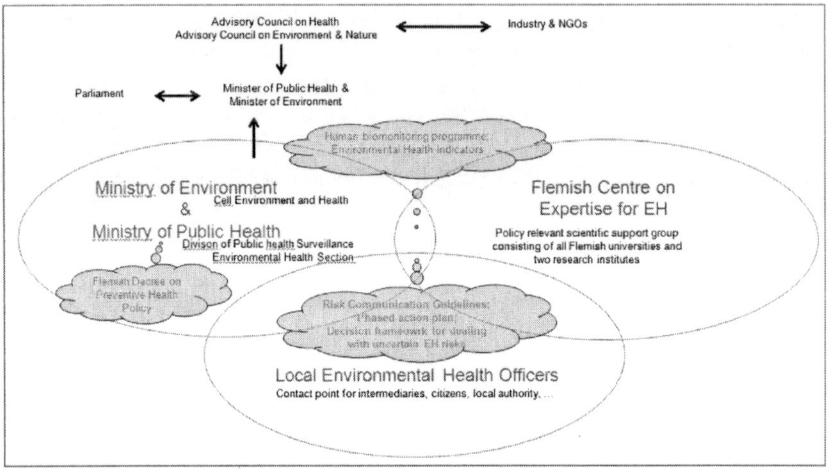

Fig. 2. Key Actors within the Flemish Environmental Health Policy Arrangement.

regards to health, to determine the origin and pathways of the pollutants into the human body and to develop policy measures. Despite its complexity, the Action Plan is presented as innovative and an example of best practice at the European level. For a more detailed description of the Phased Action Plan, read Keune et al. (2009). Comparing these discursive shifts with the converging paradigms for environmental health theories (Parkes et al., 2003), it is noteworthy that the traditional, reductionistic environmental health approach does not seem appropriate for dealing with complex environmental health risks and that more ecologically informed, upstream approaches are necessary for improving public health. In fact, the discursive shifts described above are similar to key concepts and principles underlying ecosystem based approaches to health research, such as the importance of integrating transdisciplinary, indigenous and stakeholder perspectives (Crescentia Dakubo, 2010).

At the organizational level, the Flemish Environmental Health Network was established in 2001 in order to trace and manage potential environmental health threats as expediently as possible, and to allow for greater stakeholder participation. The health network consists of three subnetworks: (1) the Local Environmental Health Officers, (2) the Administrative Services on Environment and Health within the Flemish Health Ministry and the Flemish Environment Ministry, and (3) the Flemish Centre of Expertise on Environment and Health. As is clearly illustrated in Fig. 2, the boundaries between the sub-networks overlap as each sub-network fills a combination of functions (knowledge development, policy-making and communication).

The Local Environmental Health Officers are the primary contact persons for the general public levying questions and complaints about environment and health issues. These officers have to register and analyse complaints; if necessary, they pass the problems on to the authorized government and/or the centre of expertise to find a solution. The local officers also assist the Flemish Government and the Flemish Centre of Expertise in creating social support for human biomonitoring surveys and in risk communication. Overall, they facilitate interactions between society, science and politics in order to find common ground and help to co-produce knowledge.

The Flemish environment and health policy is the result of the process and output of policy co-ordination between the environment and public health policy fields. Within each department, an Administrative Services on Environment and Health has been established. Together, both services are responsible for the preparation, execution, and evaluation of the Flemish environment and health policies. However, the development of one joint environmental health administrative service was hampered by the institutional organization of the Flemish Government and the resistance of some personalities in managerial positions. In order to avoid policy conflicts and to create a greater coherence between both services, tasks have been informally allocated and there is a bimonthly consultation. Nevertheless, friction still exists given that these two environmental health services have their own perspective on environmental health problems, their own institutional dynamics, and their own political agendas. Although efforts of increased co-operation at the administrative level have persisted, long- and medium-term environmental health objectives have yet to be developed.

There are also no agreements on well-defined environmental health indicators. Nowadays, the Flanders Environment Report uses three types of indicators: Disability Adjusted Life Years (DALYs), the concentration of polluting substances in humans measured by human biomonitoring, and the number of certain diseases and cancers which are strongly related to the environment. The multitude of environmental health indicators used reflects the complexity of environment and health problems. It is recommended to further research and to set up socio-political debates about good environmental health indicators in Flanders as well as at the European level. So far, Flanders also lacks an overarching environmental health strategy that is explicitly integrated into adjacent policy fields such as energy, transportation, agriculture, and the economy. To effectively increase public health, therefore, it is important to integrate knowledge and concerns about environment and health into cross-sectoral policies such as those pertaining to particulate matter, indoor air quality, product standards, and urban development.

The Flemish Centre of Expertise on Environment and Health fills the knowledge gaps and supports polity-oriented research. Its main goal is to set up a global human biomonitoring survey in Flanders. Within the Centre, experts from all Flemish universities and two Flemish research institutions – all with different disciplinary backgrounds – jointly investigate the complex relationship between the environment and health. The experts involved include statisticians, chemists, political and social scientists, medical experts, biologists, epidemiologists and toxicologists. The research team has evolved from a multidisciplinary into an interdisciplinary team which jointly investigates environmental health problems, using knowledge and concepts from different disciplines and integrating them into a synthesized, co-ordinated and coherent result. More recently, and influenced by the social sciences, other stakeholders are now also participating in different stages of the knowledge-development process (Keune et al., 2010). As such, the Centre has some characteristics of transdisciplinary research.

Although stakeholders' participation creates added value, there are some obstacles to putting it into practice as it is considered a time-consuming and labour-intensive process that can be hampered by participants, for example by their different levels of knowledge and power as well as their social, cultural and institutional affiliations. Possible strategies to transform the consortium of the Flemish Centre of Expertise into a real transdisciplinary research team which co-produces knowledge might be a valuable point of interest for future research. Related to the substantive focus of its work, the Flemish Centre of Expertise has given more importance to the quality of the research process in terms of open communication and dialogue about the research process and its results, as well as the management of uncertainties. However, it must be noted that in Flanders, in affiliation with the Netherlands (Van der Sluijs et al., 2003), no official procedure has been developed to assess and communicate uncertainties. Even though no standard rules for dealing with uncertainties can be given, a global reference framework has been useful. Another challenge is that the Centre of Expertise on Environment and Health is financed by the Policy Research Centres Programme in which every five years the themes are reviewed. As a consequence, the centre's research program may not be stable and durable over the long term. This is an issue in part because sustained long-term research programs enable the development of research such as the establishment of longitudinal surveys and also gives teams the time they need to build research capacity. However, advocates of the temporary character of the Centre's resource argue that this approach ensures the centre remains competitive, results-oriented, and focused on innovation.

As the findings from this Flemish case study reinforce, the Phased Action Plan and the decision framework are built on the conceptual ideas of Renn's risk governance framework in order to facilitate a transparent and balanced decision-making process in case of uncertain environmental health risks. Both tools are still in a preliminary phase and are not ready-for-use recipes; however, they do offer a promising framework which we suggest should be adapted to work in new contexts. The key feature of both tools is to enhance stakeholders' participation in order to increase the quality and legitimacy of the decision-making process, ensure more robust and well-informed decision making, expand knowledge to include both scientific as well as societal considerations, and increase public support for policy measures. These measures are needed as currently stakeholders tend to be indirectly involved in the decision-making process through traditional, formal advisory boards while local people are directly involved as participants in human biomonitoring surveys. Our research suggest that through direct involvement in the knowledge-production process, local actors could increase their knowledge about the problem and it is quite conceivable that might also want to be involved in the output phase of the decision-making process, such as helping to reach consensus on the best policy actions. In short, these are tools that put into practice Post-Normal Science epistemologies and through additional research could more robustly help to facilitate the development of multi-actor governance strategies relevant to addressing environment and health risks within a range of social-ecological health settings.

CONCLUSION

In this chapter we have made an empirical contribution to the risk governance debate by analysing the Flemish environmental health risk governance arrangement. In particular, we have focused on how to put the principles of stakeholder participation, transdisciplinarity and (policy) integration into practice. To make this case, we have studied the institutional dynamics of this arrangement and developed an evaluation framework based on a literature review to evaluate its performance. We argue that this approach represent best practices in how to identify and then learn from lessons about the use of risk governance arrangements in order to enhance mutual learning and intersectoral collaborations.

One of the strengths of the present Flemish environmental health risk governance arrangement is that it has led to increased interaction between the spheres of science, policy and society. Therefore it has clearly succeeded in increasing the exchange of information between: (1) different governmental levels; (2) scientists, policymakers and the general public; and (3) environmental and public health civil servants. Moreover, the human biomonitoring surveys, in combination with the work of local environmental health officers, has enabled the early detection and handling of concerns and problems, which has prevented them from evolving into real crises. Since the establishment of the Flemish Environmental Health Network, no environmental health disaster has occurred; however, it is difficult to determine whether this is just a coincidence or a result of a good working relationships and the formation of an early warning network.

This analysis also provides insights into some shortcomings and makes a few recommendations for how to improve the risk governance arrangement. For example, the integration of environmental health objectives in all relevant policy domains, additional research, socio-political debates about environmental health indicators, and additional research to facilitate that co-production of knowledge and multi-actor governance are all needed. From a more reflexive point of view, this Flemish case study illustrates that novel thinking on the epistemology of risks, on the one hand, and policy-making, on the other, is strengthened by a strong focus on holism. There is much to be gained, therefore, from mutual learning between the traditional environmental health field and the newly emerging field of ecohealth with its integrated, holistic approach of knowledge development and policy-making. The integration of environmental health objectives in all relevant policy fields and the integration of the whole cause–effect system within risk assessment are just two examples. Yet, the increased literature on risk assessment and risk governance and their stress on stakeholder participation and communication is an important reminder of the contemporary relevance of these issues for anyone who studies the links between health and the natural environment from a social perspective. The theoretical contribution on risk governance and the practical lessons learned in the Flemish case may also be valuable for the ecological health community as new policy domains are starting to open up through interfaces with agencies like the WHO and the International Development and Research Centre. After all, the Flemish case study presents good practices and learnings from pitfalls encountered when trying to put into practice intersectoral, transdisciplinary and multi-stakeholder knowledge development and policy-making.

Finally, a reflection is made on methodology. The combination of content analysis of scientific publications and policy documents with data gathered through in-depth interviews with the main stakeholders resulted in a detailed and balanced understanding of the Flemish environmental health risk governance arrangement. The strength of this triangulation approach is its internal validity. The external validity of this Flemish case study is rather limited as the results cannot be generalized to other regions and countries, nor to other policy sectors. Yet comparative approaches of risk governance arrangements would be interesting to investigate in further research. On balance, a longitudinal, either cross-sector or cross-regional comparison of risk governance arrangements would allow: (1) to identify good practices which can be exchanged and (2) to build a typology of risk governance arrangements in order to draw lessons from contemporary and to build future risk governance arrangements.

We hope our case approach can inform a wider debate as we are witnessing the emergence of a series of sensitizing and appealing integrative concepts, ranging from sustainable development, risk governance and environmental health to ecohealth. As these concepts are interesting, their advocates tend to believe that their application can be conceived as a univocal implementation process. In fact, as we illustrate above, we observe that these concepts need to be subjected to long and laborious processes of acceptance and application as it is through these processes that pre-existing institutional patterns gradually shift.

REFERENCES

Beck, U. (1992). *Risk society. Towards a new modernity.* London: Sage.

Briggs, D. J. (2008). A framework for integrated environmental health impact assessment of systemic risks. *Environmental Health, 7*(61). doi:10.1186/1476-069X-7-61

Dakubo, C. Y. (2010). Ecosystem approaches to human health: Key concepts and principles. In Dakubo (Ed.), *Ecosystems and human health* (pp. 37–53). New York, NY: Springer. Dakubo is one of the contributors to this book. This citation refers to this chapter http://link.springer.com/chapter/10.1007%2F978-1-4419-0206-1_3#

Douglas, M., & Wildavsky, A. (1982). *Risk and culture. An essay on the selection of technological and environmental dangers.* Berkeley, CA: University of California Press.

European Environmental Agency. (2005). *Environmental policy integration in Europe – State of play and an evaluation framework.* Copenhagen: EEA.

Gabe, J., Bury, M., & Elston, M. A. (2004). *Key concepts of medical sociology.* London: Sage.

Giddens, A. (1990). *The consequences of modernity*. Stanford, CA: Stanford University Press.

Grandjean, P. (2005). Non-precautionary aspects of toxicology. *Toxicology Applied Pharmacology*, *207*(Suppl. 2), 652–657.

Hajer, M. A. (1995). *The politics of environmental discourse*. Oxford: Oxford University press.

Jasanoff, S. (Ed.). (2004). *States of knowledge. The Co-production of science and social order*. New York, NY: Routledge.

Keune, H. (2012). Critical complexity in environmental health practice: simplify and complexify. *Environmental Health*, *11*(Suppl. 1), S19.

Keune, H., Morrens, B., Croes, K., Colles, A., Koppen, G., Springael, J., … Van Larebeke, N. (2010). Opening the research agenda for selection of hot spots for human biomonitoring research in Belgium: A participatory research project. *Environmental Health*, *9*(33). Available at http://www.ehjournal.net/content/9/1/33

Keune, H., Morrens, B., Springael, J., Koppen, G., Colles, A., Loots, I., … Van Larebeke, N. (2009). Policy interpretation of human biomonitoring research results in Belgium: Priorities and complexity, politics and science. *Environmental Policy and Governance*, *19*(2), 115–129.

Knol, A. B., Briggs, D. J., & Lebret, E. (2010). Assessment of complex environmental health problems: Framing the structures and structuring the frameworks. *Science of the Total Environment*, *408*(14), 2785–2794.

Leroy, P., Driessen, P. P. J., & Van Vierssen, W. (2010a). From climate change to social change. In P. P. J. Driessen, P. Leroy & W. Van Vierssen (Eds.), *From climate change to social change* (pp. 161–172). Utrecht: International Books Utrecht.

Leroy, P., Driessen, P. P. J., & Van Vierssen, W. (2010b). Climate, science, society, and politics: Multiple perspectives on interactions and change. In P. P. J. Driessen, P. Leroy & W. Van Vierssen (Eds.), *From climate change to social change* (pp. 15–29). Utrecht: International Books Utrecht.

Lupton, D. (1999). *Risk*. London: Routledge.

National Research Council. (2009). *Science and decisions: Advancing risk assessment*. Washington, DC: The National Academies Press.

Nowotny, H., Scott, P., & Gibbons, M. (2003). Introduction. 'Mode 2' revisited: The new production of knowledge. *Minerva*, *41*(3), 179–194.

Parkes, M., Panelli, R., & Weinstein, P. (2003). Converging paradigms for environmental health theory and practice. *Environmental Health Perspectives*, *111*(5), 669–675.

Renn, O. (2005). *White paper on risk governance: Toward an integrative framework*. Geneva: International Risk Governance Council.

Renn, O. (2008). *Risk governance: Coping with uncertainty in a complex world*. London: Earthscan.

Rosa, E. A. (1998). Metatheoretical foundations for post-normal risk. *Journal of Risk Research*, *1*(1), 15–44.

Runhaar, H. A. C., Driessen, P. P. J., & Van der Sluijs, J. P. (2009). Shifts in environmental health risk governance: An analytical framework. In S. Martorell, C. Guedos Soares & J. Barnett (Eds.), *Safety reliability and risk analysis; theory, methods and applications* (pp. 369–378). London: CRC Press.

Runhaar, H. A. C., Driessen, P. P. J., Van Bree, L., & Van der Sluijs, J. P. (2010). A meta-level analysis of major trends in environmental health risk governance. *Journal of Risk Research*, *13*(3), 319–335.

Schmidt, V. A. (2008). Discursive institutionalism: The explanatory power of ideas and discourse. *Annual Review of Political Science*, *11*, 303–326.

Scott, W. R. (2001). *Institutions and organizations* (2nd ed). Thousand Oaks, CA: Sage.

Stassen, K. R., Gislason, M., & Leroy, P. (2010). Impact of environmental discourses on public health policy arrangements: A comparative study in the UK and Flanders (Belgium). *Public Health*, *124*(10), 581–592.

Stassen, K., Smolders, R., & Leroy, P. (2013). Sensitizing events as trigger for discursive renewal and institutional change in Flanders' environmental health approach, 1970s–1990s. *Environmental Health*, *12*(46). doi:10.1186/1476-069X-12-46. Available at http://www.ehjournal.net/content/12/1/46

Thompson-Klein, J., Grossenbacher-Mansuy, W., Häberli, R., Bill, A., Scholz, R. W., & Welti, M. (Eds.). (2001). *Transdisciplinarity: Joint prolem solving among science, technology and society. An effective way for managing complexity*. Basel: Birkhäuser Verlag.

Van der Sluijs, J. P. (2007). *Post normal science: Working deliberatively within imperfections*. Studium Generale Wageningen. Science, Policy and Complex Phenomena, 21 March 2007. Retrieved from http://www.nusap.net/article.php?sid=38

Van der Sluijs, J. P., Risbey, J. S., Kloprogge, P., Ravetz, J. R., Funtowicz, S. O., et al. (2003). *RIVM/MNP guidance for uncertainty assessment and communication: Detailed guidance*. Utrecht: Utrecht University.

Van Kersbergen, K., & Van Waarden, R. (2004). 'Governance' as a bridge between disciplines: Cross-disciplinary inspiration regarding shifts in governance and problems of governability, accountability and legitimacy. *European Journal of Political Research*, *43*, 143–171.

OCEANS AND HUMAN HEALTH IN THE CARIBBEAN REGION

Alana Malinde S. N. Lancaster and
Lyndon F. Robertson

ABSTRACT

Purpose – *This chapter examines the characteristics, challenges and prospects of environmental governance and participation in issues pertaining to human health and the ocean in the CARICOM Caribbean region.*

Design/methodology – *Utilising the fisheries sector – one of the principal economic, social and environmental drivers relating to the marine environment in the Caribbean region – we discuss the concepts of hierarchical governance in contradistinction to heterarchical governance. This is done through a socio-legal analysis of the predominant top-down model of governance, a discussion of the successes and shortcomings of bottom-up governance and a proposal for more inclusive participation methodologies in the region.*

Findings – *While the paradigm of new collaborative environmental governance was birthed in the aftermath of the 1985 Brundtland Commission Report, and moreso since the 1992 Rio Conference, this analysis will show that governance of the marine resource, and consequently how the individual is juxtaposed within this matrix, has not shifted from a position of hierarchy to one of heterarchy, as prescribed*

Ecological Health: Society, Ecology and Health
Advances in Medical Sociology, Volume 15, 311–334

ISSN: 1057-6290/doi:10.1108/S1057-6290(2013)0000015018

by the governance literature. Indeed, the structures for governance remain largely top-down in nature and while many states have begun to embrace more inclusive and participatory methodologies many of these interventions will need to be bolstered if the governance of the region's marine resources is to progress from traditional top-down to more inclusive and representative typologies.

Practical implications – *These concepts, when applied to the subject of environmental governance, will demonstrate that there needs to be an improvement in participatory environmental governance in the CARICOM region if the integrity of human health and the ocean is to be maintained. Importantly, while these methodologies strive for the formulation prescribed in Principle 10 of the 1992* Rio Declaration, *and most famously exemplified in the 1998* Aarhus Convention, *issues of environmental advocacy, transparency, inequality and justice need to be reconceptualised, if the region is to see prudent governance of the interface between humans and the ocean.*

Originality/value – *This research takes established concepts on the issue of* locus standi *in the common law legal tradition and juxtaposes it within the emerging paradigm of ecohealth and environmental governance. This conceptual framework has identified both the prospects and problems of environmental governance in the Caribbean region and may provide the basis for further research as well as more inclusive and sustainable environmental governance.*

INTRODUCTION

Like emeralds scattered upon a sapphire sea, the countries of the Caribbean region glisten as do rare and precious gemstones on a jeweled belt. Situated in the region of the Americas, the Caribbean region is an archipelago that extends some 4,020 kilometres (2,500 mi) in length, 257 kilometres (160 mi) in width and covers an area of more than 2.5 million square kilometres. More than 7,000 islands, islets, reefs, and cays enclose the Caribbean Sea – the second largest sea in the world (CARSEA, 2007) – which, along with the Atlantic Ocean and the adjoining coastal areas, form a semi-enclosed sea as identified in Art. 122 of the 1982 *Montego Bay Convention*. The geographical characteristics of this enclosure means that the Caribbean Sea is extremely susceptible to social and ecological activity and therefore concerted efforts to manage and protect the integrity of the Sea as a marine and coastal resource are necessary (ECLAC, 2004). Protection of the Sea is important mainly because it is considered a biodiversity hotspot (Mittermeier et al., 2005; Myers et al., 2000) and a significant reservoir of biodiversity that is threatened with destruction. In addition to its stunning array of biodiversity, the marine environment boasts indispensable and innumerable ecological functions which are essential to the health and security of the citizens of the Caribbean region, as well as to visitors to this paradise. The location also forms part of the semi-enclosed Caribbean Large Marine Ecosystem (CLME) – a unique ecological region which is particularly important to the fisheries of the region (Fanning, Mahon, & McConney, 2009).

This chapter will focus on the CARICOM group of states within the Caribbean region, as it is within this context that governance of the marine environment and its implications for human health can be well understood. Therefore, in this chapter we first introduce the CARICOM region and describe its relevance to the study of human and marine governance. The chapter identifies the principal threats to the ocean and human health in the region and then analyses the paradigm of governance with respect to these factors. Accordingly, we approach the issues of ecological health through a legal rubric in order to draw attention to the importance of the characteristics and challenges of the current system of governance. To achieve this objective, comparisons are made links with the governance framework of the fisheries industry in the region – primarily because this industry constitutes a significant social, economic and ecological asset to the Caribbean region. Significantly, we contrast the existing hierarchical model inherent in the region with emerging approaches which promote the heterarchical model typical of more inclusive, transparent and participatory governance.

CONTEXT: EXISTING AND EMERGING ECOHEALTH ISSUES IN THE CARICOM CARIBBEAN REGION

Birthed with the Federation in 1958, CARICOM is a regional integration movement, guided by the 2001 *Revised Treaty of Chaguaramas* (*RToC*), which comprises 15 member states and 5 associate member states. This arrangement is primarily for purposes of economic integration, natural resources management, and the development and implementation of regional agreements (Map 1).

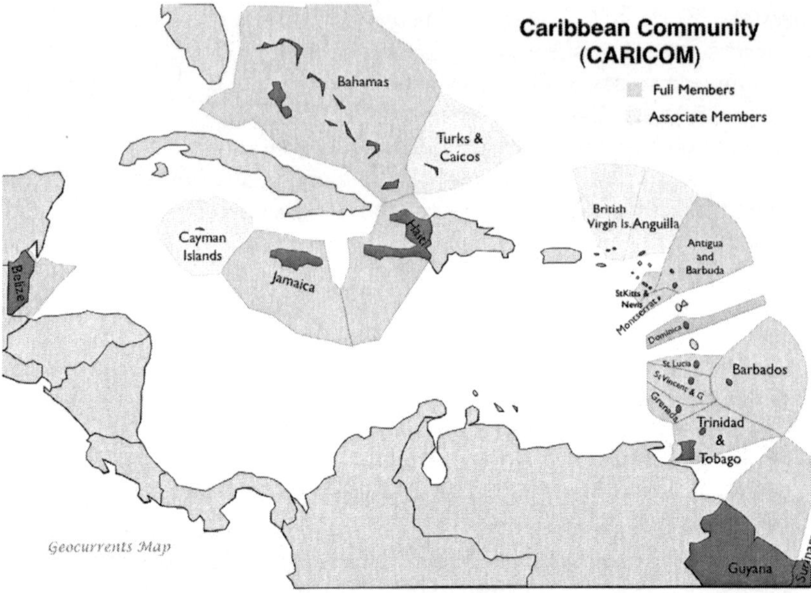

Map 1. Map of the Wider Caribbean Region with CARICOM Members Indicated (maritime boundaries are indicative only and remain largely unsettled). *Source*: http://geocurrents.info/wp-content/uploads/2011/03/Caricom44.jpg

Within the CARICOM grouping is another economic sub-union – the Organisation of Eastern Caribbean States (OECS) – which is directed by means of the 2011 *Revised Treaty of Basseterr*e (*RToB*). This sub-bloc of states currently comprises 9 members, and spreads across the Eastern Caribbean in a near-continuous archipelago which extends from the Leeward Islands to the Windward Islands. All the states are full or associate members of CARICOM.

From a socio-economic as well as an environmental perspective, the Caribbean is considered a group of Small Island Developing States (SIDS) – low-lying coastal countries and low income continental countries – that tend to share similar sustainable development challenges. This includes small but growing populations, limited resources, susceptibility to natural disasters, vulnerability to external shocks and fragile environments. Most economies are based on agriculture, tourism and the extraction of living marine resources, with the exception of the petroleum-producing state of Trinidad & Tobago (UNEP/Heileman & Walling, 2004). Many of these characteristics in themselves pose significant challenges to the functioning of effective frameworks of governance in the region; these pressures are often made more complex by poverty which is widespread throughout both the mainland and island countries. These characteristics in turn have effects on human health and the environment. It is a *sine qua non* therefore, that the aforementioned characteristics make the region vulnerable to threats to its ecosystem and health services, while at the same time underscore the urgent need to strengthen the manner in which the relationship between oceans, health and human activity is governed.

The degradation of the marine resources, pollution of the Sea principally from land-based sources, and the over-exploitation of living and non-living resources, contribute to a plethora of potentially negative effects on human health. These include *inter alia*:

- exposure to seafood-borne contaminants – chemical residues, for example heavy metals such as mercury and persistent organic pollutants (POPs), which enter the food chain where they bioaccumulate and biomagnify, predisposing humans – especially populations of coastal communities who consume relatively more seafood – to health hazards from carcinogens and heavy metals;
- increases in waterborne pathogens, microorganisms, outbreaks and other extreme water events;
- increased exposure to emerging zoonotic diseases and outbreaks, including to wildlife and domestic animals as reservoirs for human pathogens;
- exposure to skin, ear, nose and throat (ENT) and acute gastrointestinal infections (AGI) from contact with polluted and contaminated water during recreational or other activities in the Sea – the sources of these etiological agents are usually contaminants in sewage and washout from drains and main water courses after heavy rains;
- effects of urbanisation on the coastal regions, especially chemical, nutrient and pathogen loading;
- harmful algal blooms (HAB) which result in marine biotoxins synthesised by dinoflagellates which have varied effects on fish and severe health effects on humans;
- increased population migration inducing susceptibility and reducing resilience to diseases.

The sources and effects of these factors have not all been well researched, but they are to some extent quantified (UNEP/CEP, 2012); it has been recognised from as far back as 1994 that they need to be addressed by the states in the region in a holistic and coordinated manner (UNEP/CEP, 1994). As this chapter will discuss in further detail below, one such strategy for successful management is the endorsement of international and regional environmental agreements which aim at addressing these issues and another is the concomitant adoption of strategies to address them in a standard manner at the national level.

THE EMERGENCE OF A NEW PARADIGM OF ENVIRONMENTAL GOVERNANCE

Governance is a broad concept with a seemingly elusive definition. Fanning, Mahon, and McConney (2011) note that governance is not management, nor is it government, but it does encompass both. Within the context of ocean and human health, governance is taken to reflect the global shift in awareness of the increasing diversity of stakeholders involved in the use of the oceans and its resources (Johnson, 2006). In the Caribbean region, the principal model of governance has, and to a large extent still is, the formal, centralised and vertical exercise of power and authority more common to government. This is in contradistinction to governance which is horizontally decentralised and devolved to broader members of society (Harrington et al., 2008). However, as will be illustrated through examples to follow, while the practice of hierarchical management is slowly changing, empirical evidence suggests that although in some respects the mode of governance has changed the effects of these changes may not be as effective or inclusive as anticipated.

The 1992 Rio Summit, catalysed the rapid and expansive evolution of the principles of international environmental law and environmental law, which were subsequently incorporated into various instruments at the regional and national levels, with varying degrees of effectiveness. While some principles such as sustainable development, the polluter-pays and the precautionary principle are prescriptive, others give rise to environmental procedural rights. Notably, Principle 10 of the 1992 *Rio Declaration* sought to promote public participation through access to information, participation in decision making and access to judicial processes for seeking redress. This framework can feature largely in considerations for ecohealth, since its approaches are participatory, systems-based and strive for an understanding and promotion of health and wellbeing in the context of social and ecological interactions. Best elucidated in the guise of the 1998 *Aarhus Convention*, procedural rights were set about to ensure that the parity of access to information, public participation and access to justice in environmental decision-making will be incorporated into traditional approachs to decision-making and environmental management.

It is worthy to note that the main challenge to implementing environmental governance in the CARICOM Caribbean jurisdiction is the issue of the incorporation of international or regional law into a state's domestic law. The significance of the distinction between the national regimes and those at the regional and international levels is especially noteworthy in the Caribbean region – primarily those countries which are former British colonies. This conundrum arises because, with the exception of Haiti and Suriname, the legal systems of CARICOM Caribbean states subscribe to the dualist doctrine. Put simply, without proper translation and enactment of legislation to give effect to international or regional law, states can find themselves powerless to enforce or benefit from the full gamut of measures available to them within international or regional law. Examples of the former have been amply illustrated with the trio of cases – *Acting Chief of Police v. Bryan* (1987); *Natural Resources Conservation Authority v. Seafood and Ting International Ltd.*; *Natural Resources Conservation Authority v. DYC Fishing Ltd.* (1999) and *Talisman (Trinidad) Petroleum Limited v. EMA* (2002). In these cases courts in the British Virgin Islands, Jamaica and Trinidad and Tobago respectively tried unsuccessfully to implement requirements under multilateral environmental agreements (MEAs) and other instruments to protect marine resources, species or coastal resources. A key issue was that while, the respective states had signed onto and committed *internationally* to ensure this protection they did not put these commitments into *national* law.[1] The *Westerhaven* cases further serve as a cautionary tale to CARICOM Caribbean states with dualist legal traditions, since in this instance, although the relevant instrument had been incorporated – its subsequent amendment was not, even though Belize passed its legislation after the update. The effect of this lacuna was that the quantum of damages for destruction to the Belize Barrier Reef, its associated ecosystems and services, was substantially reduced which in this case proved deleterious to efforts to mitigate the damage to the reef and to restore, as much as possible, the integrity of the resource.

Environmental procedural rights, in effect, provide that anyone with a sufficient interest in environmental issues should be able to challenge the substantive or procedural rights of a decision, act or omission. This represents a shift in the modus operandi in which, through governance mechanisms, the inclusion of the public is solicited. The purpose of these rights is to foster inclusion, seek to achieve transparency, and devolve decision-making from the bottom up – that is from those who are the principal custodians and beneficiaries of the health of the [marine] environment. These persons may often be disenfranchised by the traditional command-control model or by the system of common law which pervades the Commonwealth legal system of CARICOM Caribbean states.

On this matter, English law has striven assiduously to place some restraint on the procedural rights of the citizen. Judicial authority is found in the leading House of Lords case of *Gouriet v. Union of Post Office Workers and Others*[2] where Lord Wilberforce expressed that private rights can be asserted by individuals but public rights 'can only be asserted by the Attorney General as representing the public' (1977, pp. 71–72). Issues of individual and public health have been inculcated in these kinds of debates since the Industrial Revolution when public health was a primary concern of government as well as of citizens who sought measures to secure concomitant rights[3] and do so by relying on proprietary interest,[4] or through the discretion of the Attorney General. Tensions between individual and public rights also extends to the area of administrative and public law – an area of law which is underpinned by the notion of protecting the citizen from administrative excesses of authority inherent in most traditional governance structures. As illustrated in the case *Reg. v. Inland Revenue Commissioners, ex parte National Federation of Self-Employed and Small Businesses Ltd.*,[5] speaking of the need for leave, Lord Diplock stated: '... [i]ts purpose is to prevent the time of the court being wasted by busybodies with misguided or trivial complaints of administrative error, and to remove the uncertainty in which public officers and authorities might be left as to whether they could safely proceed with administrative action...' (1981, p. 105) This restraint, according to Lord Scarman, '... is an essential protection against abuse of legal process ... [as] it enables the court to prevent abuse by busybodies, cranks, and other mischief-makers ...' (1981, p. 113). Essentially therefore, the confines of administrative and common law, have served as a significant impediment to effective environmental governance, since generally, public interest groups and environmentally driven individuals lacked 'sufficient interest' to seek redress for environmental injustices.

On the face of it, therefore, addressing issues of ecohealth which permeate society in its entirety are potentially at odds with these tenets. Consequently, there is the recognition that *locus standi* facilitates public participation in the decision making of pertinent issues which pervade the lives of the members of a society (Anderson, 1994). As stated by Lord Diplock, there is 'a grave lacuna in [the] system of public law, if a pressure group ... is prevented by outdated rules of *locus standi*, from bringing a matter to the attention of the court to vindicate the rule of law, and get unlawful conduct stopped'. Arguably, therefore, the transcendant and often non-proprietary nature of environmental issues means that the vexing question of *locus standi* persists as a significant hurdle to the successful inclusion of and access to environmental justice for the populace. The case of *R v. Secretary of State for the Environment, ex parte Rose Theatre Trust Co* (1990), perhaps marks the watershed of this restrictive approach in English common law. This restrictive approach was wholly unsatisfactory within the era of expanded environmental stewardship and inclusiveness and sadly was initially adopted in the Caribbean in cases such as *Spencer v. Canzone Del La Mere Ltd. et al.* (1993), *Scotland District Association Inc. v. Attorney General and Others* (1996) and *Spencer v. Attorney General of Antigua and Asian Village Antigua Ltd.* (1998). Since then, English law has embarked on a more relaxed approach to standing – most notably in *R v. Inspectorate of Pollution, ex parte Greenpeace Ltd. (No. 2)* (1994) and *R v. Secretary of State for Foreign Affairs, ex parte World Development Movement* (1995). These cases are perhaps reflective of the United Kingdom's movement toward signature and eventual ratification of the *Aarhus Convention*. Happily, in the Caribbean – while the issue of standing has not been specifically or substantially addressed – there is ample jurisprudence which suggests that the approach has taken root, including cases where issues affecting human health and the ocean have devolved. Arguably therefore, the new approach to standing in English law has entered the realpolitik of governance in the Caribbean region – albeit with mixed results. Further, as will be discussed later in the chapter, the Commonwealth Caribbean has also utilised command-and-control methods to circumvent the challenges of *locus standi* in administrative and common law.

What may be discerned from this analysis is that the rights of the Caribbean citizen to access and use marine resources, and enjoy the basic human right of good health, is impinged on by both internal and external factors. The external factors relate to issues of international and regional influence, such as MEAs, soft-law documents such as the 1992 *Rio Declaration* and the 1989 *Port of Spain Accord*, and principles of environmental and human rights. From an internal perspective, a citizen's rights are affected by the extent of the acceptance of international and regional obligations by their respective states and consequently by the priorities and direction of states both as individuals and as groups. The model of governance chosen by a state, therefore, dictates its policy as well as its institutional and legislative considerations in the marine environment.

Thus, while concern for environmental issues continues to be espoused in the scientific, technical, policy and academic spheres, there consistently remains great reluctance and sloth on the part of the states of the region to translate these realities into legislation and other concrete actions. This failing will have disastrous and devastating effects on the ecology, health and economy of the region in the foreseeable future. However, it is arguable that this is a situation which can change – to some extent, from the bottom up – if Caribbean states genuinely embrace or incorporate more participatory governance. This proposition will be illustrated by referencing measures in the governance of the fisheries sector and by an analysis of attempts to incorporate mechanisms aimed at participatory environmental governance into the framework of environmental management in the CARICOM Caribbean region.

GOVERNANCE IN RELATION TO THE MARINE RESOURCES AND ISSUES OF HUMAN HEALTH IN THE CARIBBEAN REGION

Kooiman, Bavinick, Jentoft, and Pullin (2005) define governance in the context of fisheries as 'the whole of public as well as private interactions taken to solve societal problems and to create opportunities within society … [g]overnance thus includes the development and application of principles, rules, norms and enabling institutions that guide public interactions' (p. 29). This definition is particularly noteworthy, especially in the context of the Caribbean. The fishing industry will be used as an example, since according to data, it has been well documented as important to the coastal communities of the Wider Caribbean region (FAO/Salas, Chuenpagdee, Charles, & Seijo, 2011). Fish is a ready source of protein, macronutrients (e.g. fatty acids such as omega-3) and micronutrients (vitamins such as selenium and minerals) for communities of the Caribbean. The fishing industry is even more important for Small Island Developing States (SIDS) such as the Caribbean, where dependence on an affordable, accessible and readily available source of protein is pronounced. The fishing industry of the Caribbean region depends on a diverse array of fish (Fanning et al., 2009) and most of these resources are considered to be overexploited (FAO, 1998; Mahon, 2002; Myers & Worm, 2003). Of the approximate 113,000 tonnes of fish caught within the exclusive economic zone (EEZ) of the CARICOM Caribbean states annually, 45% is consumed locally (by natives and tourists),

23% is shipped to European Union territories and 30% is exported to North America (Maitland & Morrison, 2010). It is therefore indisputable that the management and governance of this crucial industry should be subject to more holistic and collaborative forms of governance than the traditional command-and-control structures which were inherited from colonialism and bequeathed at independence. Further, because the industry directly pertains to concerns around issues of nutrition, public health and community wellbeing, its governance is also in direct correlation with a systematic, multi-sectoral and integrated approach to ecological health governance. The desirable mode of governance ought to involve formal and informal institutions, social groups, processes, interactions and traditions, all of which influence how power is exercised, how public decisions are taken, how citizens become engaged or disaffected, and who gains legitimacy and influence (Bakker, 2007; Olsson, 2007). The fishing industry provides an example, which will be discussed below, and which lends itself to this model of environmental governance at different scales – specifically at the national and regional levels.

Several observations may be made about the mechanisms and actors in the governance of the marine resource in the Caribbean region. Firstly, from a regulatory perspective, the command-and-control method has historically been practised and still continues to dominate and reign supreme. This is discouraging, given the shift from first generation environmental problems which lent more readily to this approach to second generation environmental issues, which are more complex, involve multiple small polluters, and do not lend themselves to traditional regulatory solutions (Gunningham & Grabosky, 1998). Nevertheless, it is arguable that within the Caribbean region this latter approach will be the ashes from which the phoenix of new environmental governance will arise. Secondly, more contemporary environmental issues and national governance strategies are often influenced by international and regional drivers – and therefore are external to the state. Thirdly, a characteristic which follows is the dominating role of governments in the governance equation. Fourthly, an element which may be discerned is the involvement of multi-lateral structures such as the United Nations Environmental Programme's Regional Seas Programme (UNEP/CEP), CARICOM and OECS arrangements, as well as NGOs and other non-traditional actors in environmental governance. However, what is woefully missing is the widescale involvement of the principal dependents/custodians of the resources – Jared Jamaica or Gitangali Guyana: the average Caribbean woman/man and the principal stakeholders in the governance process. The extent to which these stakeholders can exercise their full suite of environmental procedural rights is either a commendation or indictment of the system of governance in the region. More often than not, the average citizen is the recipient of the top-down governance structure, and their involvement is elicited when their rights are affected, often in an adverse manner. The analysis will now turn to how the region has tried to move from a command-type to a devolved method of governance and will comment upon whether these attempts have been successful.

FISHERIES GOVERNANCE: THE SAME OLD MODEL OR NEW ENVIRONMENTAL GOVERNANCE?

There is perhaps no better example of the characteristics, failings and prospects for environmental governance than the management of the fisheries sector. Along with marine delimitation, the fisheries sector in almost every Caribbean state is usually well defined and regulated – exclusively by means of command-and-control instruments. This fact belies the importance of the industry to the socio-cultural, economic and environmental identity of the Caribbean region. Additionally, in the OECS region, there has been the recognition that the sector crosses spatial and temporal borders, and therefore requires collaborative and collective governance. This is evidenced by the efforts in the 1980s by OECS states to pass almost identical fisheries legislation, aimed at addressing the transboundary nature of the fisheries resource, and by recent updates – primarily by Antigua and Barbuda (*Fisheries Act* 2006, *Fisheries Regulations* 2013) – to incorporate principles of international environmental law into fisheries management. While these laws have traditionally adopted a species approach to resource regulation, recent enactments such as Antigua and Barbuda's *Fisheries Regulations* 2013, comprise the precautionary and ecosystems approach, which are meant to address elements which pose a risk to present as well as future generations – including overfishing, introduction of invasive species and contaminants to both the freshwater and marine resource. This arguably takes into account both spatial and temporal elements (Reed & Bruyneel, 2010), and endorses the recognition of sustainable development as customary international law.[6] Further, it is a means of implanting the emerging paradigm of ecohealth firmly into the governance and management of one of the most crucial sources of protein, nutrients and income for the average Caribbean citizen.

The command-and-control structure of governance characteristic of the Caribbean fishing industry is replete with administrative support – both at the national and regional levels. Two comments are to be made: First, has the management and governance of the fisheries industry been effective with respect to the national and regional governance regimes? Arguably not, given the over-exploitation of fisheries resources outlined earlier in the chapter. Further, concern over the future of the regional fishing industry has prompted CARICOM Heads to mandate the secretariat to prepare a Common Fisheries Policy and Regime (CFPR). This is in light of deficiencies in the management and conservation of regional fish stocks due to both over-fishing in the cases of some species and the under-utilisation of other species (Stabroek News, 2009). At the time of writing, this policy, which was completed in 2011, is yet to be signed off by the Heads of Government. However, it is firmly on the agenda of the Caribbean Regional Fisheries Mechanism (CRFM) – the principal forum, apart from regional fishery [management] organisations (RF[M]Os), for the management of fisheries in CARICOM and OECS states. Accordingly, states are in the process of revamping their approaches to fisheries management which continues to be a traditional command-and-control style that is infused with more holistic and participatory features than their predecessors. These changes seek to incorporate broader principles of sustainable development, which will ensure a greater level of accountability by regional organisations and national institutions, foster community-based resource management, and consequently inspire broader engagement by key stakeholders.

A second observation is that stakeholders like Jared and Gitangali are now much more actively involved in the fisheries management structure than they were previously. The infiltration of the community-based resource management concept in the region is most evident in the fisheries sector, with an example at the regional level being the Caribbean Network of Fisherfolk Organisations (CNFO). In addition, most Caribbean states have organised and maintained active fisherfolk organisations. The attendance of the CNFO at the recent 11th Meeting of the Caribbean Fisheries Forum was praised as a step in a positive direction by at least one delegate at the Forum. This is because while governments and policymakers have featured preminately in fisheries governance in the region it is often at the exclusion of, and detriment to, the principal stakeholders: the fisherfolk of the region.

Overall, this approach largely endorses the interactive fisheries governance perspective advocated by Mahon, McConney, and Roy (2008), which involves a dynamic and complex fish chain, leading from the resource and its supporting ecosystem to the global marketplace and the local consumer. Traditionally, interactions within complex fisheries systems have tended to be ignored when fisheries resources are examined in an isolated manner, and public participation in problem solving and creating opportunities have been discouraged (FAO/Salas et al., 2011). However, it appears that there is the desire by governments of the region to depart from the old methodology, but ultimately the proof will be in the extent that these measures better address the multi-faceted challenges of the sector, and the extent to which the proletariat – Jared Jamaica, Gitangali Guyana and NGOs – can make inroads into the top-down system of the command-and-control approach.

THE NEW INSTRUMENTS OF GOVERNANCE?

While it cannot be said that the voluntary and negotiated agreements nor the regulatory flexibility initiatives envisioned within environmental governance abound in the Caribbean. Instead, states have sought to reformulate their command-and-control regulation in light of international and regional developments. Thus, states such as Belize, Guyana, Jamaica and Trinidad & Tobago have passed legislation to incorporate key principles of environmental governance and social justice into their legislative regimes by means of environmental impact assessments (EIA) as well as by introducing freedom of information legislation. Additionally, in the pursuit of broader participation and environmental justice, many jurisdictions have seen the extension – largely by statutory means – of the purview of the common law principles governing judicial review. Additionally, at the common law the approach to standing, with respect to environmental issues, has been largely expanded to afford claimants who previously could not seek leave of the court, but who were affected, the access to having their voices heard. Finally, there has been increasing reliance in the Constitution – the fundamental command-and-control-type instrument – to assert environmental, social, indigenous and human rights. It is also apposite to mention that the OECS has given indications that within the sub-group, matters relating to the governance of the marine resource will be addressed collectively and holistically under the tenets of an ocean governance regime (IISD/OECS, 2012). In sum, considerations of the governance of ocean and human health appears to have progressed from the restrictive top-down governance approach – or has it?

ENVIRONMENTAL GOVERNANCE IN THE CARIBBEAN – A PHOENIX?

> Normatively, new governance is claimed to be more responsive, legitimate, and effective from top-down approaches[,] because deliberation, cooperation and learning at [the]

> local level, may lead to responses which better take account of local circumstances [and] build upon local knowledge and capacities, and [may also] result in greater stakeholder ownership and 'buy in' [.] [New governance] is also arguably better able to transcend jurisdictional boundaries than traditional approaches[.]
>
> Gunningham, 2009a, 2009b

Ecohealth requires *inter alia*, transdisciplinarity, participation, gender and social equity, and sustainability (Charron, 2012). The Caribbean region may be considered to be attempting to perfect its own special hybrid of the shift elucidated in the governance literature. This movement from the top-down command-and-control regulation to the more desirable consensual approach seeks to coordinate at multiple levels, and is distinctly polycentric (Trubek & Trubek, 2005). While it may be true to say that the States are tentatively trying economic instruments and other inclusive forms of management, it is unquestionable that the command-and-control ethic inherent in colonial governance structures will for the next while continue to be the mainstay of the governance of ocean and marine resources, and therefore the relationship to the health of the resource and to the populace which uses and relies upon it. That having been said, it is prudent to examine, how *de jure* or *de facto* methods of decentralisation have, are, or could promote this form of neo-environmental governance for the Caribbean region.

The EIA process is perhaps the best illustration of the foray of CARICOM states away from the traditional approach and towards environmental governance. The instrument, which originated in the 1969 *National Environmental Policy Act* (United States), became more widespread with its mention in Principle 17 of the 1992 *Rio Declaration*. Today, it is implanted firmly by means of specific legislation into at least five CARICOM states,[7] while other states like Antigua & Barbuda, Barbados and St. Lucia utilise planning legislation to invoke the process as a part of development control. The instrument was a particularly attractive method of reconciling economic development with natural resources management, while incorporating the participation of the stakeholders who are the most likely to be affected significantly. The format of the legislation adopted by Belize, Guyana and Trinidad & Tobago is, for example, substantially different from that adopted in Jamaica, Antigua & Barbuda, St. Lucia and St. Christopher-Nevis, but the fundamental premise underpinning the instruments is the right of access to information, public participation in decision making and access to justice. This unquestionably expands the approach of traditional governance into the heterarchical typology of neo-environmental governance by affording the citizen's involvement in the development process.

However, despite the good intentions of these expedients, the effectiveness of the instrument in legitimate environmental governance arenas is still to materialise. Firstly, concern for the public interest in environmental litigation has been rare, with only a handful of cases taken to the courts from Jamaica,[8] Belize,[9] Trinidad & Tobago,[10] the British Virgin Islands[11] and the Bahamas,[12] in the last 10 years (Andrade, Excell, & Gonzalez, 2011). This trend has to be viewed from several perspectives. On a positive note, it is a marked improvement that cases on issues which have potential far-reaching effects on the populace of the Caribbean have been able to reach the courts – without the traditional impediments to the annals of environmental justice. However, it is here the good news ends, and the shortcomings of this process begin. Firstly, an analysis of the operation of the EIA process, as well as the case law emanating from challenges to the legitimacy of the process, show that as a tool of more participatory and transparent governance, it has largely failed in the Commonwealth Caribbean. Bynoe (2006) underscores that the weaknesses of the EIA process in Guyana lies with the method of engagement and involvement of the public, the lack of consideration of indigenous rights, and questions as to whether trust, awareness, transparency and accountability are truly present in the participatory process. Many of these concerns arguably also apply to the EIA experience in Belize, which formed the basis of the first distinctive environmental litigation to reach the courts of the region. In the *Belize Alliance of Conservation NGOs* case, many of the issues raised by Bynoe surfaced in the two-part case which ended up at the Judicial Committee of the Privy Council for consideration. In what would become the trend in the subsequent cases of *Fisherman and Friends* and *Save Guana Cay* that arrived at its doorstep, the Privy Council in the *Belize Alliance of Conservation NGOs* case was careful to enunciate that its task was to decide on the procedural, as opposed to substantive, requirements of the EIA process. Accordingly, for essentially this reason, all three cases which arrived at the Privy Council failed on the grounds that the procedures prescribed under the relevant statute were adequately followed. However, each case had significantly alarming substantive facts – including issues of transparency and accountability which escaped the examination of the court. However, the blame cannot be placed solely with the Judicial Committee of the Privy Council, since judicial review actions are fundamentally based on the breaches in procedure, not substance, of the actions of the public authorities.

The fault appears to lie in the substantive formulation of the EIA process in the region. As Ramlogan (2010) opines in reference to the *Talisman*, *ALUTRINT* and *Fisherman and Friends* cases, there are several critical factors militating against effective and substantial public participation in the decision making process. These include the state as an active economic player; the role of multi-national corporations; and perhaps most cogently, the absence of a culture of environmental advocacy and democracy in the Caribbean region. The combination, as seen in the *Virgin Islands Environmental Council* case, often means that environment and development are at opposite ends of the continuum – a situation which defeats the purpose and objectives of neo-environmental governance. However, it is our considered assessment that all is not lost, since litigation with respect to EIA has largely expanded standing in the region and ushered in the era of involvement by NGOs and other private interest groups. What is lacking from the equation, is the increased awareness of issues, as well as of their collective power on behalf of the citizenry. In this regard, it is hoped that the recent profusion of [proposed] freedom of information legislation surfacing around the region will serve to make information more accessible, and consequently lead to a more aware public. When this awareness and a culture of environmental democracy take root, the EIA process will take on some semblance of legitimacy. Perhaps this culture is slowly but surely growing, as the public's consciousness is gripped by examples such as the recent hunger-strike demonstration by Highway Re-Route Movement activist, Dr Wayne Kublalsingh,[13] and the challenge by Oceana in Belize to the validity of the offshore oil contracts issued by the Government of Belize.[14]

According to Upson and Hughes (1994), the public nature of environmental law uses judicial review as a mechanism to regulate decisions. However, because of the increased complexity of such issues, the regulator mechanisms aimed to protect the public are fading, as they are no longer efficient in protecting the environment. This is however not the case with Jamaica, which has seen substantial success in the quest for environmental justice through judicial review at the national level. In the leading case of *Pear Tree Bottom*, the Court found in favor of the applicants – two NGOs and four local residents – and quashed the permit granted by the National Resources Conservation Authority for the construction of a hotel. In a judgment which upholds the tenets of the public participation process, it was found that the Agency had not acted transparently and that its failure deprived the public of the information necessary to make a fully informed, intelligent decision. The dereliction constituted a breach of the public's legitimate expectation of fair and meaningful participation. *Pear Tree* was followed in similar spirit by the *Harbour View* and *Palisadoes* cases, and what is perhaps interesting is that of the CARICOM states with EIA legislation, Jamaica's is the least detailed legislation with respect to the conduct and review of EIAs. Barbados,[15] Belize and Trinidad & Tobago have also taken steps to expand the scope of the principles of judicial review, by the adoption of legislation which makes provision for individuals to bring judicial review proceedings in the public interest.

Finally, perhaps an ironic use of the command-and-control-type instrument is to expand procedural environmental rights, through recourse to the supreme law in Commonwealth Caribbean states, which prescribes the fundamental rights of the citizenry. Most CARICOM Constitutions do not contain provisions which guarantee environmental rights, but the respective constitutions of Guyana and Jamaica contain provisions which can be construed to engender environmental rights. Guyana, on the face of it, has the more extensive provisions[16] but Jamaica's provisions[17] resonate similar sentiments. One observation is that with respect to health-related considerations, Guyana's constitution seems to place a fetter akin to the hypersensitive claimant in nuisance law under subsection 3, which Jamaica's does not appear to do. This may constitute a limitation to the guarantee to every citizen, but the interpretation of this is subject to speculation. Despite these provisions, whether under constitutional provisions whether an individual citizen is able to bring a constitutional motion to protect the environment, or is able to argue for the striking down of legislation inconsistent with these provisions. However, the generous interpretation given to the constitutional provisions concerning the human rights of indigenous people in the recent case of *Aurelio Cal*, provides a beacon of hope (Anderson, 2012). Additionally, Caribbean courts have derived environmental rights from other fundamental rights in the *Constitution* – for example, in *Soodeen v. Attorney General of Trinidad & Tobago*, lead waste caused injury to residents; the court held that a failure or omission by the government to safeguard citizens would be considered an infringement of their right to life. Accordingly, the Court took cognisance of the fact that a healthy environment is paramount to the enjoyment of the right to life. Utilising the Constitution in this manner would therefore seem to be a means of fostering more inclusive participation, imposing some level of accountability on the government, and affording the citizens a means of ecological justice.

CONCLUSION

From the foregoing discussion, it is possible to draw some general conclusions. Firstly, at all levels the approach to governance has devolved from one where the power was centralised to government into, a participatory model which requires greater access to information and environmental justice. Through examples, it can be illustrated that the model of governance for oceans, health and humans, tends to involve a broad gamut of stakeholders and the basic framework is a hierarchical model of governance. However, the CARICOM Caribbean region has been responsive to the new paradigm for environmental governance and a heterarchical structure is emerging, both at the regional level and at the national level. In addition, states are incorporating spatial and temporal considerations into their decision making; thereby extending governance across borders and between generations.

Much of the fundamental changes to governance will need to come from the national level. However, the state's monopoly on decision making needs to be tempered by the input of those at the grassroots, such as by principal stakeholders like Jared Jamaica and Gitangali Guyana, to yield a form of more collaborative governance. Additionally, the state needs to make information more available, foster the trust of the populace, promote awareness, participation and access to justice, and most importantly be held accountable for its actions. While the region in general has not devolved to a level of neo-environmental governance, there are several illustrations of systems and other tools, the majority gently couched in a command-and-control setting, which are fostering more participative governance strategies. These mechanisms should be viewed as a mixed bag in terms of their success in effecting environmental governance. Nevertheless they represent the growing strength of the phoenix of neo-governance for the CARICOM Caribbean region.

NOTES

1. For further discussion on this see Winston Anderson, *Principles of Caribbean Environmental Law* (2013), pp. 36–37 and Winston Anderson. Implementing MEAs in the Caribbean: Hard Lessons from *Seafood and Ting*. (Anderson, 2001), 10(2), *Review of European Community & International Environmental Law*, 227.
2. [1977] 3 All ER 70.
3. See for example *Boyce v Paddington Borough Council (1903), Robinson v Kilvert* (1889) LR 41 ChD 88 and *McKinnon Industries v Walker* [1951] WN 401.
4. See for example *Hunter v Canary Wharf Ltd.* [1997] UKHL 14.
5. [1981] 2 W.L.R. 722.
6. See for example the *Gabčíkovo–Nagymaros (Hungary v. Slovakia)* 25 September 1997 and *Pulp Mills On The River Uruguay (Argentina v. Uruguay)* 13 July 2006 (see also Lee, 2006).

7. Belize, Guyana, Jamaica, Trinidad & Tobago and St. Christopher-Nevis.

8. *Northern Jamaica Conservation Association et al. v. Natural Resources Conservation Authority and National Environmental Planning Agency (No. 1)* J.M. (2006) S.C. 49.

9. *Belize Alliance of Conservation Non-Governmental Organizations v. DOE and Belize Electricity Company Limited (No. 2)* [2004] UKPC 6; *Alliance of Conservation Non-Governmental Organizations v DOE and Belize Electricity Company Limited No. 1)* Privy Council Appeal No. 47 of 2003 (13 August 2003); *Belize Institute for Environmental Law and Policy v Chief Officer of the Department of Environment* Claim No. 302 of 2007 (Belize).

10. *Fishermen and Friends of the Sea v. The Environmental Management Authority & Anor* [2005] UKPC 32.

11. *Virgin Islands Environmental Council v. Attorney General and Quorum Island BVI Limited*, BVIHCV 185 of 2007 (High Court of the BVI) (7 March 2008).

12. *Save Guana Cay Reef Association Ltd. and others v. The Queen and others* [2009] UKPC 44.

13. Trinidad Tobago's First Hunger Strike Grips Nation, Trinidad Express, 27 December 2012, available at http://www.trinidadexpress.com/news/T_T_s_first_hunger_strike_grips_nation-184996731.html.

14. *Oceana in Belize; Citizens Organised for Liberty Through Action (COLA) and the Belize Coalition to Save Our Natural Heritage v Minister Of Natural Resources* (Claim No. 810 of 2011), 23 April, 2013 (Supreme Court of Belize, 2013).

15. See for example *Scotland District Association Inc. v. Attorney General and Others*, 1996 (Supreme Court of Barbados).

16. Art 149 J of the 2003 *Constitution (Amendment) (No. 2)* (Guyana).

17. Art. 13(3)(l) of the 2011 *Charter of Fundamental Rights and Freedoms (Constitutional Amendment) Act* (Jamaica).

REFERENCES

Anderson, W. (1994). Locus standi in Commonwealth Environmental Law – Caribbean perspectives. *Caribbean Law Review*, 379.

Anderson, W. (2001). Implementing MEAs in the caribbean: Hard lessons from seafood and ting. *Review of European Community & International Environmental Law*, *10*(2), 227.

Anderson, W. C. (2012). *Principles of Caribbean environmental law*. Washington, DC: Environmental Law Institute.

Andrade, D. E., Excell, C., & Gonzalez, C. (2011). *Citizen enforcement of procedural rights in the. environmental impact assessment process in Belize and Jamaica*. Paper Presented at the Ninth International Conference on Environmental Compliance and Enforcement 2011. Retrieved from http://inece.org/conference/9/proceedings/49_Andrade.pdf

Bakker, K. (Ed.). (2007). *Eau Canada: The future of Canada's water* (pp. 1–22). Vancouver: UBC Press.

Bynoe, M. (2006). Citizen participation in the environmental impact assessment process in guyana: Reality or fallacy? *Law, Environment and Development Journal*, *2*(1), 34. Retrieved from http://www.lead-journal.org/content/06034.pdf

CARSEA. (2007). *Caribbean sea ecosystem assessment (CARSEA): A sub-global component of the millennium ecosystem assessment (MA)*. Caribbean marine studies, special edition (Port-of-Spain: The Cropper Foundation). Retrieved from http://www.acs-aec.org/purple_beauty/Documents/Legal/CSC/CARSEA%20Document.pdf

Charron, D. F. (2012). *Ecohealth research in practice innovative applications of an ecosystem approach to health*. Ottawa: Springer, IDRC. Retrieved from http://idl-bnc.idrc.ca/dspace/bitstream/10625/47809/1/IDL-47809.pdf

ECLAC. (2004). *Major issues in the management of enclosed or semi-enclosed Seas, with particular reference to the caribbean Sea*. Retrieved from http://www.eclac.org/publicaciones/xml/1/20811/L0024.pdf

Fanning, L., Mahon, R., & McConney, P. (2009). Focusing on living marine resource governance: The Caribbean large marine ecosystem and adjacent areas project. *Coastal Management*, *37*(3), 222.

Fanning, L., Mahon, R., & McConney, P. (2011). *Towards ecosystem-based management in the wider caribbean*. Amsterdam: Amsterdam University Press.

FAO. (1998). Report of the Seventh Session of the Working Party on the Assessment of Marine Fishery Resources, Belize City, Belize, December 2–5, 1997. FAO Fish. Report No. 576, FAO, Rome.

FAO/Salas, S., Chuenpagdee, R., Charles, A., & Seijo, J. C. (Eds.). (2011). *Coastal fisheries of Latin America and the Caribbean*. FAO Fisheries and Aquaculture Technical Paper No. 544, FAO, Rome. Retrieved from http://www.fao.org/docrep/014/i1926e/i1926e.pdf

Gunningham, N. (2009a). Environment law, regulation and governance: Shifting architectures. *Journal of Environmental Law*, *21*(2), 179.

Gunningham, N. (2009b). The new collaborative environmental governance: The localization of regulation. *Journal of Law and Society*, *36*(1), 145.

Gunningham, N., & Grabosky, P. (1998). Smart regulation: Designing environmental policy. (Oxford Socio-legal Studies). Oxford, Clarendon Press.

Harrington, C., et al. (2008). Locating communities in natural resource management. *Journal of Environmental Policy and Planning*, *10*(2), 199.

IISD/OCES. (2012). *Workshop contemplates OECS Ocean Resource Management Policy*. OECS Press Release. Retrieved from http://sids-l.iisd.org/news/workshop-contemplates-oecs-ocean-resource-management-policy/

Johnson, D. M. (2006). The challenge of international Ocean governance: Institutional, ethical and conceptual dilemmas. In D. R. Rothwell & D. L. VanderZwaag (Eds.), *Towads principled Ocean governance: Australian and Canadian approaches and challenges* (pp. 349–399). New York, NY: Routledge.

Kooiman, J., Bavinick, M., Jentoft, S., & Pullin, R. (Eds.). (2005). *Fish for life: Interactive governance for fisheries*. Amsterdam: Amsterdam University Press.

Lee, M. K. (2006). *The Uruguay Paper Pulp Mill Dispute: Highlighting the Growing Importance of NGOs and Public Protest in the Enforcement of International Environmental Law. Sustainable Development Law & Policy*, *71*. Retreived from http://digitalcommons.wcl.american.edu/cgi/viewcontent.cgi?article=1215&context=sdlp

Mahon, R. (2002). Living aquatic resource management. In I. Goodbody & E. Thomas-Hope (Eds.), *Natural resource management for sustainable development in the Caribbean* (pp. 143–218). Kingston, Jamaica: Canoe Press, UWI.

Mahon, R., McConney, P., & Roy, R. N. (2008). Governing fisheries as complex systems. *Marine Policy*, *31*(1), 104.

Maitland, T., & Morrison, K. (2010). *Oceans & human health training course*. Barbados: UWI CaveHill Campus.

Mittermeier, R. A., et al. (2005). *Hotspots revisited: Earth's biologically richest and most threatened terrestrial*. Mexico City: CEMEX.

Myers, N., et al. (2000). Biodiversity hotspots for conservation priorities. *Nature*, *403*, 853. Retrieved from http://www.nature.com/nature/journal/v403/n6772/full/403853a0.html

Myers, R. A., & Worm, B. (2003). Rapid worldwide depletion of predatory fish communities. *Nature*, *423*, 280–283.

Olsson, P. (2007). The role of vision in framing adaptive co-management processes: Lessons from Kristianstads Vattenrike, Southern Sweden. In D. Armitage, F. Berkes & N. Doubleday (Eds.), *Adaptive co-management: Collaboration, learning and multilevel governance* (pp. 268–285). Vancouver: UBC Press.

Ramlogan, R. (2010). Using the law to achieve environmental democracy and sustainable development: An elusive dream for trinidad and tobago. *Electronic Green Journal*, *1*(30). Retrieved from http://www.escholarship.org/uc/item/11p9f74m#page-2

Reed, M. G., & Bruyneel, J. (2010). Rescaling environmental governance, rethinking the state: A three dimensional view. *Progress in Human Geography Online*, 26 January 2010.

Stabroek News. (2009). Caribbean fisheries sector must improve management, conservation or face trouble. *Stabroek News*, April 16. Retrieved from http://www.stabroeknews.com/2009/business/04/16/caribbean-fisheries-sector-must-improve-management-conservation-or-face-trouble/

Trubek, D., & Trubek, L. (2005). The co-existence of new governance and legal regulation: Complementarity or rivalry? 542. Paper presented at Annual Meeting of Research Committee on Sociology of Law, Madison, WI.

Upson, J., & Hughes, D. (1994). Locus standi: An essential hurdle in judicial review of environmental action [1994] ELM 136.

UNEP. (1994). *Regional Overview of Land-Based Sources of Pollution in the Wider Caribbean Region*. CEP Technical Report No. 33. UNEP Caribbean Environment Programme, Kingston, UK. Retrieved from http://www.cep.unep.org/publications-and-resources/technical-reports/tr33en.pdf

UNEP. (2012). *Atlas of the Assessment and Management of Environmental Pollution in the Wider Caribbean Region*. CEP Technical Report No. 53. UNEP Caribbean Environment Programme, Kingston, UK. Retrieved from http://www.cep.unep.org/publications-and-resources/technical-reports/AMEP%20ATLAS%202007-2009.pdf

UNEP/Heileman, S., & Walling, L. J. (Eds.). (2004). *Caribbean environment outlook*. Nairobi: UNEP/Earthprint. Retrieved from http://www.unep.org/geo/pdfs/Caribbean_EO_final.pdf.

FURTHER READING

Acting Chief of Police v. Bryan 1987, 1985 (High Court).

Attorney General of Belize v MS Westerhaven Schiffahrts and anor (Supreme Court of Belize), Claim No. 45 of 2009. Retreived from http://www.elaw.org/system/files/westerhaven.26.4.10.pdf

Aurelio Cal et al. v. Attorney General of Belize, 2007, Supreme Court of Belize, 171 and 172.

Belize Alliance of Conservation Non-Governmental Organizations v. DOE and Belize Electricity Company Limited, (No. 1), 2003 (Judicial Committee of the Privy Council) 47. Retrieved from http://www.bailii.org/uk/cases/UKPC/2003/63.html

Belize Alliance of Conservation Non-Governmental Organizations v. DOE and Belize Electricity Company Limited (No. 2), 2004 (Judicial Committee of the Privy Council) 6. Retrieved from http://www.bailii.org/uk/cases/UKPC/2004/6.html

Belize Institute for Environmental Law and Policy v. Chief Officer of the Department of Environment, Claim, 2007 (Supreme Court of Belize) 302. Retrieved from http://www.elaw.org/node/3607

Charter of Fundamental Rights and Freedoms (Constitutional Amendment) Act (Jamaica). 2011.

Constitution (Amendment) (No. 2) (Guyana), 2003.

Fisheries Act (Antigua), 2006. Retrieved from http://laws.gov.ag/acts/2006/a2006-22.pdf

Fisheries Regulations (Antigua), 2013. Retrieved from http://faolex.fao.org/docs/pdf/ant122461.pdf

Fishermen and Friends of the Sea v. The Environmental Management Authority & Anor., 2005 (Judicial Committee of the Privy Council) 32. Retrieved from http://www.bailii.org/uk/cases/UKPC/2005/32.html

Jamaica Environment Trust and Another v. National Water Commission and Others, 2010 (Supreme Court of Judicature of Jamaica), Claim No. HCV 00114/2010. Retrieved from http://www.jamentrust.org/advocacy-a-law/legal/cases/126-judicial-review-harbour-view-sewage-treatment-plant.html

Jamaica Environment Trust v. Natural Resources Conservation Authority and National Environmental Planning Agency (Supreme Court of Judicature of Jamaica), Claim No. HCV 5674/2010. Retrieved from http://www.jamentrust.org/advocacy-a-law/legal/cases/142-judicial-review-roadworks-on-the-palisadoes-strip.html

Montego Bay Convention (United Nations Convention on the Law of the Sea), Montego Bay, Jamaica, 10 December 1982, 21 I.L.M. 1261. Retrieved from http://www.un.org/Depts/los/convention_agreements/texts/unclos/closindx.htm

MS Westerhaven Schiffahrts v. Attorney General of Belize (Court of Appeal, Belize) Civil Appeal No. 19 of 2010, Claim Retreived from http://www.ecolex.org/server2.php/libcat/docs/COU/Full/En/COU-159759.pdf

Natural Resources Conservation Authority v. Seafood and Ting International Ltd.; Natural Resources Conservation Authority v. DYC Fishing Ltd., 1999 (Supreme Court of Judicature of Jamaica) 269.

Northern Jamaica Conservation Association et al. v. Natural Resources Conservation Authority and National Environmental Planning Agency (No. 1), 2006 (Supreme Court of Judicature of Jamaica) Claim No. HCV 3022 of 2005. Retrieved from http://www.jamentrust.org/advocacy-a-law/legal/cases/99-pear-tree-bottom-papers.html

People United Respecting the Environment (PURE) and Rights Action Group (RAG) v. Environmental Management Authority and ALUTRINT Ltd. (High Court, Trinidad & Tobago), HC – CV 2007 – 02263. Retrieved from http://www.elaw.org/system/files/smelter+judgement.pdf

R v. Inspectorate of Pollution ex parte Greenpeace, 1994 (Court of Appeal) 329.

R v. Panel on Take-overs and Mergers, ex parte Datafin plc and another, 1987 (Court of Appeal of England and Wales) 815.

R v. Secretary of State for the Environment ex parte Rose Theatre Trust, 1990 (Queen's Bench Division) 754.

R v. Secretary of State for Foreign and Commonwealth Affairs Ex parte. World Development Movement Ltd., 1995 (Divisional Court of England and Wales) 386.

Revised Treaty of Chaguaramas establishing the Caribbean Community including the CARICOM Single Market and Economy, 5 July, 2001, 32 ILM 289 (1993). Retrieved from http://www.sice.oas.org/trade/caricom/caricind.asp

Save Guana Cay Reef Association Ltd. and others v. The Queen and others, 2009 (Judicial Committee of the Privy Council) 44. Retrieved from http://www.jcpc.gov.uk/decided-cases/docs/JCPC_2009_0013_Judgment.pdf

Soodeen v. Attorney-General of Trinidad and Tobago (High Court of Trinidad and Tobago), No S-839 of 1996 (Unreported), Retrieved from http://webopac.ttlawcourts.org/LibraryJud/Judgments/HC/rlee/1996/cv_96_839DD16jul2012.pdf. No. S-839 of 1996 (Unreported).

Talisman (Trinidad) Petroleum Limited v. EMA, 2002 (Environmental Tribunal of Trinidad and Tobago) EA 003 of 2002, Retreived from http://www.ecolegal2001.com/resource/ea3_2002.pdf

Virgin Islands Environmental Council v. Attorney General and Quorum Island BVI Limited, 2008 (Eastern Caribbean Supreme Court (Civil)) BVIHCV2007/0185. Retrieved from http://www.eccourts.org/wp-content/files_mf/1359390929_magicfields_pdf_file_upload_1_1.pdf